Theory of Elastic Wave Propagation and its Application to Scattering Problems

Elastic wave propagation applies to a wide variety of fields, including seismology, non-destructive testing, energy resource exploration, and site characterization. New applications for elastic waves are still being discovered. *Theory of Elastic Wave Propagation and its Application to Scattering Problems* starts from the standpoint of continuum mechanics, explaining stress and strain tensors in terms of mathematics and physics, and showing the derivation of equations for elastic wave motions, to give readers a stronger foundation. It emphasizes the importance of Green's function for applications of the elastic wave equation to practical engineering problems and covers elastic wave propagation in a half-space, in addition to the spectral representation of Green's function. Finally, the MUSIC algorithm is used to address inverse scattering problems.

- Offers comprehensive coverage of fundamental concepts through to contemporary applications of elastic wave propagation
- Bridges the gap between theoretical principles and practical engineering solutions

The book's website provides the author's software for analyzing elastic wave propagations, along with detailed answers to the problems presented, to suit graduate students across engineering and applied mathematics.

Terumi Touhei is a Professor at the Tokyo University of Science, with extensive experience of teaching graduate students.

Theory of Elastic Wave Propagation and its Application to Scattering Problems

Terumi Touhei

CRC Press
Taylor & Francis Group
Boca Raton London New York

CRC Press is an imprint of the
Taylor & Francis Group, an **informa** business

First edition published 2024
by CRC Press
2385 NW Executive Center Drive, Suite 320, Boca Raton FL 33431

and by CRC Press
4 Park Square, Milton Park, Abingdon, Oxon, OX14 4RN

CRC Press is an imprint of Taylor & Francis Group, LLC

ISBN: 978-1-032-17077-0 (hbk)
ISBN: 978-1-032-17078-7 (pbk)
ISBN: 978-1-003-25172-9 (ebk)

DOI: 10.1201/9781003251729

Typeset in Nimbus font
by KnowledgeWorks Global Ltd.

Publisher's note: This book has been prepared from camera-ready copy provided by the author.

Access the Support Material: www.routledge.com/9781032170770

To Kinuko

Contents

Preface xiii

Acknowledgment xvii

Author xix

Chapter 1 Introduction 1

1.1 Basic concept of continuum mechanics 1
1.2 Strain tensor 2
1.2.1 Deformation of continuum body 2
1.2.2 Definition of strain tensor 4
1.2.3 Characterization of infinitesimal strain tensor 7
1.2.4 Volumetric strain based on the infinitesimal strain tensor 9
1.3 Equation of motion for continuum body and concept of stress tensor 11
1.3.1 Material derivative and Reynolds transport theorem 11
1.3.2 Cauchy's relation and definition of stress tensor 15
1.3.3 Equilibrium equation in terms of stress tensor and symmetry of stress tensor 19
1.4 Elastic wave equation and its reciprocity 22
1.4.1 Elastic wave equation 22
1.4.2 Elastic constants 24
1.4.3 Reciprocity of elastic wavefield 26
1.5 Problems 29

Chapter 2 Basic properties of solution for elastic wave equation and representation theorem 31

2.1 Solution for scalar wave equation for unbounded media 31
2.2 Helmholtz decomposition of vector field 35
2.2.1 Gradient and rotation operators 35
2.2.2 Decomposition of vector fields into irrotational and divergence-free vector fields 38
2.3 P and S wave solutions derived from elastic wave equation 40
2.3.1 Scalar potentials for elastic wave equation 40
2.3.2 Plane wave solutions for elastic wave equation 41
2.4 Green's function for elastic wave equation 44

2.4.1 Definition of Green's function for elastic wave equation ... 44
2.4.2 Green's function for scalar wave equation ... 46
2.4.3 Derivation of Green's function for elastic wave equation ... 52
2.5 Representation theorem for solutions for elastic wavefield .. 57
2.5.1 Overview of considered problem ... 57
2.5.2 Representation of solutions for elastic wavefield in the frequency domain ... 58
2.5.3 Representation of solution for elastic wavefield in the time domain ... 64
2.6 Problems ... 68

Chapter 3 Elastic wave propagation in 3D elastic half-space ... 71

3.1 Plane wave propagation ... 71
3.1.1 Coordinate system, governing equation, free boundary conditions, and scalar potentials ... 71
3.1.2 Properties of scalar potentials defined by Eq. (3.1.7) ... 74
3.1.3 Reflection and polarization of plane waves at free surface ... 77
3.1.4 Propagation of the Rayleigh wave ... 85
3.2 Green's function for elastic half-space ... 88
3.2.1 Weyl and Sommerfeld integrals ... 88
3.2.2 Representation of Green's function in terms of superposition of plane waves ... 95
3.2.3 Representation of Green's function in terms of superposition of cylindrical waves ... 99
3.3 Evaluation of Green's function in the complex wavenumber plane by use of branch line integrals ... 109
3.3.1 General remarks ... 109
3.3.2 Method based on branch line integral for Sommerfeld integral ... 109
3.3.3 Approximation of Green's function for elastic half-space by method based on branch line integral ... 115
3.4 Evaluation of Green's function in complex wavenumber plane using steepest descent path method ... 123
3.4.1 Concept of steepest descent path ... 123
3.4.2 Application of steepest descent path method to Sommerfeld integral ... 126
3.4.3 Application of steepest descent path method to Green's function for elastic half-space ... 130

3.5 Spectral representation of Green's function for elastic half-space ... 138
3.5.1 Review of elastic wave equation and boundary conditions ... 139
3.5.2 Self-adjointness of operator $\mathscr{A}_{k'l'}$ ... 141
3.5.3 Relationship between resolvent and eigenfunction for operator $\mathscr{A}_{i'j'}$ ... 146
3.5.4 Eigenfunction expansion for solution for elastic half-space ... 151
3.5.5 Spectral representation of Green's function ... 153
3.6 Problems ... 155

Chapter 4 Analysis of scattering problems by means of Green's functions 157

4.1 Application of representation theorem to solid-fluid interaction problem ... 157
4.1.1 Definition of problem and basic equations ... 157
4.1.2 Representation theorem and boundary integral equation ... 159
4.1.3 Coupling of boundary integral equations via discretization method ... 163
4.1.4 Analysis of sound propagation in virtual LPG tank ... 165
4.2 Application of the spectral representation of Green's function to scattering problem ... 167
4.2.1 Overview of problem ... 167
4.2.2 Derivation of Lippmann-Schwinger equation for scattering problem ... 169
4.2.3 Application of generalized Fourier transform to Lippmann-Schwinger equation ... 171
4.2.4 Numerical method for Lippmann-Schwinger equation in wavenumber domain ... 172
4.2.5 Numerical examples ... 179
4.3 Inversion of point-like scatterers in elastic half-space ... 185
4.3.1 Definition of problem and basic notation ... 185
4.3.2 Method for the identification of the location of point-like scatterers ... 188
4.3.3 Numerical examples ... 199

Appendix A Tensor algebra for continuum mechanics ... 202

A.1 General remarks ... 202
A.2 Transformation rules for vector components ... 202
A.3 Definition and basic properties of tensors ... 205
A.4 Metric tensor ... 206

A.5 Triple scalar product ... 207
A.6 Isotropic rank-4 tensor ... 208

Appendix B Fourier transform, Fourier-Hankel transform, and Dirac delta function ... 214

B.1 Fourier integral transform in $\mathbb{R}^n$ ... 214
B.2 Hankel transform and Fourier-Hankel transform ... 215
B.3 Dirac delta function ... 216

Appendix C Green's function in the wavenumber domain ... 219

C.1 Overview of task ... 219
C.2 Construction of Green's function in the wavenumber domain ... 219
C.2.1 Homogeneous solution for elastic wave equation in the wavenumber domain ... 219
C.2.2 Green's function in the wavenumber domain expressed as undetermined coefficients ... 220
C.2.3 Conditions for determining coefficients of Green's function in the wavenumber domain ... 221
C.2.4 Homogeneous solution at the root of the Rayleigh function ... 223
C.3 Resolvent kernel, spectrum, and eigenfunctions ... 224
C.3.1 Resolvent kernel ... 224
C.3.2 Eigenfunction for point spectrum ... 227
C.3.3 Eigenfunction for continuous spectrum ... 228
C.3.4 Explicit form of the definition function ... 231

Appendix D Comparison of Green's function obtained using various computational methods ... 233

D.1 Computational methods for Green's function for comparisons and analyzed model ... 233
D.1.1 Formulations for Green's function for comparisons ... 233
D.1.2 Fourier-Hankel transform for Green's function ... 234
D.1.3 Steepest descent path method for Green's function ... 235
D.1.4 Spectral form of Green's function ... 235
D.2 Numerical results ... 236

Appendix E Music algorithm for detecting location of point-like scatterers ... 238

Answers **243**

References **257**

Index 261

Preface

The theory of elastic wave propagation is open to a large variety of areas of applications such as seismology, site characterization, nondestructive testing, medical image processing, and among others. These vast areas of applications, which are sometimes connected with scattering problems, are the result of its long history and growing theory. The purpose of the book is to describe the theory and applications for graduate students, researchers and engineers interested in elastic wave propagation.

Chapter 1 serves as an introduction to the theory of elastic wave propagation. We first clarify the basic concept of the continuum mechanics together with the definition of the Euler and Lagrange approaches. In addition, we define the strain and stress tensors using the modern style found in textbooks (Schutz, 1990). We try to connect the physical understandings with mathematical definitions for these tensors. Even today, it seems that there are not so many textbooks for the continuum mechanics concerned with the modern mathematical viewpoint of tensors. That is the reason why we chose the above description of the strain and stress tensors. The tensor algebra needed to understand the above content is provided in Appendix A.

We also introduce the Reynolds transport theorem to derive the equation of motion for a continuum medium. We linearize the equation of motion assuming that the motions in continuum media exhibit small-amplitude vibrations. The elastic wave equation, the treatments and/or the investigations of which are the theme of the book, is derived from the linearized equation above under the assumption that the medium is isotropic.

At the end of this chapter, the reciprocity of the elastic wavefield is derived using the Gauss divergence theorem. It may be true that the contents of this chapter seem to be complex and difficult, readers are, however, just required to have the knowledge of the strength of material, general mechanics, fluid mechanics, elementary of calculus and linear algebra, which should be taught at the undergraduate level in the engineering course.

In Chapter 2, we discuss the solutions for the elastic wave equation for a 3-D full space. As preparations for the discussions in this chapter, we also derived solutions for the scalar wave and Helmholtz equations. Solutions of the elastic wave equation are discussed by decomposing the elastic wave equation into scalar wave equations for P and S waves. The presence of the P and S waves for the elastic wavefield is an important fact, which is verified by the properties of a vector field consisting of irrotational and divergence-free components.

Green's functions for the scalar wave, as well as the elastic wave equations, have special standpoints for theoretical and numerical analyses since Green's function not only reveals the properties of wave propagation but also transforms the partial differential equation into an integral equation. We present the explicit forms as well as the derivation processes of Green's functions for a 3-D full space in detail in §2.4. The derivation of Green's function by the use of the Dirac delta function and Fourier transform can be thought of as a basic skill in mathematical physics for graduate engineering students. The basics for the Dirac delta function as well as the Fourier transform to understand the above procedures are explained in Appendix B.

The coupling of the reciprocity of the wavefield obtained in Chapter 1 and Green's function yields a representation theorem for the solution of the elastic wavefield. The representation theorem clarifies the relationship among the solution of the elastic wave equation, the boundary values and body forces. A practical application of the representation theorem for an engineering problem is presented in §4.1.

In Chapter 3, we discuss the solutions of the elastic wave equation for a 3D half-space. We show the presence of three types of waves: P, SV, and the SH waves. The interaction between P and SV waves is caused at the free boundary, whereas SH wave exists independently from the P and SV waves. The presence of the Rayleigh wave can be recognized based on the interaction of P and SV waves for a special case.

We also derived Green's function for an elastic half-space. We find that derivation of the closed form of Green's function for an elastic half-space is impossible. In stead of the closed form, we derive Green's function for an elastic half-space in the form of a Fourier-Hankel transform. We introduced two types of historical approximation methods for computing Green's function. The approximation methods we introduce are branch line integral and steepest descent path methods. Nowadays, it is not difficult to compute Green's function for an elastic half-space without using approximation methods. In this sense, the approximation methods for Green's function are sometimes thought to only have historical importance. We will see, however, that the historical approximation approach has the potential to improve modern computational methods. This is discussed in the next chapter.

At the end of this chapter in §3.5, we introduce the modern viewpoint of mathematics, which is the spectral theory of the operator to Green's function for an elastic half-space. According to §3.3, Green's function can be expressed by the residue term and branch-line integrals, respectively. These terms are unified in terms of the eigenfunctions of the point and continuous spectra using the spectral theory. We seek the representation of Green's function in terms of eigenfunctions. The discussions seem to be in a rather abstract manner. The representation of Green's function, however, enables us to formulate an efficient method for the scattering problem, which is also discussed in the next chapter.

The derivation of Green's function in the wavenumber domain, which is used in §3.2–§3.4, is very complicated. In addition, Green's function in the wavenumber domain yields the resolvent kernel used in §3.5, whose properties are very important for the spectral representation of Green's function. Therefore, the derivation of Green's function in the wavenumber domain as well as of the resolvent kernel together with its properties are separated from Chapter 3, and summarized in Appendix C. The comparisons of Green's functions obtained from various computational methods are summarized in Appendix D.

In Chapter 4, we develop numerical methods for the scattering of the elastic waves using Green's functions derived in Chapters 2 and 3. In §4.1, we apply the representation theorem to a solid-fluid interaction problem. Discretization of the representation theorem yields a boundary-element technique. We analyze the vibration of a virtual underground energy storage system and examine its properties.

In §4.2, we apply the spectral representation of Green's function and the generalized Fourier transform obtained in §3.5 to the equation of the type of the Lippmann-Schwinger equation. We compute the scattering wavefield caused by an underground fluctuation of the wavefield. It is true that the spectral representation of Green's function as well as the generalized Fourier transforms obtained in §3.5 are rather abstract. In spite of these mathematical forms, we show that the spectral representation of Green's function and the generalized Fourier transform can provide an efficient tool for analyzing engineering problems.

In §4.3, we compute the inversion of the point-like scatterers by means of the pseudo-projection and MUSIC algorithms. The MUSIC algorithm is presented in Appendix E. The pseudo-projection method is developed by the steepest descent path method, which is considered to be of historical importance. The discussions in §4.3 show a case in which the historical method can also improve the efficiency of modern computational methods.

In this book, we could not include the discussions for wave propagation in layered media, dispersion of guided waves, the time domain solutions via the Cagniard-de Hoop method and among others. The author, however, would like to disclose every computer program used for Chapter 4, so that the readers of this book can reproduce the same results. The author would appreciate it if many engineers were interested in the methods described in Chapter 4. The software is disclosed at `https://www.rs.tus.ac.jp/~ttouhei/main.pdf`.

Acknowledgment

First of all, I would like to thank all of the faculty members of the Department of Civil Engineering, Tokyo University of Science for their support and time in completing the book. For the description of the concept of continuum mechanics, I would like to thank many discussions made by Professor Hiroshi Okada and Professor Takahiro Tsukahara, Department of Mechanical and Aerospace Engineering, Tokyo University of Science. I would like to thank Professor Keiichi Kato, Department of Mathematics, Tokyo University of Science. He invited me to participate in the Division of Research Alliance for Mathematical Analysis at the Research Institute for Science and Technology. The discussions in the Research Alliance were important for the spectral representation of Green's function for an elastic half-space and its application in scattering problems. I appreciate Professor Masayuki Nagano and Professor Yoshifumi Ohmiya, Department of Architecture, Tokyo University Science. We worked together for the Disaster Risk Management Course and conducted many discussions, which were also important for this study.

I would like to thank JOGMEC for the use of the analysis model shown in Fig. 4.1.3 in Chapter 4, as a commissioned project by the Agency for Natural Resources and Energy. I would appreciate Tokyo Electric Power Services Co., Ltd. and Hakusan Corporation in cooperation with the numerical results presented in §4.1.

I enjoyed the discussions with Professor Emeritus, Sohichi Hirose, Tokyo Institute of Technology, Professor Kazuhisa Abe, Professor Kazuhiro Koro, Niigata University, Dr. Takahiro Saitoh, Gunma University, Dr. Taizo Maruyama, Tokyo Institute of Technology, and Dr. Akira Furukawa, Hokkaido University in joint seminars on applied mechanics.

I am grateful to Mr. Tony Moore, editor, as well as Mr. Frazer Merritt and Ms. Aimee Wragg, editorial assistants at Taylor & Francis Group, for their invaluable support in initiating the book project and for their valuable advice regarding the manuscript.

Special thanks I would like to show to Professor Emeritus, Tatsuo Ohmachi, Tokyo Institute of Technology. I really appreciate his encouragement, careful reading of the draft of the manuscript, many suggestions and more.

I acknowledge that any remaining errors are solely my responsibility. I would greatly appreciate it if those are kindly pointed out.

Finally, I would like to thank my wife Kinuko. With her understanding and support, I could complete the book.

Author

Terumi Touhei received a B.S. in 1981 and an M.Eng. in 1983 from Waseda University, Japan. Following that, he served as an engineer at Sato Kogyo. Co., Ltd. until 1992. In 1991, he earned a Ph.D. from the Tokyo Institute of Technology in Japan. Since 1992, he has been a faculty member at Tokyo University of Science, currently holding the position of a professor. Professor Touhei's research interests primarily focus on solid mechanics and the representation of Green's function in an elastic wave field, particularly applied to scattering problems.

1 Introduction

1.1 BASIC CONCEPT OF CONTINUUM MECHANICS

Materials can take the form of solids, liquids, or gases. Liquids and gases, the shapes of which are easily deformed, are considered to be fluids. Gases, however, are different from liquids in that their volumes can easily change. We encounter many types of solids and fluids in our daily lives. However, the distinction is not always so obvious. For example, the Earth's mantle and glaciers are considered to be solids. However, both behave like fluids if they are observed over a long time period. Therefore, in order to investigate different phenomena, not only objects to be observed, but also the temporal and spatial scales need to be considered.

In general, the distance between molecules or atoms is about 1 nm. As a result, an enormous number of molecules or atoms are present, even in a cubic volume with side lengths of 1 μm. Therefore, a macroscopic approach to investigating deformation phenomena for solid or fluid flow is possible, in which the solid or fluid body can be assumed to be continuous. This is a starting point of continuum mechanics, and we refer to the investigated body as a *continuum body*. We also define a particle or spatial point in a continuum body as it still contains enormous number of molecules or atoms. As a consequence, the motion of a particle has to be understood as an averaged value for an enormous number of molecules or atoms in a particle under certain temporal and spatial scales.

In the context of continuum mechanics, we define physical quantities in a spatial and/or temporal coordinate system. For example, let ΔV and ΔM be the volume of a small element and the mass of the element, respectively. Then, the relationship between ΔV and ΔM can be expressed as

$$\Delta M = \rho(\boldsymbol{x})\Delta V \tag{1.1.1}$$

where ρ is the mass density at a spatial point $\boldsymbol{x}$.

In this textbook, three-dimensional (3D) solid media are considered, and a spatial point is expressed in a Cartesian coordinate system, unless otherwise stated. We generally express the components of the position vector in the following form:

$$\boldsymbol{x} = (x_1, x_2, x_3) = (x_j), \; (j = 1, 2, 3) \tag{1.1.2}$$

where the subscripted index describes the component of the coordinate system. In this chapter, however, we express the components of a vector and clarify the chosen coordinate system. For example, we use the following expression for the components of a vector:

$$\boldsymbol{x} \underset{e}{\rightarrow} (x_1, x_2, x_3) = (x_j), \; (j = 1, 2, 3) \tag{1.1.3}$$

DOI: 10.1201/9781003251729-1

where $\mathscr{O}$ denotes the coordinate system, which defines the location of the origin and the direction of the base vectors [1]. Note that a vector itself represents a physical quantity that is independent of the coordinate system. Equation (1.1.3) is convenient for determining the transformation rule for the components between two different coordinate systems, i.e., different locations of the origins as well as different directions of the base vectors. The tensor algebra necessary for continuum mechanics in this textbook is explained in Appendix A.

Here, we define the term "particle" in a continuum body. Based on the concept of a particle, we can formulate two different approaches in continuum mechanics, which are the Lagrange approach and the Euler approach. The Lagrange approach considers the trajectory of a particle in a continuum body, whereas the Euler approach considers the motion of a particle at its current coordinates. The Lagrange approach is used when the current state of a continuum body is strongly affected by its initial state, whereas, in the Euler approach, the effects of the initial state can be ignored. For wave problems in elastic solids, where the vibration amplitude is small compared to the wavelength, the difference between the Lagrange and Euler approaches is very small, which leads to a linear equation for elastic wave motion, which is the theme of this text book. This introduction describes the linear equation for elastic wave motion and its reciprocity, after the description of the strain and stress tensors. In addition, the Euler and Lagrange approaches, as well as the Reynolds transport theorem, are also discussed.

1.2 STRAIN TENSOR

1.2.1 DEFORMATION OF CONTINUUM BODY

Figure 1.2.1 shows a continuum body, which occupies a region V_0 at time $t = 0$. The body undergoes deformation and moves to a region V at time t. In this section, we deal with a finite region of the continuum body to investigate the deformation.

With respect to the Lagrange and Euler approaches, we use two coordinate systems to investigate the deformation, which are the original coordinate system denoted by $(X_i), (i = 1,2,3)$, and the current coordinate system denoted by $(x_i), (i = 1,2,3)$. We also use the following notations:

$$\boldsymbol{X} \underset{\mathscr{O}}{\rightarrow} (X_1, X_2, X_3) \tag{1.2.1}$$

$$\boldsymbol{x} \underset{\mathscr{O}}{\rightarrow} (x_1, x_2, x_3) \tag{1.2.2}$$

The original coordinate system is used to distinguish particles in the original state, whereas the current coordinate system is used to describe the location of particles at the current time. Note that the location of the origin as well as the direction of the

---

[1] We use the notation in Eq. (1.1.3) following Schutz, B.F. (1990). We discuss the concept of the tensor algebra by means of the above notation in Appendix A. We also try to provide the physical meaning of the strain and stress tensors in this chapter from the mathematical definition of tensors given in Appendix A.

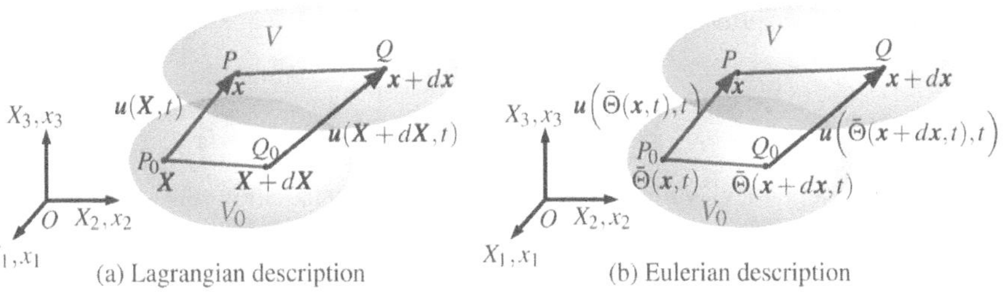

Figure 1.2.1 Deformation of a continuum body. Region V_0 undergoes deformation and moves to V. Consequently, the line element $\overline{P_0Q_0}$ in V_0 shifts to $\overline{PQ}$ in V. The original coordinate system X_i is used to distinguish particles in V_0, and the current coordinate system x_i is used to describe the location of the particle at the current time. We show the Lagrangian and Eulerian descriptions for the displacement field in (a) and (b), respectively.

base vectors for both the original and current coordinate systems can be identical, as we used the notations for Eqs. (1.2.1) and (1.2.2).

The relationship between the original and current coordinate systems is described by a map defined as:

$$\boldsymbol{x} = \Theta(\boldsymbol{X}, t) \tag{1.2.3}$$

where t is the current time. We call Eq. (1.2.3) the *Lagrangian description*, since the map Θ enables us to trace the trajectory of a particle in a continuum body. Alternatively, we can also define a map from $(\boldsymbol{X}, t)$ to its original position of a particle as

$$\boldsymbol{X} = \bar{\Theta}(\boldsymbol{x}, t) \tag{1.2.4}$$

which is called the *Eulerian description*. Equations (1.2.3) and (1.2.4) have to be consistent each other in the sense that Eq. (1.2.4) can be derived by solving Eq. (1.2.3).

As shown in Fig. 1.2.1, a line element $\overline{P_0Q_0} \in V_0$ shifts to $\overline{PQ} \in V$ due to the deformation. The position vectors for $P_0, Q_0 \in V_0$ are denoted by $\boldsymbol{X}$ and $\boldsymbol{X} + d\boldsymbol{X}$, whereas $P, Q \in V$ are $\boldsymbol{x}$ and $\boldsymbol{x} + d\boldsymbol{x}$, respectively. Vectors $\overrightarrow{P_0P}$ and $\overrightarrow{Q_0Q}$ are the displacement vectors defined by the endpoints of the line elements. We introduce a displacement field to express these vectors such that

$$\begin{aligned}
\overrightarrow{P_0P} &= \boldsymbol{u}(\boldsymbol{X}, t) = \boldsymbol{x} - \boldsymbol{X} \\
&= \Theta(\boldsymbol{X}, t) - \boldsymbol{X} \\
\overrightarrow{Q_0Q} &= \boldsymbol{u}(\boldsymbol{X} + d\boldsymbol{X}, t) = \boldsymbol{x} + d\boldsymbol{x} - (\boldsymbol{X} + d\boldsymbol{X}) \\
&= \Theta(\boldsymbol{X} + d\boldsymbol{X}, t) - (\boldsymbol{X} + d\boldsymbol{X})
\end{aligned} \tag{1.2.5}$$

where $\boldsymbol{u}$ denotes the displacement field. We applied the Lagrangian description to the displacement field, where the independent variables $(\boldsymbol{X}, t)$ are used. We can also use the Eulerian description to the displacement field by

$$\begin{aligned} \overrightarrow{P_0P} &= \boldsymbol{u}\left(\bar{\Theta}(\boldsymbol{x},t),t\right) = \boldsymbol{x} - \bar{\Theta}(\boldsymbol{x},t) \\ \overrightarrow{Q_0Q} &= \boldsymbol{u}\left(\bar{\Theta}(\boldsymbol{x}+d\boldsymbol{x},t),t\right) = \boldsymbol{x}+d\boldsymbol{x} - \bar{\Theta}(\boldsymbol{x}+d\boldsymbol{x},t) \end{aligned} \tag{1.2.6}$$

where the independent variables $(\boldsymbol{x},t)$ are used. We also show the concept of the Lagrangian and Eulerian descriptions of the displacement field in Figs. 1.2.1 (a) and (b), respectively.

1.2.2 DEFINITION OF STRAIN TENSOR

It is necessary to have a quantitative measure of the degree of deformation that enables us to describe the internal states of the body. This is expressed in the following form:

$$\eta := \frac{|\overline{PQ}| - |\overline{P_0Q_0}|}{|\overline{P_0Q_0}|} = \frac{|d\boldsymbol{x}| - |d\boldsymbol{X}|}{|d\boldsymbol{X}|} \tag{1.2.7}$$

which is the ratio of the change in length to the original length of the line element (see Fig. 1.2.1). We will derive the expression for η in terms of the components of the coordinate system. By means of the properties of η, which is a physical quantity independent of the coordinate system, we will proceed to the concept of the strain tensor.

We start with an evaluation of the difference between the squared lengths of the line elements before and after deformation:

$$|d\boldsymbol{x}|^2 - |d\boldsymbol{X}|^2$$

because this is related to η by the following equation:

$$\begin{aligned} |d\boldsymbol{x}|^2 - |d\boldsymbol{X}|^2 &= \frac{|d\boldsymbol{x}| - |d\boldsymbol{X}|}{|d\boldsymbol{X}|} \frac{|d\boldsymbol{x}| - |d\boldsymbol{X}| + 2|d\boldsymbol{X}|}{|d\boldsymbol{X}|} |d\boldsymbol{X}|^2 \\ &= \eta(\eta+2)\,|d\boldsymbol{X}|^2 \end{aligned} \tag{1.2.8}$$

For $|\eta| \ll 1$, Eq. (1.2.8) can be simplified as

$$\eta = (1/2)\,\frac{|d\boldsymbol{x}|^2 - |d\boldsymbol{X}|^2}{|d\boldsymbol{X}|^2} \tag{1.2.9}$$

Now, let the components of $d\boldsymbol{x}$ and $d\boldsymbol{X}$ be expressed by

$$\begin{aligned} d\boldsymbol{x} &\underset{\mathcal{O}}{\rightarrow} (dx_1, dx_2, dx_3) \\ d\boldsymbol{X} &\underset{\mathcal{O}}{\rightarrow} (dX_1, dX_2, dX_3) \end{aligned} \tag{1.2.10}$$

Then, the difference in the squared lengths can be expressed as

$$\begin{aligned} |d\boldsymbol{x}|^2 - |d\boldsymbol{X}|^2 &= dx_i dx_i - dX_i dX_i \\ &= \left(\frac{\partial x_i}{\partial X_k}\frac{\partial x_i}{\partial X_l} - \delta_{kl}\right) dX_k dX_l \\ &= 2L_{kl}\,dX_k dX_l \end{aligned} \tag{1.2.11}$$

where

$$L_{kl} = (1/2)\left(\frac{\partial x_i}{\partial X_k}\frac{\partial x_i}{\partial X_l} - \delta_{kl}\right) \tag{1.2.12}$$

Note that the summation convention for the subscripted index (see Appendix A) is applied to Eq. (1.2.11). In the following, the summation convention is also applied to subscripted indices for all of the equations, unless otherwise stated. For the derivation of Eq. (1.2.11), we use the following equation:

$$dx_i = \frac{\partial x_i}{\partial X_k} dX_k \tag{1.2.13}$$

Equation (1.2.11) can be interpreted as a linear mapping of two vectors into a real number, if we distinguish dX_k and dX_l as different vectors. In this sense, L_{kl} is a component of a rank-2 tensor. We call L_{kl} the Lagrangian (Green) finite strain tensor. Alternatively, if we apply the following equation:

$$dX_i = \frac{\partial X_i}{\partial x_k} dx_k \tag{1.2.14}$$

to the difference between the squared lengths, then we have

$$\begin{aligned} |d\boldsymbol{x}|^2 - |d\boldsymbol{X}|^2 &= dx_i dx_i - dX_i dX_i \\ &= \left(\delta_{kl} - \frac{\partial X_i}{\partial x_k}\frac{\partial X_i}{\partial x_l}\right) dx_k dx_l \\ &= 2E_{kl} dx_k dx_l \end{aligned} \tag{1.2.15}$$

where E_{kl} is defined by

$$E_{kl} = (1/2)\left(\delta_{kl} - \frac{\partial X_i}{\partial x_k}\frac{\partial X_i}{\partial x_l}\right) \tag{1.2.16}$$

As in the case for L_{kl} in Eq. (1.2.11), E_{kl} can be regarded as a component of a rank-2 tensor. We refer to E_{kl} as the Eulerian (Almansi) finite strain tensor. The difference between Eqs. (1.2.12) and (1.2.16) is in the use of independent variables. Namely, the original coordinate system $(\boldsymbol{X}, t)$ is used for Eq. (1.2.12), whereas the current coordinate system $(\boldsymbol{x}, t)$ is used for Eq. (1.2.16), yielding the strain tensors based on the Lagrange and the Euler approaches, respectively.

In order to analyze wave propagation in a continuum body, the governing equation uses the displacement field as an unknown function to be solved. Therefore, it is necessary to determine the relationship between the strain tensor and the displacement field. According to Eq. (1.2.5), we have

$$x_i = u_i + X_i \tag{1.2.17}$$

and as a result,

$$\frac{\partial x_i}{\partial X_k} = \frac{\partial u_i}{\partial X_k} + \delta_{ik} \tag{1.2.18}$$

Therefore, the Green finite strain tensor can be expressed as

$$L_{kl} = (1/2)\left(\frac{\partial u_k}{\partial X_l} + \frac{\partial u_l}{\partial X_k} + \frac{\partial u_i}{\partial X_l}\frac{\partial u_i}{\partial X_k}\right) \tag{1.2.19}$$

Likewise, from

$$X_i = x_i - u_i \tag{1.2.20}$$

we obtain the expression for the Eulerian strain tensor in the following form:

$$E_{kl} = (1/2)\left(\frac{\partial u_k}{\partial x_l} + \frac{\partial u_l}{\partial x_k} - \frac{\partial u_i}{\partial x_l}\frac{\partial u_i}{\partial x_k}\right) \tag{1.2.21}$$

The relationship of the gradient of the displacement field between the Lagrange and Euler approaches is expressed as

$$\begin{aligned}\frac{\partial u_k}{\partial x_l} &= \frac{\partial u_k}{\partial X_j}\frac{\partial X_j}{\partial x_l} \\ &= \frac{\partial u_k}{\partial X_j}\left(\delta_{jl} - \frac{\partial u_j}{\partial x_l}\right) \\ &= \frac{\partial u_k}{\partial X_l} - \frac{\partial u_k}{\partial X_j}\frac{\partial u_j}{\partial x_l}\end{aligned} \tag{1.2.22}$$

Therefore, for a case that the gradient of the displacement field is very small, namely

$$\left|\frac{\partial u_k}{\partial X_j}\right| \ll 1, \quad \left|\frac{\partial u_k}{\partial x_l}\right| \ll 1 \tag{1.2.23}$$

we can proceed our discussions by assuming that

$$\frac{\partial u_k}{\partial x_l} = \frac{\partial u_k}{\partial X_l} \tag{1.2.24}$$

Equation (1.2.24) shows that the differences between the Lagrange and Euler approaches are very small if Eq. (1.2.23) holds. For this case, Eqs. (1.2.19) and (1.2.21) yield:

$$\varepsilon_{ij} = (1/2)\left(\frac{\partial u_j}{\partial x_i} + \frac{\partial u_i}{\partial x_j}\right) \tag{1.2.25}$$

which is referred to as the infinitesimal strain tensor. As for the Green and Almansi strain tensors, ε_{ij} can be also regarded as a rank-2 tensor. The difference between the squared lengths of the line elements in terms of the infinitesimal strain tensor is expressed as

$$|d\mathbf{x}|^2 - |d\mathbf{X}|^2 = 2\varepsilon_{ij}dx_i dx_j \tag{1.2.26}$$

1.2.3 CHARACTERIZATION OF INFINITESIMAL STRAIN TENSOR

Now, let us discuss the properties of the infinitesimal strain tensor. For the discussion, we do not distinguish between the original and current coordinate systems, because we consider a situation in which the difference between the results of the Euler and Lagrange approaches is very small. Under these circumstances, we simply use the spatial derivative $\partial/\partial x_j$. According to Eqs. (1.2.9) and (1.2.26), the relationship between the parameter η and the infinitesimal strain tensor is

$$\eta = \frac{\varepsilon_{ij} dx_i dx_j}{|\boldsymbol{dx}|^2} \tag{1.2.27}$$

At this point, we have two tasks related to Eq. (1.2.27). One is to connect the tensor algebra discussed in Appendix A with Eq. (1.2.27). We have to determine the transformation rule for the components of the infinitesimal strain tensor for the different coordinate systems. Note that we do not distinguish between the original and current coordinate systems for the discussion of the infinitesimal strain tensor. At this point, the transformation rule is considered for different base vectors that span the continuum body. The second task is to connect the infinitesimal strain tensor with the concept of strain learned in undergraduate engineering mechanics.

The starting point of the discussion is Eq. (1.2.27). We consider two coordinate systems $O\{\boldsymbol{e}_1, \boldsymbol{e}_2, \boldsymbol{e}_3\}$, $O'\{\boldsymbol{e}'_1, \boldsymbol{e}'_2, \boldsymbol{e}'_3\}$ and we express the components of the direction vector in the following form:

$$\begin{aligned} \frac{\boldsymbol{dx}}{|\boldsymbol{dx}|} &\underset{\mathscr{O}}{\rightarrow} (d_1, d_2, d_3) \\ \frac{\boldsymbol{dx}}{|\boldsymbol{dx}|} &\underset{\mathscr{O}'}{\rightarrow} (d'_1, d'_2, d'_3) \end{aligned} \tag{1.2.28}$$

Note that the origin of the two coordinate systems $\mathscr{O}$ and $\mathscr{O}'$ is identical for the discussion.

As shown in Appendix A, the transformation rule for the components of the direction vector is

$$d_j = a_{ij} d'_i \tag{1.2.29}$$

where a_{ij} is defined by

$$a_{ij} = \boldsymbol{e}'_i \cdot \boldsymbol{e}_j \tag{1.2.30}$$

In addition, since the left-hand side of Eq. (1.2.27) is independent of the coordinate system, the parameter η can also be expressed as

$$\eta = \varepsilon'_{ij} d'_i d'_j \tag{1.2.31}$$

As a result, we have

$$\eta = \varepsilon'_{ij} d'_i d'_j = \varepsilon_{kl} a_{ik} a_{lj} d'_i d'_j \tag{1.2.32}$$

which yields

$$d'_i d'_j (\varepsilon'_{ij} - \varepsilon_{kl} a_{ik} a_{jl}) = 0 \Longleftrightarrow \varepsilon'_{ij} = a_{ik} a_{jl} \varepsilon_{kl} \tag{1.2.33}$$

from which we find that ε_{ij} follows the transformation for a rank-2 tensor. Therefore, we can define a linear mapping of two vectors $\boldsymbol{\varepsilon}(\cdot,\cdot)$, the components of which are defined by

$$\boldsymbol{\varepsilon}(\boldsymbol{e}_i, \boldsymbol{e}_j) = \varepsilon_{ij} \tag{1.2.34}$$

which is called the strain tensor. For the case in which $\boldsymbol{\varepsilon}$ takes the arguments of direction vectors, $\boldsymbol{\varepsilon}$ provides a clear physical meaning of

$$\boldsymbol{\varepsilon}\left(\frac{d\boldsymbol{x}}{|d\boldsymbol{x}|}, \frac{d\boldsymbol{x}}{|d\boldsymbol{x}|}\right) = \eta \tag{1.2.35}$$

Here, we consider the components of the strain tensor in more detail. At an undergraduate level, normal strain is defined as the ratio of the change in length to the original length, whereas shear strain is defined as the decrease in angle from $\pi/2$. Figure 1.2.2 shows the concept of normal and shear strains in the $x_1 - x_2$ plane. We now connect these concepts with the components of the strain tensor using ε_{11} and ε_{12}. According to Eq. (1.2.25), the strain tensor component ε_{11} is

$$\varepsilon_{11} = \lim_{\Delta x_1 \to 0} \frac{u_1(\boldsymbol{x} + \Delta x_1 \boldsymbol{e}_1) - u_1(\boldsymbol{x})}{\Delta x_1} \tag{1.2.36}$$

We see that $u_1(\boldsymbol{x} + \Delta x_1 \boldsymbol{e}_1) - u_1(\boldsymbol{x})$ in Eq. (1.2.36) is the change in length, and Δx_1 is the original length of the line element, which are δ and L, respectively, in Fig. 1.2.2. Therefore, ε_{11} corresponds to the normal strain for the x_1 direction.

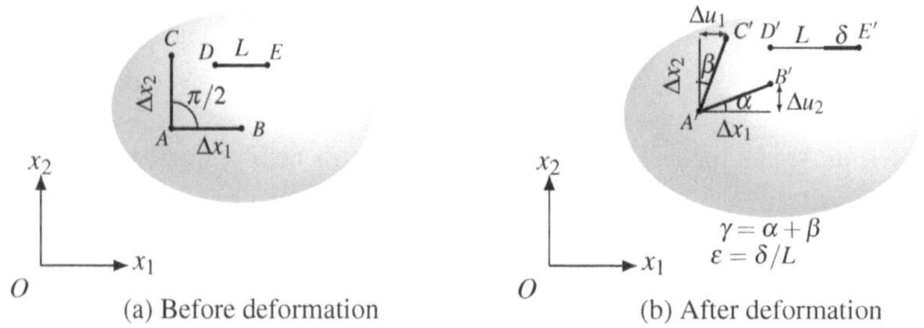

Figure 1.2.2 Normal and shear strains commonly taught at the undergraduate level. Normal strain ε is defined as the ratio of the change in length δ to the original length L. Shear strain γ is defined as the decrease in angle from $\pi/2$. Note that points from A to E shift to A' to E', respectively, due to deformation.

The component ε_{12} is described by

$$\begin{aligned}\varepsilon_{12} &= (1/2)\Big(\lim_{\Delta x_1\to 0}\frac{\Delta u_2(\boldsymbol{x})}{\Delta x_1}+\lim_{\Delta x_2\to 0}\frac{\Delta u_1(\boldsymbol{x})}{\Delta x_2}\Big)\\ &= (1/2)\Big(\lim_{\Delta x_1\to 0}\frac{u_2(\boldsymbol{x}+\Delta x_1\boldsymbol{e}_1)-u_2(\boldsymbol{x})}{\Delta x_1}\\ &\quad+\lim_{\Delta x_2\to 0}\frac{u_1(\boldsymbol{x}+\Delta x_2\boldsymbol{e}_2)-u_1(\boldsymbol{x})}{\Delta x_2}\Big)\end{aligned} \tag{1.2.37}$$

According to Fig. 1.2.2, we find that

$$\begin{aligned}\tan\alpha &= \frac{\Delta u_2(\boldsymbol{x})}{\Delta x_1}\\ \tan\beta &= \frac{\Delta u_1(\boldsymbol{x})}{\Delta x_2}\end{aligned} \tag{1.2.38}$$

For $|\alpha| << 1$ and $|\beta| << 1$, we have

$$\alpha \sim \tan\alpha,\ \beta \sim \tan\beta \tag{1.2.39}$$

which indicates that ε_{12} represents half of the shear strain γ in the $x_1 - x_2$ plane. In general, we see that the diagonal component of the strain tensor ε_{ii} (summation convention is not applied) denotes the normal strain in the ith direction, whereas the off-diagonal element ε_{ij}, $(i \neq j)$ corresponds to the shear strain in the $x_i - x_j$ plane.

1.2.4 VOLUMETRIC STRAIN BASED ON THE INFINITESIMAL STRAIN TENSOR

The next example describes the use of the infinitesimal strain tensor to investigate changes in a volume element due to deformation. In Fig. 1.2.3(a), there is a parallelepiped spanned by $\overrightarrow{AB}$, $\overrightarrow{AC}$, and $\overrightarrow{AD}$, which is a volume element before deformation. Due to deformation, points A, B, C, and D shift to A', B', C', and D', respectively,

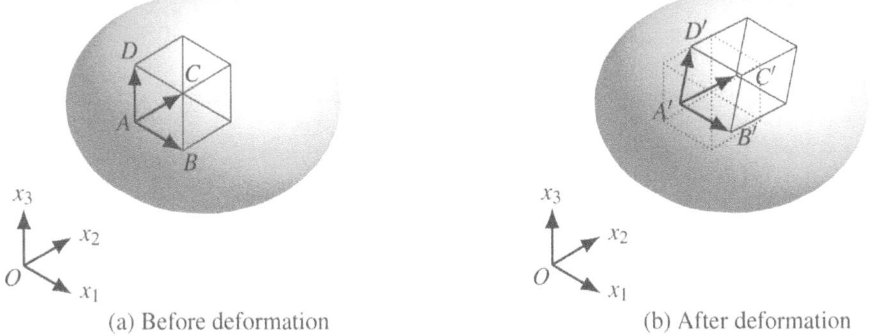

Figure 1.2.3 Deformation of volume element in continuum body. Note that points from A to D shift to A' to D' due to deformation.

and vectors $\overrightarrow{A'B'}$, $\overrightarrow{A'C'}$, and $\overrightarrow{A'D'}$ span the parallelepiped after deformation (see Fig. 1.2.3). Let the position vectors for A, B, C and D be $\boldsymbol{x}$, $\boldsymbol{x}+d\boldsymbol{x}_1$, $\boldsymbol{x}+d\boldsymbol{x}_2$, and $\boldsymbol{x}+d\boldsymbol{x}_3$, respectively. After deformation, position vectors A', B', C', and $'D$ shift to

$$\begin{aligned} A' &: \boldsymbol{x}+\boldsymbol{u}(\boldsymbol{x}) \\ B' &: \boldsymbol{x}+d\boldsymbol{x}_1+\boldsymbol{u}(\boldsymbol{x}+d\boldsymbol{x}_1) \\ C' &: \boldsymbol{x}+d\boldsymbol{x}_2+\boldsymbol{u}(\boldsymbol{x}+d\boldsymbol{x}_2) \\ D' &: \boldsymbol{x}+d\boldsymbol{x}_3+\boldsymbol{u}(\boldsymbol{x}+d\boldsymbol{x}_3) \end{aligned} \tag{1.2.40}$$

where $\boldsymbol{u}$ is the displacement field. The volume of the parallelepiped dV before deformation is

$$dV = \left| \begin{array}{ccc} d\boldsymbol{x}_1 & d\boldsymbol{x}_2 & d\boldsymbol{x}_3 \end{array} \right| = dx_1 dx_2 dx_3 \tag{1.2.41}$$

where the components $d\boldsymbol{x}_j$, $(j=1,2,3)$ are

$$\begin{aligned} d\boldsymbol{x}_1 &\underset{\mathcal{O}}{\rightarrow} \left(\begin{array}{ccc} dx_1 & 0 & 0 \end{array} \right) \\ d\boldsymbol{x}_2 &\underset{\mathcal{O}}{\rightarrow} \left(\begin{array}{ccc} 0 & dx_2 & 0 \end{array} \right) \\ d\boldsymbol{x}_3 &\underset{\mathcal{O}}{\rightarrow} \left(\begin{array}{ccc} 0 & 0 & dx_3 \end{array} \right) \end{aligned} \tag{1.2.42}$$

Note that $\left| \begin{array}{ccc} d\boldsymbol{x}_1 & d\boldsymbol{x}_2 & d\boldsymbol{x}_3 \end{array} \right|$ is the triple scalar product defined by these three vectors (Appendix-A). After deformation, the parallelepiped is spanned by $d\boldsymbol{x}'_j$, $(j=1,2,3)$ the components of which are

$$\begin{aligned} d\boldsymbol{x}'_1 &= d\boldsymbol{x}_1+\boldsymbol{u}(\boldsymbol{x}+d\boldsymbol{x}_1)-\boldsymbol{u}(\boldsymbol{x}) \\ &\underset{\mathcal{O}}{\rightarrow} dx_1 \frac{\partial}{\partial x_1}\left(\begin{array}{ccc} u_1(\boldsymbol{x}) & u_2(\boldsymbol{x}) & u_3(\boldsymbol{x}) \end{array} \right) + dx_1 \left(\begin{array}{ccc} 1 & 0 & 0 \end{array} \right) \\ d\boldsymbol{x}'_2 &= d\boldsymbol{x}_2+\boldsymbol{u}(\boldsymbol{x}+d\boldsymbol{x}_2)-\boldsymbol{u}(\boldsymbol{x}) \\ &\underset{\mathcal{O}}{\rightarrow} dx_2 \frac{\partial}{\partial x_2}\left(\begin{array}{ccc} u_1(\boldsymbol{x}) & u_2(\boldsymbol{x}) & u_3(\boldsymbol{x}) \end{array} \right) + dx_2 \left(\begin{array}{ccc} 0 & 1 & 0 \end{array} \right) \\ d\boldsymbol{x}'_3 &= d\boldsymbol{x}_3+\boldsymbol{u}(\boldsymbol{x}+d\boldsymbol{x}_3)-\boldsymbol{u}(\boldsymbol{x}) \\ &\underset{\mathcal{O}}{\rightarrow} dx_3 \frac{\partial}{\partial x_3}\left(\begin{array}{ccc} u_1(\boldsymbol{x}) & u_2(\boldsymbol{x}) & u_3(\boldsymbol{x}) \end{array} \right) + dx_3 \left(\begin{array}{ccc} 0 & 0 & 1 \end{array} \right) \end{aligned} \tag{1.2.43}$$

Therefore, the volume of the parallelepiped after deformation is

$$\begin{aligned} dV' &= \left| \begin{array}{ccc} d\boldsymbol{x}'_1 & d\boldsymbol{x}'_2 & d\boldsymbol{x}'_3 \end{array} \right| \\ &= dx_1 dx_2 dx_3 \left(1 + \frac{\partial}{\partial x_i} u_i(\boldsymbol{x}) \right) \end{aligned} \tag{1.2.44}$$

Now, we can define the volumetric strain, which is the ratio of the change in volume to the original volume, as follows:

$$\begin{aligned} \eta_v(\boldsymbol{x}) &= \frac{dV' - dV}{dV} \\ &= \frac{\partial}{\partial x_i} u_i(\boldsymbol{x}) = \varepsilon_{ii}(\boldsymbol{x}) \end{aligned} \tag{1.2.45}$$

Note that we apply the summation convention to ε_{ii}, namely

$$\varepsilon_{ii} = \varepsilon_{11} + \varepsilon_{22} + \varepsilon_{33} \tag{1.2.46}$$

According to the definition, the volumetric strain should be independent of the coordinate system. Let us check this fact by means of the transformation rule for a component of the rank-2 tensor:

$$\varepsilon'_{ij} = a_{ik} a_{jl} \varepsilon_{kl}$$

Then, we have

$$\begin{aligned} \varepsilon'_{ii} &= a_{ik} a_{il} \varepsilon_{kl} \\ &= \delta_{kl} \varepsilon_{kl} = \varepsilon_{ll} \end{aligned} \tag{1.2.47}$$

For the derivation of Eq. (1.2.47), we use Eq. (A.2.11).

1.3 EQUATION OF MOTION FOR CONTINUUM BODY AND CONCEPT OF STRESS TENSOR

1.3.1 MATERIAL DERIVATIVE AND REYNOLDS TRANSPORT THEOREM

In this section, we start with the material derivative and the Reynolds transport theorem, which are applied to the equation of motion for a continuum body. As we will see later, we will proceed to the definition of the stress tensor by investigating the internal fore for the equation of motion for a continuum body. We return to a situation again where we have to distinguish between the original and the current coordinate systems.

Figure 1.3.1 shows an example of a continuum body, the volume and boundary of which are denoted by V and S, respectively. Likewise, as in the discussion for

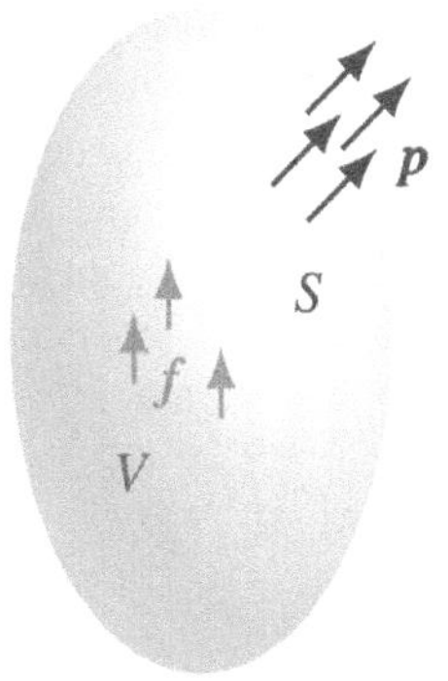

Figure 1.3.1 Volume V and its surface S to establish equilibrium equation. The body force f and traction $\boldsymbol{p}$ act on V and S, respectively.

the strain tensor, we also consider a finite region of the continuum body. We assume that the body force $\boldsymbol{f}$ and traction $\boldsymbol{p}$, which are external forces, act on V and S, respectively. The body force corresponds to action at a distance, such as that due to gravity, whereas traction is caused by contact with the surface of the body. The body force is expressed per unit volume and the traction is expressed per unit area. Both are described in the form of vectors. The equilibrium equation for the continuum body due to these forces is

$$\frac{d}{dt}\int_V \rho \boldsymbol{v}\, dV = \int_S \boldsymbol{p}\, dS + \int_V \boldsymbol{f}\, dV \tag{1.3.1}$$

where ρ is the mass density, and $\boldsymbol{v}$ is the velocity of a particle. Note that the left-hand side of Eq. (1.3.1) shows the rate of change in momentum with respect to time, which corresponds to the actions on the body due to external forces. We see that $\int_V \rho \boldsymbol{v} dV$ is a function of time as a result of the spatial integral. Therefore, we apply an ordinary differential operator d/dt to the result of the volume integral. The equilibrium equation can also be expressed by vector components in the following form:

$$\frac{d}{dt}\int_V \rho v_i dV = \int_S p_i dS + \int_V f_i dV \tag{1.3.2}$$

where the particle velocity and forces are expressed by

$$\begin{aligned} \boldsymbol{v} &\underset{\mathscr{O}}{\rightarrow} \begin{pmatrix} v_1 & v_2 & v_3 \end{pmatrix} \\ \boldsymbol{p} &\underset{\mathscr{O}}{\rightarrow} \begin{pmatrix} p_1 & p_2 & p_3 \end{pmatrix} \\ \boldsymbol{f} &\underset{\mathscr{O}}{\rightarrow} \begin{pmatrix} f_1 & f_2 & f_3 \end{pmatrix} \end{aligned} \tag{1.3.3}$$

Note that we assume the Eulerian description for Eqs. (1.3.1) and (1.3.2).

We have to be aware that the volume V changes its configuration with time. In this sense, distinction of the Lagrange and Euler approaches is also necessary to treat the rate of change in momentum exactly for Eqs. (1.3.1) and (1.3.2). We also have to be careful about interchanging the order of the integral and the operator (d/dt). Here, we resolve the problem of interchanging the order of the integral and the operator (d/dt) by introducing the material derivative and the Reynolds transport theorem.

Let χ be a physical quantity, such as the temperature or mass density in a continuum body. We use the current coordinate system $(\boldsymbol{x},t)$ for the independent variables for χ, which is the Euler approach. If the particle velocity is expressed as

$$\boldsymbol{v} \underset{\mathscr{O}}{\rightarrow} \begin{pmatrix} v_1 & v_2 & v_3 \end{pmatrix} \tag{1.3.4}$$

then the derivative of χ with respect to time is expressed as

$$\lim_{\Delta t \to 0} \frac{\chi(\boldsymbol{x}+\boldsymbol{v}\Delta t, t+\Delta t) - \chi(\boldsymbol{v},t)}{\Delta t} = \left.\frac{\partial \chi(\boldsymbol{x},t)}{\partial t}\right|_{\boldsymbol{x}\,\text{fixed}} + v_k \left.\frac{\partial \chi(\boldsymbol{x},t)}{\partial x_k}\right|_{t:\text{fixed}} \tag{1.3.5}$$

by means of the chain rule. According to Eq. (1.3.5), we define the derivative of χ as a material derivative of the following form:

$$\frac{D\chi(\boldsymbol{x},t)}{Dt} := \frac{\partial\chi(\boldsymbol{x},t)}{\partial t} + v_k\frac{\partial\chi(\boldsymbol{x},t)}{\partial x_k} \tag{1.3.6}$$

where D/Dt is the operator for the material derivative. It is possible to express the material derivative in terms of the original coordinate system. Application of the Lagrangian description shown in Eq. (1.2.3) to the material derivative yields

$$\frac{D}{Dt}\chi\big(\Theta(\boldsymbol{X},t),t\big) = \frac{\partial}{\partial t}\chi\big(\Theta(\boldsymbol{X},t),t\big)\bigg|_{\boldsymbol{X}:\text{fixed}} \tag{1.3.7}$$

As an example of the material derivative, let us investigate $D\boldsymbol{x}/Dt$. In terms of the current coordinate system, we have

$$\frac{D\boldsymbol{x}}{Dt} = \frac{\partial\boldsymbol{x}}{\partial t}\bigg|_{\boldsymbol{x}:\text{fixed}} + v_k\frac{\partial\boldsymbol{x}}{\partial x_k}\bigg|_{t:\text{fixed}} = \boldsymbol{v} \tag{1.3.8}$$

since

$$\frac{\partial\boldsymbol{x}}{\partial t}\bigg|_{\boldsymbol{x}:\text{fixed}} = \boldsymbol{0} \tag{1.3.9}$$

$$\frac{\partial\boldsymbol{x}}{\partial x_k}\bigg|_{t:\text{fixed}} \underset{\mathcal{C}}{\rightarrow} \begin{pmatrix} \delta_{1k} & \delta_{2k} & \delta_{3k} \end{pmatrix} \tag{1.3.10}$$

In terms of the original coordinate system

$$\frac{D\boldsymbol{x}}{Dt} = \frac{\partial\boldsymbol{x}}{\partial t}\bigg|_{\boldsymbol{X}:\text{fixed}} \tag{1.3.11}$$

Equations (1.3.8) and (1.3.11) show the meanings of the material derivative and the particle velocity, respectively.

Based on the above discussion, let us consider the following integral

$$F(t) = \frac{d}{dt}\int_V \chi(\boldsymbol{x},t)dV \tag{1.3.12}$$

for the interchange of the order of the differential and integral operators. We will see that this interchange is resolved by the following Reynolds transport theorem:

Theorem 1.1 *The Reynolds transport theorem ensures the interchange of the differential and integral operators by*

$$\frac{d}{dt}\int_V \chi(\boldsymbol{x},t)dV = \int_V\Big(\frac{D}{Dt}\chi(\boldsymbol{x},t) + \chi(\boldsymbol{x},t)\,div\,\boldsymbol{v}\Big)dV \tag{1.3.13}$$

where

$$div\,\boldsymbol{v} = \frac{\partial v_k}{\partial x_k}\Big(= \frac{\partial v_1}{\partial x_1} + \frac{\partial v_2}{\partial x_2} + \frac{\partial v_3}{\partial x_3}\Big) \tag{1.3.14}$$

[Proof]: We express the integral shown in Eq. (1.3.12) in terms of the original coordinate system by means of

$$dV = JdV_0 \tag{1.3.15}$$

where J is the Jacobian

$$J = \begin{vmatrix} \frac{\partial x_1}{\partial X_1} & \frac{\partial x_1}{\partial X_2} & \frac{\partial x_1}{\partial X_3} \\ \frac{\partial x_2}{\partial X_1} & \frac{\partial x_2}{\partial X_2} & \frac{\partial x_2}{\partial X_3} \\ \frac{\partial x_3}{\partial X_1} & \frac{\partial x_3}{\partial X_2} & \frac{\partial x_3}{\partial X_3} \end{vmatrix} \tag{1.3.16}$$

and V_0 is the original configuration of the volume at time $t = 0$. Then, Eq. (1.3.12) can be modified to

$$\begin{aligned} \frac{d}{dt}\int_V \chi(\boldsymbol{x},t)dV &= \frac{d}{dt}\int_{V_0} \chi(\Theta(\boldsymbol{X},t),t)JdV_0 \\ &= \int_{V_0} \frac{D}{Dt}\Big(\chi(\Theta(\boldsymbol{X},t),t)J\Big)dV_0 \\ &= \int_{V_0} \Big(J\frac{D}{Dt}\chi(\Theta(\boldsymbol{X},t),t) + \chi(\Theta(\boldsymbol{X},t),t)\frac{D}{Dt}J\Big)dV_0 \end{aligned} \tag{1.3.17}$$

For the derivation of Eq. (1.3.17), we use

$$\frac{D}{Dt}(\chi J) = J\frac{D\chi}{Dt} + \chi\frac{DJ}{Dt} \tag{1.3.18}$$

We still have to express Eq. (1.3.17) in terms of the current coordinate system by introducing the following[2]

$$\frac{DJ}{Dt} = (\operatorname{div}\boldsymbol{v})J \tag{1.3.19}$$

The derivation of Eq. (1.3.19) uses the following:

$$\frac{D}{Dt}\frac{\partial x_i}{\partial X_j} = \frac{\partial}{\partial t}\left(\frac{\partial x_i}{\partial X_j}\right)\Bigg|_{\boldsymbol{X}:\text{fixed}} = \frac{\partial v_i}{\partial X_j} = \frac{\partial v_i}{\partial x_k}\frac{\partial x_k}{\partial X_j} \tag{1.3.20}$$

Therefore, the right-hand side of Eq. (1.3.17) can be expressed in the current coordinate system as:

$$\frac{d}{dt}\int_V \chi(\boldsymbol{x},t)dV = \int_V \Big(\frac{D}{Dt}\chi(\boldsymbol{x},t) + \chi(\boldsymbol{x},t)\operatorname{div}\boldsymbol{v}\Big)dV \tag{1.3.21}$$

which is the Reynolds transport theorem. □

---

[2]Verify Eq. (1.3.19). Problem §1.5.3.

Note that the Reynolds transport theorem yields the well-known continuity equation. The mass conservation law tells us that

$$\int_V \rho(\boldsymbol{x},t)dV = \text{Const.} \tag{1.3.22}$$

regardless of time. Therefore, according to the Reynolds transport theorem, we have

$$0 = \frac{d}{dt}\int_V \rho(\boldsymbol{x},t)dV = \int_V \left(\frac{D\rho}{Dt} + \rho\,\text{div}\boldsymbol{v}\right)dV \tag{1.3.23}$$

Since V can be arbitrary, we have

$$\frac{D\rho}{Dt} + \rho\,\text{div}\boldsymbol{v} = 0 \tag{1.3.24}$$

which is known as the continuity equation. Using the continuity equation and the Reynolds transport theorem, the left-hand side of Eq. (1.3.1) becomes

$$\begin{aligned} \frac{d}{dt}\int_V \rho\boldsymbol{v}dV &= \int_V \left(\rho\frac{D\boldsymbol{v}}{Dt} + \boldsymbol{v}\frac{D\rho}{Dt} + \rho\boldsymbol{v}\,\text{div}\boldsymbol{v}\right)dV \\ &= \int_V \rho\frac{D\boldsymbol{v}}{Dt}dV \end{aligned} \tag{1.3.25}$$

Therefore, the equations of motion shown in Eqs. (1.3.1) and (1.3.2) are also expressed as

$$\int_V \rho\frac{D\boldsymbol{v}}{Dt}dV = \int_S \boldsymbol{p}\,dS + \int_V \boldsymbol{f}\,dV \tag{1.3.26}$$

and

$$\int_V \rho\frac{Dv_i}{Dt}dV = \int_S p_i dS + \int_V f_i dV \tag{1.3.27}$$

respectively.

1.3.2 CAUCHY'S RELATION AND DEFINITION OF STRESS TENSOR

Often, it is necessary to investigate the inside of a continuum body by computing the internal forces. For this purpose, the body is virtually divided into pieces. Figure 1.3.2(b) shows an example of dividing the body using a cross section Π. As shown in Fig. 1.3.2(b), traction $\boldsymbol{p}_+(\boldsymbol{x},t)$ and $\boldsymbol{p}_-(\boldsymbol{x},t)$, $\boldsymbol{x}\in\Pi$ are required for V_+ and V_- to have divided bodies that satisfy the following equilibrium equations:

$$\begin{aligned} \int_{V_+} \rho\frac{D\boldsymbol{v}(\boldsymbol{x},t)}{Dt}dV &= \int_{S_+} \boldsymbol{p}(\boldsymbol{x},t)dS + \int_\Pi \boldsymbol{p}_+(\boldsymbol{x},t)d\Pi + \int_{V_+} \boldsymbol{f}(\boldsymbol{x},t)dV \\ \int_{V_-} \rho\frac{D\boldsymbol{v}(\boldsymbol{x},t)}{Dt}dV &= \int_{S_-} \boldsymbol{p}(\boldsymbol{x},t)dS + \int_\Pi \boldsymbol{p}_-(\boldsymbol{x},t)d\Pi + \int_{V_-} \boldsymbol{f}(\boldsymbol{x},t)dV \end{aligned} \tag{1.3.28}$$

where

$$V = V_+ \cup V_-, \quad S = S_+ \cup S_- \tag{1.3.29}$$

Traction $\boldsymbol{p}_+(\boldsymbol{x},t)$ can be recognized as an action from the volume V_-, and $\boldsymbol{p}_-(\boldsymbol{x},t)$ as an action from V_+. Based on Newton's third law of motion, which describes action and reaction, the traction $\boldsymbol{p}_+(\boldsymbol{x},t)$ and $\boldsymbol{p}_-(\boldsymbol{x},t)$, as an action and reaction, have to satisfy

$$\boldsymbol{p}_+(\boldsymbol{x},t) + \boldsymbol{p}_-(\boldsymbol{x},t) = \boldsymbol{0}, \ \boldsymbol{x} \in \Pi \tag{1.3.30}$$

It is evident that the traction vector defined on the cross section depends on the normal vector for the cross section. In other words, if we set another cross section with a different normal vector, then we obtain a different traction vector. Therefore, in order to describe the internal force in a continuum body, the relationship between the normal vector and the traction vector at the cross section is necessary.

Figure 1.3.3 shows the volume element in the form of a tetrahedron taken from the continuum body shown in Fig. 1.3.2. We set a coordinate system for the tetrahedron such that the coordinate axes coincide with the ridge lines. The coordinates of the vertices of the tetrahedron are A$(a,0,0)$, B$(0,b,0)$, and C$(0,0,c)$, where a, b, and c are positive real numbers. In addition, internal forces acting on each side of the tetrahedron, the positive direction of which is the tensile side, are shown in Fig. 1.3.3. The notation σ_{ij} indicates the force per unit area acting on the x_i plane in the

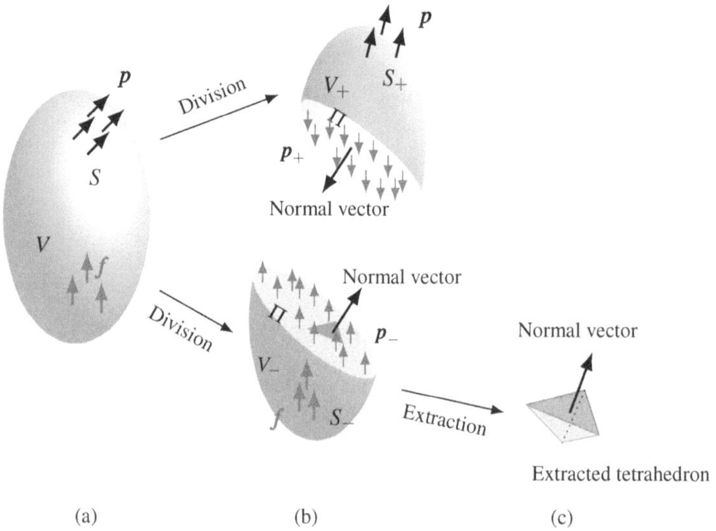

Figure 1.3.2 (a) Volume V and its surface S to establish the equilibrium equation. (b) In order to determine the internal force, the volume is divided into two parts by a cross section denoted by Π. The two volumes are referred to as V_+ and V_-. Traction $\boldsymbol{p}_+$ and $\boldsymbol{p}_-$ are required for V_+ and V_-. (c) Tetrahedral volume element separated in order to discuss the Cauchy relation.

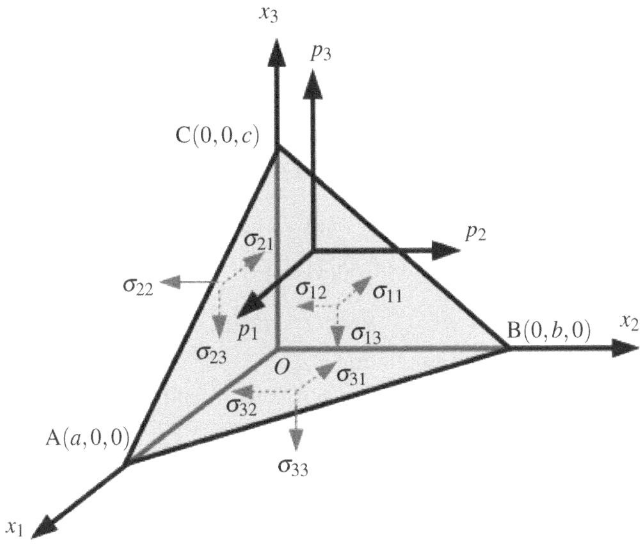

Figure 1.3.3 Equilibrium of internal forces acting on surface of tetrahedron extracted from continuum body shown in Fig.1.3.2.

jth direction. Therefore, the equilibrium equation for the internal forces acting on each side of the tetrahedron is

$$\begin{aligned} \int_{V_E} \rho \frac{Dv_i}{Dt} dV_E &= \int_{S_n} p_i dS_n + \int_{V_E} f_i dV_E - \int_{S_1} \sigma_{1i} dS_1 \\ &\quad - \int_{S_2} \sigma_{2i} dS_2 - \int_{S_3} \sigma_{3i} dS_3, \\ &\quad (i = 1,2,3) \end{aligned} \tag{1.3.31}$$

where V_E is the volume of the tetrahedron, and S_n, S_1, S_2, and S_3 are the areas of $\triangle$ABC, $\triangle$OBC, $\triangle$OAC, and $\triangle$OAB, respectively. The components of the normal vector for $\triangle$ABC are expressed as

$$\boldsymbol{n} \underset{O}{\rightarrow} \begin{pmatrix} n_1 & n_2 & n_3 \end{pmatrix} \tag{1.3.32}$$

and represent the ratio of the area[3]

$$n_j = S_j / S_n, \quad (j = 1,2,3) \tag{1.3.33}$$

[3] Verify Eq. (1.3.33). Problem §1.5.4.

Therefore, Eq. (1.3.31) can be written as

$$\int_{V_E} \rho \frac{Dv_i}{Dt} dV_E = \int_{V_E} f_i dV_E + \int_{S_n} (p_i - \sigma_{ji} n_j) dS_n \quad (i = 1,2,3) \tag{1.3.34}$$

We see that

$$V_E / S_n \to 0, \quad (S_n \to 0) \tag{1.3.35}$$

Therefore, division of both sides of Eq. (1.3.34) by S_n and $S_n \to 0$ yields

$$p_i = \sigma_{ji} n_j, \quad (i = 1,2,3) \tag{1.3.36}$$

which is known as the Cauchy relation.

Characterization of σ_{ji} as a rank-2 tensor is possible via Eq. (1.3.36). We now determine the transformation rule for σ_{ji} for the different coordinate systems. Let O' be a coordinate system for which the base vectors are denoted by $\{\boldsymbol{e}'_1, \boldsymbol{e}'_2, \boldsymbol{e}'_3\}$. The normal vector and the traction vector for Eq. (1.3.36) in this coordinate system are expressed as

$$p_i = a_{ki} p'_k, \quad n_j = a_{lj} n'_l \tag{1.3.37}$$

Therefore, Eq. (1.3.36) becomes

$$a_{ki} p'_k = \sigma_{ji} a_{lj} n'_l \iff p'_k = a_{ki} \sigma_{ji} a_{lj} n'_l \tag{1.3.38}$$

As a result, we have

$$p'_k = \sigma'_{lk} n'_l \tag{1.3.39}$$

where

$$\sigma'_{lk} = a_{lj} a_{ki} \sigma_{ji} \tag{1.3.40}$$

from which we find that σ_{ji} follows the transformation rule for the components of a rank-2 tensor. Therefore, we can define a linear mapping $\boldsymbol{\sigma}$ from two vectors into a real number such that

$$\boldsymbol{\sigma}(\boldsymbol{e}_i, \boldsymbol{e}_j) = \sigma_{ji} \tag{1.3.41}$$

which is referred to as the *stress tensor*. In this sense, σ_{ji} is a component of the stress tensor [4]. If we take the normal vector

$$\boldsymbol{n} = n_j \boldsymbol{e}_j \tag{1.3.42}$$

for one of the arguments of the stress tensor, we have

$$\begin{aligned} \boldsymbol{\sigma}(\boldsymbol{e}_i, \boldsymbol{n}) &= \boldsymbol{\sigma}(\boldsymbol{e}_i, \boldsymbol{e}_j n_j) \\ &= \sigma_{ji} n_j = p_i \end{aligned} \tag{1.3.43}$$

which is the i-th component of the traction. The physical meaning of the stress tensor as a linear mapping from two vectors into real numbers becomes clear once we take the normal vector at the cross section.

[4]The definition of the components of the stress tensor shown in Eq. (1.3.41) is different from Eq. (A.3.2), in the sense that the order of the subscripts is reversed. We employ σ_{ji} for Eq. (1.3.41) according to its definition provided in Fig. 1.3.3. Later, the symmetry of the stress tensor will resolve the problem.

1.3.3 EQUILIBRIUM EQUATION IN TERMS OF STRESS TENSOR AND SYMMETRY OF STRESS TENSOR

Here, we derive the equilibrium equation for elastic wave motion in terms of the stress tensor. In this derivation, the Gauss divergence theorem is important. We first describe the Gauss divergence theorem together with its proof. Before the description, we need to refer to scalar, vector and tensor fields:

Definition 1.1 *A scalar-valued function $f(\boldsymbol{x}), (\boldsymbol{x} \in V)$ that maps a spatial point $\boldsymbol{x} \in V$ to a scalar f is referred to as a scalar field in V. Similarly, a vector-valued function $f_i(\boldsymbol{x})$ or a tensor-valued function (for example, a rank-2 tensor) $f_{ij}(\boldsymbol{x})$, $(\boldsymbol{x} \in V)$ that map a spatial point $\boldsymbol{x} \in V$ to a vector or tensor is also known as a vector or tensor field in V.*

Lemma 1.1 *Let V and S be a region and its boundary for a closed surface, respectively, and let f be the scalar field defined in V. We express the normal vector for the boundary by*

$$\boldsymbol{n} \underset{\theta}{\rightarrow} (n_1,\ n_2,\ n_3) \tag{1.3.44}$$

the direction of which is outward from region V. Then, we have the following equation:

$$\int_V \frac{\partial f}{\partial x_i} dV = \int_S f n_i dS \tag{1.3.45}$$

[Proof]

We now provide the proof for Eq. (1.3.45) for the case of $i = 3$. Let surface S be expressed by h_+ and h_- such that:

$$x_{3+} = h_+(x_1, x_2) \geq x_{3-} = h_-(x_1, x_2), \quad (x_1, x_2) \in D \tag{1.3.46}$$

where D is a projection of S on the $x_1 - x_2$ plane, as shown in Fig. 1.3.4, and (x_1, x_2, x_{3+}) and (x_1, x_2, x_{3-}) are position vectors on the surface S, which are characterized by

$$(x_1, x_2, x_{3+}) \in S_+, \quad (x_1, x_2, x_{3-}) \in S_- \tag{1.3.47}$$

Note that S_+ and S_- is a division of S such that

$$S = S_+ \cup S_-, \quad S_+ \cap S_- = \emptyset, \tag{1.3.48}$$

According to the separation of S, we have following:

$$\begin{aligned}
\int_V \frac{\partial f}{\partial x_3} dV &= \int_D \int_{x_{3-}}^{x_{3+}} \frac{\partial f}{\partial x_3} dx_3 dx_1 dx_2 \\
&= \int_D \Big(f(x_1, x_2, x_{3+}) - f(x_1, x_2, x_{3-}) \Big) dx_1 dx_2 \\
&= \int_{S_+} n_3 f(x_1, x_2, x_{3+}) dS_+ + \int_{S_-} n_3 f(x_1, x_2, x_{3-}) dS_- \\
&= \int_S n_3 f dS
\end{aligned} \tag{1.3.49}$$

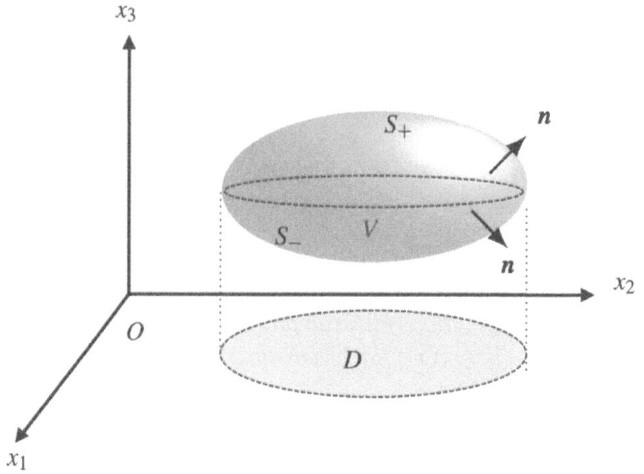

Figure 1.3.4 Region V and its closed surface $S = S_+ \cup S_-$ used to prove Gauss divergence theorem. Region D on the $x_1 - x_2$ plane is a projection of region V.

For Eq. (1.3.49), we use

$$dS_+ = dx_1 dx_2 n_3, \quad dS_- = -dx_1 dx_2 n_3 \tag{1.3.50}$$

□

According to the above lemma, we obtain the Gauss divergence theorem directly:

Theorem 1.2 *(The Gauss divergence theorem)*
Let V and S be a region and its boundary for a closed surface, respectively. Then, we have the following:

$$\int_V \frac{\partial f_i}{\partial x_i} dV = \int_S n_i f_i dS \tag{1.3.51}$$

[Remark 1] The Gauss divergence theorem can also be used for a tensor field. For example, the following equation can be established:

$$\int_V \frac{\partial \sigma_{ji}}{\partial x_j} dV = \int_S \sigma_{ji} n_j dS \tag{1.3.52}$$

[Remark 2] Let

$$f_i = u \frac{\partial v}{\partial x_i} - v \frac{\partial u}{\partial x_i} \tag{1.3.53}$$

for Eq. (1.3.51). Then, Eq. (1.3.51) becomes[5]

$$\int_V \left[u\nabla^2 v - v\nabla^2 u\right] dV = \int_S \left[u\frac{\partial v}{\partial n} - v\frac{\partial u}{\partial n}\right] dS \tag{1.3.54}$$

which is known as the Green theorem, where ∇^2 is the Laplace operator defined by

$$\nabla^2 = \frac{\partial^2}{\partial x_1^2} + \frac{\partial^2}{\partial x_2^2} + \frac{\partial^2}{\partial x_3^2} \tag{1.3.55}$$

Now, let us return to Eq. (1.3.27). According to the Cauchy relation shown in Eq. (1.3.36), Eq. (1.3.27) can be rewritten as

$$\int_V \rho \frac{Dv_i}{Dt} dV = \int_S \sigma_{ji} n_j dS + \int_V f_i dV \tag{1.3.56}$$

Application of the Gauss divergence theorem to Eq. (1.3.56) yields

$$\int_V \rho \frac{Dv_i}{Dt} dV = \int_V \left(\frac{\partial \sigma_{ji}}{\partial x_j} + f_i\right) dV \tag{1.3.57}$$

The equilibrium equation in terms of the stress tensor can be expressed in the following form:

$$\rho \frac{Dv_i}{Dt} = \frac{\partial \sigma_{ji}}{\partial x_j} + f_i \tag{1.3.58}$$

Finally, in this section, we examine the symmetry of σ_{ij}. Using Eq. (1.3.56), the equation for the equilibrium of the moment can be derived:

$$\int_V \boldsymbol{r} \times \rho \frac{D}{Dt}(v_i) dV = \int_S \boldsymbol{r} \times (\sigma_{ji} n_j) dS + \int_V \boldsymbol{r} \times (f_i) dV \tag{1.3.59}$$

where

$$\boldsymbol{r} \underset{\mathcal{O}}{\rightarrow} (x_1, x_2, x_3) \tag{1.3.60}$$

Equation (1.3.59) can also be expressed as

$$\int_V \rho\, e_{ijk} x_j \frac{Dv_k}{Dt} dV = \int_S e_{ijk} x_j \sigma_{lk} n_l dS + \int_V e_{ijk} x_j f_k dV \tag{1.3.61}$$

[5] See problem §1.5.2. For Eq. (1.3.54), we have used

$$\frac{\partial u}{\partial n} = \frac{\partial u}{\partial x_i} n_i, \quad \frac{\partial v}{\partial n} = \frac{\partial v}{\partial x_i} n_i$$

According to the Gauss divergence theorem

$$\begin{aligned}\int_S e_{ijk}x_j\sigma_{lk}n_l dS &= \int_V \frac{\partial}{\partial x_l}(e_{ijk}x_j\sigma_{lk})dV \\ &= \int_V e_{ijk}\left(\delta_{lj}\sigma_{lk}+x_j\frac{\partial \sigma_{lk}}{\partial x_l}\right)dV \end{aligned} \tag{1.3.62}$$

Equation (1.3.61) then becomes

$$\int_V \rho\, e_{ijk}x_j\frac{Dv_k}{Dt}dV = \int_V e_{ijk}\left(\delta_{lj}\sigma_{lk}+x_j\left(\frac{\partial \sigma_{lk}}{\partial x_l}+f_k\right)\right)dV \tag{1.3.63}$$

Incorporating the equilibrium equation:

$$\rho\frac{Dv_k}{Dt} = \frac{\partial \sigma_{lk}}{\partial x_l}+f_k \tag{1.3.64}$$

into Eq. (1.3.63) yields

$$\int_V e_{ijk}\delta_{lj}\sigma_{lk}dV = 0 \tag{1.3.65}$$

Therefore, we find

$$e_{ijk}\sigma_{jk} = 0 \tag{1.3.66}$$

and the components of the stress tensor are symmetric, namely, $\sigma_{ij} = \sigma_{ji}$. Consequently, we express the equilibrium equation in the following form:

$$\rho\frac{Dv_i}{Dt} = \frac{\partial \sigma_{ij}}{\partial x_j}+f_i \tag{1.3.67}$$

1.4 ELASTIC WAVE EQUATION AND ITS RECIPROCITY

1.4.1 ELASTIC WAVE EQUATION

Remember that the material derivative of the particle velocity, as shown on the left-hand side of Eq. (1.3.67), is expressed as:

$$\frac{Dv_i}{Dt} = \frac{\partial v_i}{\partial t}+v_k\partial_k v_i \tag{1.4.1}$$

according to Eq. (1.3.6). For the material derivative of the particle velocity, we assume a situation in which the amplitude of the particle velocity is much smaller than the wave velocity in the continuum body, and it is in small vibrational motion. Namely, we assume:

$$|v_i| \ll c \tag{1.4.2}$$

where c is the wave velocity.

Suppose that the particle velocity has the following form [6]:

$$v_i(\boldsymbol{x},t) = v_i(t - \boldsymbol{s}\cdot\boldsymbol{x}/c) \tag{1.4.3}$$

where $\boldsymbol{s}$ is given as

$$\boldsymbol{s} \rightarrow (s_1, s_2, s_3), \quad (|\boldsymbol{s}| = 1) \tag{1.4.4}$$

describing the direction of wave propagation. For this case, the material derivative for the particle velocity becomes

$$\begin{aligned} \frac{Dv_i}{Dt} &= v_i'\left(1 - v_k s_k/c\right) \\ &\sim \frac{\partial v_i}{\partial t} \end{aligned} \tag{1.4.5}$$

which shows that we can linearize the material derivative and we do not have to distinguish the current and original coordinate system due to the assumption of the small amplitude vibration. Therefore, using Eq. (1.3.11), the expression of the material derivative of the particle velocity in terms of the displacement field can be expressed as

$$\frac{Dv_i}{Dt} \sim \frac{\partial^2 u_i}{\partial t^2} \tag{1.4.6}$$

where u_i is the displacement field. As a result, the equilibrium equation given in Eq. (1.3.67) becomes

$$\rho \frac{\partial^2 u_i}{\partial t^2} = \partial_j \sigma_{ij} + f_i \tag{1.4.7}$$

where $\partial_j = \partial/\partial x_j$.

Our task at this point is to rewrite the right-hand side of Eq. (1.4.7) in terms of the displacement field. In order to carry out our task, we have to clarify the relationship between the stress tensor and the displacement field. For this purpose, we need to have the relationship between the infinitesimal strain tensor and the displacement shown in Eq. (1.2.25), as well as the relationship between the stress and the strain tensors. For the relationship between the stress and strain tensors, we use Hooke's law, which indicates that stress and strain are proportional. The generalized form of Hooke's law is described by a linear transformation from the strain tensor to the stress tensor in the following form:

$$\sigma_{ij} = C_{ijkl}\varepsilon_{kl} \tag{1.4.8}$$

where C_{ijkl} is referred to as the *elastic tensor*. The elastic tensor is rank 4, which has the following properties:

$$\begin{aligned} C_{ijkl} &= C_{jikl} \\ C_{ijkl} &= C_{ijlk} \end{aligned} \tag{1.4.9}$$

due to the symmetry of the stress and strain tensors.

[6] The form of the function, such as Eq. (1.4.3), is referred to as the wavefunction, and will be discussed in detail in the next chapter.

Throughout this textbook, the elastic medium is assumed to be isotropic, which means that the physical properties of the medium are invariant with respect to any coordinate system. Therefore, the elastic tensor is isotropic, having the properties of

$$C_{pqrs} = a_{pi}a_{qj}a_{rk}a_{sl}C_{ijkl} \tag{1.4.10}$$

where a_{pi}, a_{qj}, a_{rk}, and a_{sl} are the components of the transformation matrix defined by Eq. (A.2.4). As described in Appendix A, the isotropic rank-4 tensor satisfying the first expression of Eq. (1.4.9) has the following form:

$$C_{ijkl} = \lambda\delta_{ij}\delta_{kl} + \mu\delta_{ik}\delta_{jl} + \mu\delta_{il}\delta_{jk} \tag{1.4.11}$$

where the parameters λ and μ for the elastic tensor are referred to as the Lamé constants. The Lamé constants are elastic constants that can be related to Young's modulus and the Poisson ratio. We will discuss the physical meaning of the Lamé constants later.

Substitution of Eqs. (1.4.11) and (1.2.25) into Eq. (1.4.8) yields

$$\begin{aligned} \sigma_{ij} &= \lambda\delta_{ij}\varepsilon_{ll} + 2\mu\varepsilon_{ij} \\ &= \lambda\delta_{ij}\partial_l u_l + \mu(\partial_i u_j + \partial_j u_i) \end{aligned} \tag{1.4.12}$$

which is referred to as the *constitutive equation*. By means of Eq. (1.4.12), we can rewrite Eq. (1.4.7) in terms of the displacement field as

$$(\lambda+\mu)\partial_i\partial_j u_j(\boldsymbol{x},t) + \mu\partial_k^2 u_i(\boldsymbol{x},t) = \rho\frac{\partial^2}{\partial t^2}u_i(\boldsymbol{x},t) - f_i(\boldsymbol{x},t) \tag{1.4.13}$$

We refer to Eq. (1.4.13) as the elastic wave equation. Alternatively, we also refer to Eq. (1.4.13) as the governing equation for the elastic wave motion. We will sometimes express Eq. (1.4.13) in the following form:

$$L_{ij}(\partial_1,\partial_2,\partial_3)u_j(\boldsymbol{x},t) = \rho\frac{\partial^2}{\partial t^2}u_i(\boldsymbol{x},t) - f_i(\boldsymbol{x},t) \tag{1.4.14}$$

where L_{ij} is referred to as the *Navier operator* and is defined by

$$L_{ij}(\partial_1,\partial_2,\partial_3) = (\lambda+\mu)\partial_i\partial_j + \mu\delta_{ij}\nabla^2 \tag{1.4.15}$$

1.4.2 ELASTIC CONSTANTS

Undergraduate students often learn about Young's modulus E, the Poisson ratio ν, and the shear modulus G as elastic constants. It is possible to show the relationship between the Lamé constants and the elastic constants taught at the undergraduate level. Figure 1.4.1 shows the tension of a rectangular parallelepiped element, where $\boldsymbol{p}$ is the traction acting on the x_1 plane. According to the constitutive equation shown in Eq. (1.4.12), we see that

$$\begin{aligned} \sigma_{11} &= \lambda(\varepsilon_{11}+\varepsilon_{22}+\varepsilon_{33}) + 2\mu\varepsilon_{11} = p \\ \sigma_{22} &= \lambda(\varepsilon_{11}+\varepsilon_{22}+\varepsilon_{33}) + 2\mu\varepsilon_{22} = 0 \\ \sigma_{33} &= \lambda(\varepsilon_{11}+\varepsilon_{22}+\varepsilon_{33}) + 2\mu\varepsilon_{33} = 0 \end{aligned} \tag{1.4.16}$$

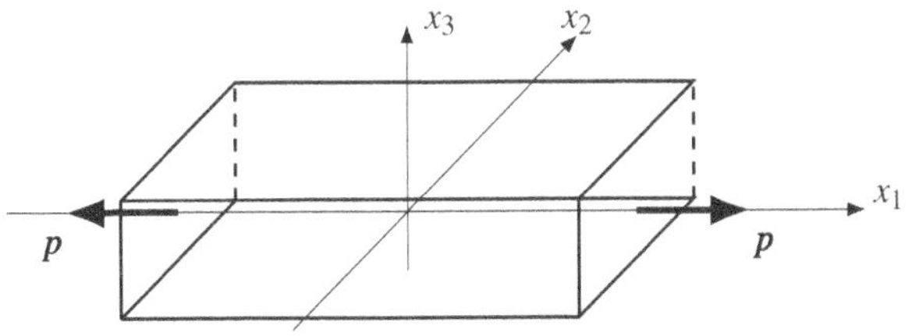

Figure 1.4.1 Tension of rectangular parallelepiped element to show the relationship between Lamé parameters and elastic constants learned at undergraduate level.

where $p = |\boldsymbol{p}|$. Equation (1.4.16) yields

$$\varepsilon_{11} + \varepsilon_{22} + \varepsilon_{33} = \frac{p}{3\lambda + 2\mu} \tag{1.4.17}$$

As a result, we have

$$\begin{aligned} \varepsilon_{11} &= \frac{\lambda + \mu}{\mu(3\lambda + 2\mu)} p \\ \varepsilon_{22} &= -\frac{\lambda}{2\mu(3\lambda + 2\mu)} p (= \varepsilon_{33}) \end{aligned} \tag{1.4.18}$$

Young's modulus E and Poisson's ratio ν can be derived from Eq. (1.4.18) such that

$$\begin{aligned} E &= \frac{p}{\varepsilon_{11}} = \frac{\mu(3\lambda + 2\mu)}{\lambda + \mu} \\ \nu &= -\frac{\varepsilon_{22}}{\varepsilon_{11}} = \frac{\lambda}{2(\lambda + \mu)} \end{aligned} \tag{1.4.19}$$

Alternatively, we also have

$$\begin{aligned} \lambda &= \frac{\nu}{(1 - 2\nu)(1 + \nu)} E \\ \mu &= \frac{E}{2(1 + \nu)} \end{aligned} \tag{1.4.20}$$

According to the constitutive equation shown in Eq. (1.4.12), we see that

$$\sigma_{ij} = 2\mu\varepsilon_{ij}, \quad (i \neq j) \tag{1.4.21}$$

Therefore, we have

$$\mu = G \tag{1.4.22}$$

We summarize examples of the elastic constants for some materials in Table 1.1.

Table 1.1
Examples of elastic constants and mass densities (Sato, Y. 1978).

	E (GPa)	μ (GPa)	ρ (kg/m^3)	ν
Zinc	1.08×10^2	4.34×10	7.12×10^3	0.249
Aluminum	7.03×10^2	2.61×10	2.69×10^3	0.345
Steel	2.08×10^2	8.1×10	7.85×10^3	0.30
Cast iron	1.52×10^2	6.08×10	7.8×10^3	0.27
Glass	7.31×10	2.40×10	2.94×10^3	0.258

1.4.3 RECIPROCITY OF ELASTIC WAVEFIELD

We have derived the elastic wave equation as the governing equation in the form of a partial differential equation. As a consequence, it is necessary to clarify what kind of boundary conditions are required to express the solution of the elastic wave equation. In order to resolve this problem, we derive the reciprocity of the elastic wavefield. At first, for simplicity, we assume that the elastic wavefield in the steady state with the time factor of $\exp(i\omega t)$, where ω denotes the circular frequency.

As shown in Fig. 1.4.2, we consider two different states inside the same region of the continuum body, the volume and boundary of which are denoted by V and S, respectively. The Lamé parameters and mass density for the region are denoted by λ, μ and ρ, respectively. The different states are caused by different body forces f_i and $f_i^\star$. We use the superscript $\star$ to distinguish the different states.

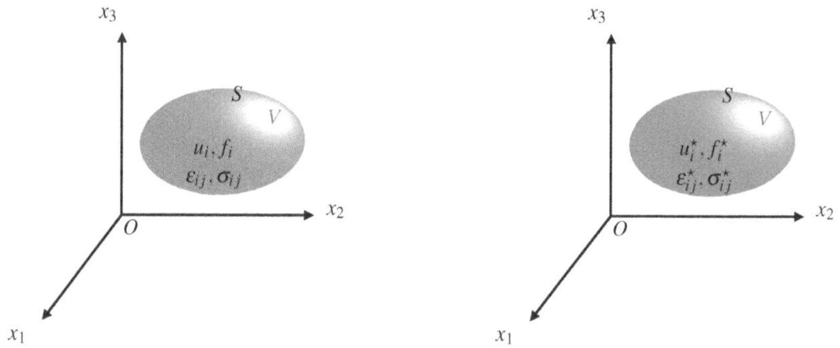

Figure 1.4.2 Two different states for same region V and its boundary S illustrating reciprocity.

These two different states satisfy the following equations:

$$\begin{aligned}\left(L_{ij}(\partial_1,\partial_2,\partial_3)+\rho\omega^2\delta_{ij}\right)u_j(\boldsymbol{x}) &= -f_i(\boldsymbol{x})\\ \left(L_{ij}(\partial_1,\partial_2,\partial_3)+\rho\omega^2\delta_{ij}\right)u_j^\star(\boldsymbol{x}) &= -f_i^\star(\boldsymbol{x})\end{aligned} \tag{1.4.23}$$

We also find that

$$L_{ij}(\partial_1,\partial_2,\partial_3)u_j(\boldsymbol{x}) = \frac{\partial}{\partial x_j}\sigma_{ij}(\boldsymbol{x}) \tag{1.4.24}$$

according to Eq. (1.4.12). In addition, the Gauss divergence theorem yields

$$\begin{aligned}&\int_S n_j\sigma_{ij}(\boldsymbol{x})u_i^\star(\boldsymbol{x})dS\\ &= \int_V \frac{\partial}{\partial x_j}\left(u_i^*(\boldsymbol{x})\,\sigma_{ij}(\boldsymbol{x})\right)dV\\ &= \int_V\left(\frac{\partial}{\partial x_j}u_i^\star(\boldsymbol{x})\right)\sigma_{ij}(\boldsymbol{x})dV+\int_V u_i^\star(\boldsymbol{x})\left(\frac{\partial}{\partial x_j}\sigma_{ij}(\boldsymbol{x})\right)dV\\ &= \int_V \varepsilon_{ij}^\star(\boldsymbol{x})\sigma_{ij}(\boldsymbol{x})dV+\int_V u_i^\star(\boldsymbol{x})L_{ij}(\partial_1,\partial_2,\partial_3)u_j(\boldsymbol{x})dV\end{aligned} \tag{1.4.25}$$

Likewise, we have

$$\begin{aligned}&\int_S n_j\sigma_{ij}^\star(\boldsymbol{x})u_i(\boldsymbol{x})dS\\ &= \int_V \frac{\partial}{\partial x_j}\left(u_i(\boldsymbol{x})\,\sigma_{ij}^\star(\boldsymbol{x})\right)dV\\ &= \int_V \varepsilon_{ij}(\boldsymbol{x})\sigma_{ij}^\star(\boldsymbol{x})dV+\int_V u_i(\boldsymbol{x})L_{ij}(\partial_1,\partial_2,\partial_3)u_j^\star(\boldsymbol{x})dV\end{aligned} \tag{1.4.26}$$

Subtracting Eq. (1.4.26) from Eq. (1.4.25) yields

$$\begin{aligned}&\int_S\left[u_i^\star(\boldsymbol{x})p_i(\boldsymbol{x})-u_i(\boldsymbol{x})p_i^\star(\boldsymbol{x})\right]dS\\ &= \int_V\Big[u_i^\star(\boldsymbol{x})\left(L_{ij}(\partial_1,\partial_2,\partial_3)+\rho\omega^2\delta_{ij}\right)u_j(\boldsymbol{x})\\ &\qquad -u_i(\boldsymbol{x})\left(L_{ij}(\partial_1,\partial_2,\partial_2)+\rho\omega^2\delta_{ij}\right)u_j^\star(\boldsymbol{x})\Big]dV\\ &= \int_V\left[-u_i^\star(\boldsymbol{x})f_i(\boldsymbol{x})+u_i(\boldsymbol{x})f_i^\star(\boldsymbol{x})\right]dV\end{aligned} \tag{1.4.27}$$

because

$$\begin{aligned}\varepsilon_{ij}^\star\sigma_{ij} &= \varepsilon_{ij}^\star\left(\lambda\delta_{ij}\varepsilon_{kk}+2\mu\varepsilon_{ij}\right)\\ &= \lambda\varepsilon_{ii}^\star\varepsilon_{kk}+2\mu\varepsilon_{ij}^\star\varepsilon_{ij}=\sigma_{ij}^\star\varepsilon_{ij}\end{aligned} \tag{1.4.28}$$

For Eq. (1.4.27), we also use

$$
\begin{aligned}
p_i^{\star} &= \sigma_{ij}^{\star} n_j \\
p_i &= \sigma_{ij} n_j
\end{aligned} \tag{1.4.29}
$$

Equation (1.4.27) shows the reciprocity of the elastic wavefield in the steady state. As can be seen, the boundary conditions for S are found to be expressed by the displacement and traction. The surface integral with respect to the displacement and traction equates to the right-hand side of the volume integral.

Next, we consider the two different states satisfying the following equations:

$$
\begin{aligned}
\left(L_{ij}(\partial_1,\partial_2,\partial_3) - \rho\delta_{ij}\frac{\partial^2}{\partial t^2}\right) u_j(\boldsymbol{x},t) &= -f_i(\boldsymbol{x},t) \\
\left(L_{ij}(\partial_1,\partial_2,\partial_3) - \rho\delta_{ij}\frac{\partial^2}{\partial t^2}\right) u_j^{\star}(\boldsymbol{x},t) &= -f_i^{\star}(\boldsymbol{x},t)
\end{aligned} \tag{1.4.30}
$$

where external body forces are defined in the range of $0 \le t < \infty$. Based on the same procedure for obtaining the reciprocity of the elastic wavefield in a steady state, we have the following reciprocity in a transient state:

$$
\begin{aligned}
&\int_0^{\infty} d\tau \int_S \left[u_i^{\star}(\boldsymbol{x},\tau)p_i(\boldsymbol{x},\tau) - u_i(\boldsymbol{x},\tau)p_i^{\star}(\boldsymbol{x},\tau)\right] dS \\
&\quad - \int_V \rho \left[u_i^{\star}(\boldsymbol{x},\tau)\frac{\partial}{\partial \tau}u_i(\boldsymbol{x},\tau) - u_i(\boldsymbol{x},\tau)\frac{\partial}{\partial \tau}u_i^{\star}(\boldsymbol{x},\tau)\right]_{\tau=0}^{\tau=\infty} dV \\
&= \int_0^{\infty} d\tau \int_V \left[u_i^{\star}(\boldsymbol{x},\tau)\left(L_{ij}(\partial_1,\partial_2,\partial_3) - \delta_{ij}\rho\frac{\partial^2}{\partial \tau^2}\right)u_j(\boldsymbol{x},\tau)\right. \\
&\quad \left. - u_i(\boldsymbol{x},\tau)\left(L_{ij}(\partial_1,\partial_2,\partial_3) - \delta_{ij}\rho\frac{\partial^2}{\partial \tau^2}\right)u_j^{\star}(\boldsymbol{x},\tau)\right] dV \\
&= \int_0^{\infty} d\tau \int_V \left[-u_i^{\star}(\boldsymbol{x},\tau)f_i(\boldsymbol{x},\tau) + u_i(\boldsymbol{x},\tau)f_i^{\star}(\boldsymbol{x},\tau)\right] dV
\end{aligned} \tag{1.4.31}
$$

Equation (1.4.31) shows that the boundary conditions for S are also displacement, and traction, even in a transient state. In addition, we notice that the initial and terminal conditions with respect to the displacement as well as the velocity in the volume are required to equate to the right-hand side of the volume integral. It is of course that we need to introduce the integration with respect to times for the reciprocity expression in the transient state.

Remember that the starting point of deriving the elastic wave equation was Eqs. (1.3.1) and/or (1.3.2), where we used the traction acting on the boundary S. At the end of this chapter, we derived the reciprocity of the elastic wavefield, from which we found the boundary conditions required for the elastic wave equation were traction and displacement. We have to emphasize that discussions carried out for the elastic wave equation are consistent from the construction of the equations of motion to the derivation of reciprocity of the equations. Furthermore, it's important to note that the reciprocity of the elastic wavefield was established without the need for knowledge about solutions to the elastic wave equation. Instead, we relied on the Gauss divergence theorem, the constitutive equation, and the equilibrium equation to establish the reciprocity of the elastic wavefield.

1.5 PROBLEMS

1. Consider 2D plane as shown in Fig. 1.5.1, in which the base vectors for the two different Cartesian coordinate systems are denoted by $\mathcal{O}\{\boldsymbol{e}_1, \boldsymbol{e}_2\}$ and $\mathcal{O}'\{\boldsymbol{e}'_1, \boldsymbol{e}'_2\}$, respectively. The angle between x_1 and x'_1 axes is set by θ.
   a. Let the transformation rule for the base vectors be expressed by $\boldsymbol{e}'_k = a_{kj}\boldsymbol{e}_j$. Clarify a_{kj} in terms of θ.
   b. Let the expressions of a vector be $\boldsymbol{v} = v_i\boldsymbol{e}_i = v'_i\boldsymbol{e}'_i$. Derive the relation ship between v_i and v'_i in terms of θ.
   c. Let a rank-2 tensor be denoted by $\boldsymbol{T}$ and define its components by

$$\begin{aligned} T_{ij} &= \boldsymbol{T}(\boldsymbol{e}_i, \boldsymbol{e}_j) \\ T'_{ij} &= \boldsymbol{T}(\boldsymbol{e}'_i, \boldsymbol{e}'_j) \end{aligned} \tag{1.5.1}$$

   Clarify the relationship between T_{ij} and T'_{ij} in terms of the product of matrices.
   d. Suppose $T_{ij} = T_{ji}$. Express the each component T'_{ij} in terms of T_{ij} and θ.
2. a. Let $\boldsymbol{n}$ be a unit vector whose components are

$$\boldsymbol{n} \underset{\mathcal{O}}{\rightarrow} (n_1, n_2, n_3) \tag{1.5.2}$$

   Show the directional derivative of a scalar function $\varphi(\boldsymbol{x})$, $\boldsymbol{x} \in \mathbb{R}^3$ with respect to the direction $\boldsymbol{n}$ is expressed by

$$\frac{\partial \varphi(\boldsymbol{x})}{\partial n} = \nabla\varphi(\boldsymbol{x}) \cdot \boldsymbol{n} \tag{1.5.3}$$

   where $\nabla\varphi$ is defined by

$$\nabla\varphi(\boldsymbol{x}) \underset{\mathcal{O}}{\rightarrow} (\partial_1\varphi(\boldsymbol{x}), \partial_2\varphi(\boldsymbol{x}), \partial_3\varphi(\boldsymbol{x})) \tag{1.5.4}$$

   b. Show the following identity

$$\int_V \nabla^2\varphi(\boldsymbol{x})dV = \int_S \frac{\partial \varphi(\boldsymbol{x})}{\partial n} dS \tag{1.5.5}$$

   by means of the Gauss divergence theorem, where V and S are a region and its boundary of closed surface.

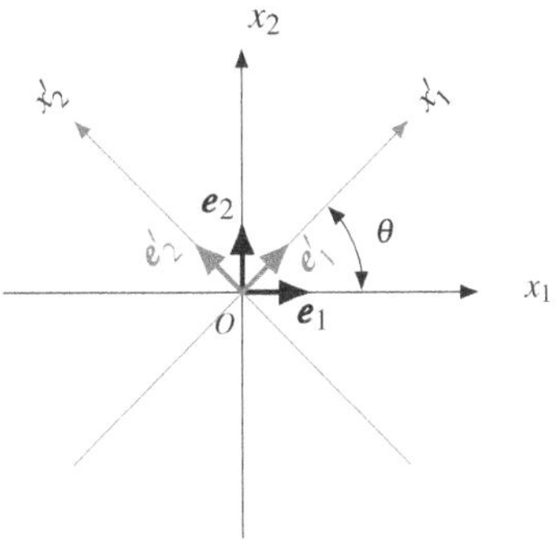

Figure 1.5.1 2D plane spanned two different coordinate systems, which are $x_1 - x_2$ and $x'_1 - x'_2$ coordinate systems. Base vectors for the coordinate systems are denoted by $\mathcal{O}\{\boldsymbol{e}_1, \boldsymbol{e}_2\}$ and $\mathcal{O}'\{\boldsymbol{e}'_1, \boldsymbol{e}'_2\}$

c. Consider the two states of the wavefield such that

$$\begin{aligned}(\nabla^2 + k^2)u(\boldsymbol{x}) &= -f(\boldsymbol{x}) \\ (\nabla^2 + k^2)u^\star(\boldsymbol{x}) &= -f^\star(\boldsymbol{x})\end{aligned} \tag{1.5.6}$$

Then, prove the following reciprocity relationship

$$\begin{aligned}&\int_V \left[u^\star(\boldsymbol{x})(\nabla^2 + k^2)u(\boldsymbol{x}) - u(\boldsymbol{x})(\nabla^2 + k^2)u^\star(\boldsymbol{x})\right] dV \\ &= \int_V \left[-u^\star(\boldsymbol{x})f(\boldsymbol{x}) + u(\boldsymbol{x})f^\star(\boldsymbol{x})\right] dV \\ &= \int_S \left[u^\star(\boldsymbol{x})\frac{\partial u(\boldsymbol{x})}{\partial n} - u(\boldsymbol{x})\frac{\partial u^\star(\boldsymbol{x})}{\partial n}\right] dS\end{aligned} \tag{1.5.7}$$

3. Verify Eq. (1.3.19).
4. Verify Eq. (1.3.33).

2 Basic properties of solution for elastic wave equation and representation theorem

Our task in this chapter is to find solutions for the elastic wave equation shown in Eq. (1.4.13) and to derive the representation theorem for the solutions. Wavefield we deal with in this chapter is 3-D full space. The derivation of the representation theorem is based on the reciprocity of the elastic wave equation presented in Eqs. (1.4.27) and (1.4.31). The search for a homogeneous solution for the elastic wave equation is carried out by decomposing the equation into scalar wave equations for the P and S waves. This decomposition is based on a Helmholtz decomposition of vector field. Therefore, in this chapter, we start with scalar wave equations and the Helmholtz decomposition. Green's function is important in the representation theorem. We present a detailed derivation of Green's function because it aids in understanding the properties of this function.

2.1 SOLUTION FOR SCALAR WAVE EQUATION FOR UNBOUNDED MEDIA

We call the following equation a *scalar wave equation*:

$$\nabla^2 u(\boldsymbol{x},t) = \frac{1}{c^2}\frac{\partial^2 u(\boldsymbol{x},t)}{\partial t^2}, \quad \left((\boldsymbol{x},t) \in \mathbb{R}^3 \times \mathbb{R}\right) \tag{2.1.1}$$

where $c \in \mathbb{R}$ is a constant. Later, we will see that c is the wave velocity. When a solution for Eq. (2.1.1) can be expressed in the form

$$u(\boldsymbol{x},t) = u_0(\boldsymbol{x})e^{i\omega t} \tag{2.1.2}$$

where ω is the circular frequency, Eq. (2.1.1) is modified as

$$\left(\nabla^2 + \frac{\omega^2}{c^2}\right)u_0(\boldsymbol{x}) = 0, \quad \left(\boldsymbol{x} \in \mathbb{R}^3\right) \tag{2.1.3}$$

We call Eq. (2.1.3) the *Helmholtz equation*.

To find solutions for Eq. (2.1.3), we focus on functions φ_α that satisfy

$$\nabla^2 \varphi_\alpha(\boldsymbol{x}) = \alpha^2 \varphi_\alpha(\boldsymbol{x}), \quad (\alpha \in \mathbb{C}) \tag{2.1.4}$$

DOI: 10.1201/9781003251729-2

The application of the Laplace operator to φ_α is equivalent to the multiplication of a scalar α^2 to the function. It is not very difficult to exemplify such functions. For example, the function

$$\varphi_\alpha(\boldsymbol{x}) = \exp\left(\alpha \boldsymbol{p} \cdot \boldsymbol{x}\right), \quad (\boldsymbol{x}, \boldsymbol{p} \in \mathbb{R}^3, \ |\boldsymbol{p}| = 1) \tag{2.1.5}$$

satisfies Eq. (2.1.4), where $\boldsymbol{p} \cdot \boldsymbol{x}$ is the scalar product of $\boldsymbol{p}$ and $\boldsymbol{x}$. It can be simply verified that Eq. (2.1.5) satisfies Eq. (2.1.4) by assuming that the components of the vectors $\boldsymbol{p}$ and $\boldsymbol{x}$ are given as

$$\begin{aligned} \boldsymbol{x} &= \begin{pmatrix} x_1 & x_2 & x_3 \end{pmatrix} \\ \boldsymbol{p} &= \begin{pmatrix} p_1 & p_2 & p_3 \end{pmatrix} \end{aligned} \tag{2.1.6}$$

It is readily seen from Eq. (2.1.4) that $\alpha = \pm ik$ solves Eq. (2.1.3). Therefore, the two functions $\exp(+ik\boldsymbol{p} \cdot \boldsymbol{x})$ and $\exp(-ik\boldsymbol{p} \cdot \boldsymbol{x})$ are solutions for Eq. (2.1.3). The linear combination of these two functions

$$u_0(\boldsymbol{x}) = C_1 \exp(-ik\, \boldsymbol{p} \cdot \boldsymbol{x}) + C_2 \exp(ik\, \boldsymbol{p} \cdot \boldsymbol{x}) \tag{2.1.7}$$

where C_1 and C_2 are arbitrary constants, is also a solution for Eq. (2.1.7).

Equation (2.1.7) yields a solution for Eq. (2.1.1) in the form

$$u(\boldsymbol{x}, t) = C_1 \exp\left(i\omega\left(t - \frac{\boldsymbol{p} \cdot \boldsymbol{x}}{c}\right)\right) + C_2 \exp\left(i\omega\left(t + \frac{\boldsymbol{p} \cdot \boldsymbol{x}}{c}\right)\right) \tag{2.1.8}$$

In order to have physical interpretations of Eq. (2.1.8), let us define the set of $\boldsymbol{x}$ with respect to the real parameter s as:

$$M(s) = \{\boldsymbol{x} \in \mathbb{R}^3 \,|\, \boldsymbol{p} \cdot \boldsymbol{x} = s, \ |\boldsymbol{p}| = 1, \ s \in \mathbb{R}\} \tag{2.1.9}$$

For a fixed s, $M(s)$ forms a plane whose normal vector is $\boldsymbol{p}$ and whose distance from the origin of the coordinate system is $|s|$. On the plane $M(s)$, the two functions

$$\exp\left(i\omega\left(t - \frac{\boldsymbol{p} \cdot \boldsymbol{x}}{c}\right)\right), \ \exp\left(i\omega\left(t + \frac{\boldsymbol{p} \cdot \boldsymbol{x}}{c}\right)\right) \tag{2.1.10}$$

each have a constant value for a fixed time t. Now, consider the time period from t to $t + \Delta t$. On the plane $M(s + \Delta s)$, we have

$$\exp\left(i\omega\left(t + \Delta t - \frac{s + \Delta s}{c}\right)\right) = \exp\left(i\omega\left(t - \frac{s}{c}\right)\right) \tag{2.1.11}$$

if Δs satisfies $\Delta s = c\Delta t$. The value of the function at $\boldsymbol{x} \in M(s)$ appears again at a point $\boldsymbol{x} \in M(s + c\Delta t)$ after a time Δt. This means that a constant value of the function on the plane $M(s)$ moves with a velocity of c toward the direction of the normal vector $\boldsymbol{p}$ (see Fig. 2.1.1). Likewise, a constant value of the function

$$\exp\left(i\omega\left(t + \frac{\boldsymbol{p} \cdot \boldsymbol{x}}{c}\right)\right)$$

on the plane $M(s)$ moves with a velocity of c toward the direction opposite to the normal vector of the plane. As a result, the function $u(\boldsymbol{x},t)$ in Eq. (2.1.8) describes plane waves with a velocity of c; that is, it describes the superposition of waves propagating in the forward and backward directions with respect to the vector $\boldsymbol{p}$.

Now, we can determine the physical meaning of ω/c in Eq. (2.1.3). Remember that the relationship between the wavelength λ and frequency f is $c = f\lambda$. Therefore,

$$\omega/c = 2\pi/\lambda \tag{2.1.12}$$

which means that ω/c denotes the wavenumber.

In general, the function

$$u(\boldsymbol{x},t) = f\left(t - \frac{\boldsymbol{p}\cdot\boldsymbol{x}}{c}\right) + g\left(t + \frac{\boldsymbol{p}\cdot\boldsymbol{x}}{c}\right) \tag{2.1.13}$$

where f and g are arbitrary twice-differentiable functions, is also a solution for Eq. (2.1.1) in the form of plane waves. It can be simply verified that Eq. (2.1.13) is a solution for Eq. (2.1.1) by substituting Eq. (2.1.13) into Eq. (2.1.1). A solution in the form of Eq. (2.1.13) is known as the *d'Alembert formula*. The possibility of

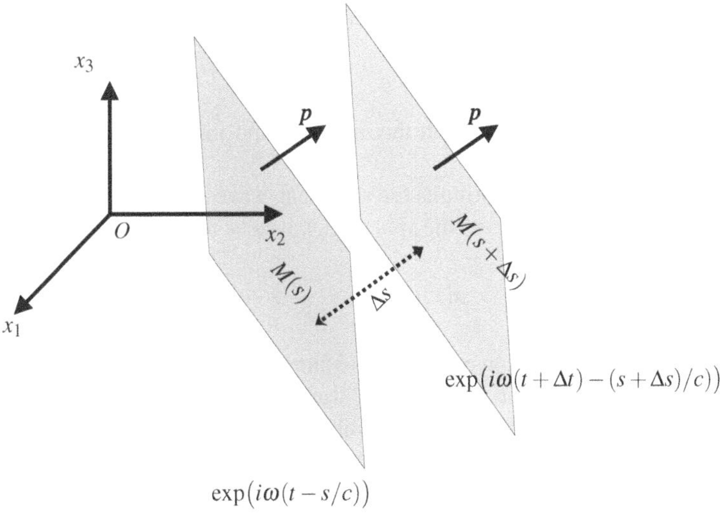

Figure 2.1.1 Concept of plane waves. $M(s)$ and $M(s+\Delta s)$ form planes whose normal vector is $\boldsymbol{p}$. The distances of the planes from the origin of the coordinate system are $|s|$ and $|s+\Delta s|$, respectively. Equation (2.1.11) is established for the case where $\Delta s = c\Delta t$.

using arbitrary functions for the solutions for Eq. (2.1.1) shows the complexity and diversity of solutions for partial differential equations.

Plane waves are not the only solutions for Eq. (2.1.1). The function

$$\varphi_\alpha(\boldsymbol{x}) = \frac{\exp(\alpha|\boldsymbol{x}|)}{|\boldsymbol{x}|}, \quad (\boldsymbol{x} \neq 0, \ \alpha \in \mathbb{C}) \tag{2.1.14}$$

also satisfies Eq. (2.1.4). That is, we have

$$\nabla^2 \frac{\exp(\alpha|\boldsymbol{x}|)}{|\boldsymbol{x}|} = \alpha^2 \frac{\exp(\alpha|\boldsymbol{x}|)}{|\boldsymbol{x}|} \tag{2.1.15}$$

The verification of Eq. (2.1.15) is left as an exercise.

Since $\alpha = \pm ik$ for Eq. (2.1.14) solves Eq. (2.1.3), a solution for Eq. (2.1.1) can be written in the form

$$u(\boldsymbol{x},t) = C_1 \frac{1}{|\boldsymbol{x}|} \exp\Big(i\omega\big(t - \frac{|\boldsymbol{x}|}{c}\big)\Big) + C_2 \frac{1}{|\boldsymbol{x}|} \exp\Big(i\omega\big(t + \frac{|\boldsymbol{x}|}{c}\big)\Big), \quad (\boldsymbol{x} \neq 0) \tag{2.1.16}$$

Equation (2.1.16) shows that the waves have spherical symmetry. Therefore, Eq. (2.1.16) describes spherical waves. The first term on the right-hand side of the equation describes the outgoing wave. As time $t \to \infty$, the wave propagates toward $|\boldsymbol{x}| \to \infty$ with a geometric decay factor $|\boldsymbol{x}|^{-1}$. The second term on the right-hand side of Eq. (2.1.16) describes the incoming wave, which propagates in the direction $|\boldsymbol{x}| \to 0$ as $t \to \infty$. The amplitude increases with the factor $|\boldsymbol{x}|^{-1}$. Note that a spherical incoming wave is physically unnatural and is not acceptable in the sense of physical phenomena. The presence of the incoming wave as a solution for the wave equation is due to the operator for the wave equation being invariant to time reversal for

$$t' = -t \tag{2.1.17}$$

In this chapter, we use both the outgoing and incoming spherical waves to discuss Green's function.

The d'Alembert formula for spherical waves is also possible. Referring to Eq. (2.1.16), we can express the spherical waves as follows:

$$u(\boldsymbol{x},t) = \frac{f(t - |\boldsymbol{x}|/c)}{|\boldsymbol{x}|} + \frac{g(t + |\boldsymbol{x}|/c)}{|\boldsymbol{x}|} \tag{2.1.18}$$

The direct substitution of Eq. (2.1.18) into Eq. (2.1.1) clarifies the justification of Eq. (2.1.18) as a solution for the scalar wave equation.

Various types of wave are possible for Eq. (2.1.1). The precise discussions are the representation of the Laplace operator in terms of the various coordinate systems and solving Eq. (2.1.4) using the separation of variables. A solution for cylindrical waves can be derived from Eq. (2.1.1) by employing a cylindrical coordinate system and the separation of variables, which are left as exercises[1].

[1] See problems shown in § 2.6.1.

2.2 HELMHOLTZ DECOMPOSITION OF VECTOR FIELD

2.2.1 GRADIENT AND ROTATION OPERATORS

Now, let $\varphi(\boldsymbol{x})$, $(\boldsymbol{x} \in V \subset \mathbb{R}^3)$, which is differentiable, be a scalar field in V. The gradient of the scalar field is defined as

$$\nabla\varphi(\boldsymbol{x}) = \begin{pmatrix} \partial_1\varphi(\boldsymbol{x}) & \partial_2\varphi(\boldsymbol{x}) & \partial_3\varphi(\boldsymbol{x}) \end{pmatrix}, \quad (\boldsymbol{x} \in V \subset \mathbb{R}^3) \tag{2.2.1}$$

in a Cartesian coordinate system, where ∇ is referred to as the gradient operator. $\nabla\varphi(\boldsymbol{x})$ defines a vector field in V.

According to Eq. (1.5.3)[2], the directional derivative of $\varphi(\boldsymbol{x})$ with respect to the direction vector $\boldsymbol{s}$ is expressed by

$$\frac{\partial\varphi(\boldsymbol{x})}{\partial s} = \nabla\varphi(\boldsymbol{x}) \cdot \boldsymbol{s}, \quad (|\boldsymbol{s}| = 1) \tag{2.2.2}$$

Therefore, for the case where $\boldsymbol{s}_t$ is the tangent vector for a surface with a constant value of $\varphi(\boldsymbol{x})$, we have

$$\frac{\partial\varphi(\boldsymbol{x})}{\partial s_t} = \nabla\varphi(\boldsymbol{x}) \cdot \boldsymbol{s}_t = 0 \tag{2.2.3}$$

which shows

$$\nabla\varphi(\boldsymbol{x}) \perp \boldsymbol{s}_t \tag{2.2.4}$$

For a case where the direction vector is given by

$$\boldsymbol{s}_n = \frac{\nabla\varphi(\boldsymbol{x})}{|\nabla\varphi(\boldsymbol{x})|} \tag{2.2.5}$$

the directional derivative of φ becomes

$$\frac{\partial\varphi(\boldsymbol{x})}{\partial s_n} = |\nabla\varphi(\boldsymbol{x})| \geq |\nabla\varphi(\boldsymbol{x}) \cdot \boldsymbol{s}|, \quad (\forall \boldsymbol{s} \in \mathbb{R}^3,\ |\boldsymbol{s}| = 1) \tag{2.2.6}$$

Equation (2.2.6) shows that the directional derivative of φ achieves its maximum value along the direction of $\nabla\varphi$, and the direction of $\nabla\varphi$ points towards the increase of $\varphi(\boldsymbol{x})$. Equations (2.2.4) and (2.2.6) characterize the properties of the ∇ operation.

As an example of $\nabla\varphi(\boldsymbol{x})$, we define a scalar field as

$$\varphi(\boldsymbol{x}) = r \left(= \sqrt{x_1^2 + x_2^2 + x_3^2} \right), \quad (\boldsymbol{x} \in \mathbb{R}^3) \tag{2.2.7}$$

Then, the gradient of the scalar wavefield becomes

$$\nabla\varphi(\boldsymbol{x}) = \frac{1}{r}\begin{pmatrix} x_1 & x_2 & x_3 \end{pmatrix}, \quad (\boldsymbol{x} \in \mathbb{R}^3 \setminus \{0\}) \tag{2.2.8}$$

Figure 2.2.1 shows the vector field of ∇r, in which vectors defined by ∇r are orthogonal to the surface of a sphere (i.e., constant value of r) and the direction of ∇r is also found to be increase of r.

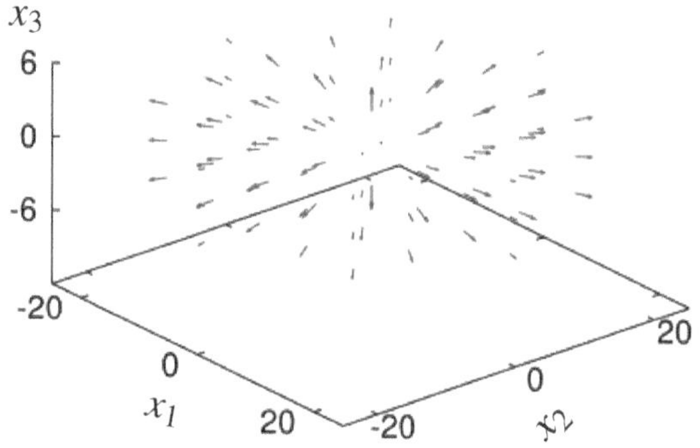

Figure 2.2.1 Vector field defined by ∇r. The vectors are orthogonal to the surface of a sphere of $r = \text{const}$ and directions of the vectors are toward the increase of r.

Next, let $\boldsymbol{u}(\boldsymbol{x})$, $(\boldsymbol{x} \in V \subset \mathbb{R}^3)$, which is also differentiable, be a vector field. The rotation of the vector field is denoted by $\nabla \times \boldsymbol{u}$, whose components in a Cartesian coordinate system are defined as

$$\left(\nabla \times \boldsymbol{u}(\boldsymbol{x})\right)_i = e_{ijk}\partial_j u_k(\boldsymbol{x}), \quad (\boldsymbol{x} \in V) \tag{2.2.9}$$

where e_{ijk} in Eq. (2.2.9) is the Eddington symbol defined in Appendix A. In the following, the rotation of a vector field is sometimes expressed as

$$\text{rot}\,\boldsymbol{u}\,(= \nabla \times \boldsymbol{u}) \tag{2.2.10}$$

To investigate the results of applying the rotation operator to a vector field, let

$$\boldsymbol{u}(\boldsymbol{x}) = \begin{pmatrix} 0 & 0 & u_3(\boldsymbol{x}) \end{pmatrix}, \quad (\boldsymbol{x} \in \mathbb{R}^3) \tag{2.2.11}$$

Then, we find that

$$\text{rot}\,\boldsymbol{u}(\boldsymbol{x}) = \begin{pmatrix} \partial_2 u_3(\boldsymbol{x}) & -\partial_1 u_3(\boldsymbol{x}) & 0 \end{pmatrix} \tag{2.2.12}$$

and as a result, we have

$$\text{rot}\,\boldsymbol{u}(\boldsymbol{x}) \perp \nabla u_3(\boldsymbol{x}) \tag{2.2.13}$$

For this case, rot $\boldsymbol{u}$ is embedded in a plane that is tangential to a surface with a constant value of $u_3(\boldsymbol{x})$, which yields a rotational vector field.

As an example, we examine the vector field

$$\boldsymbol{u}(\boldsymbol{x}) = \begin{pmatrix} 0 & 0 & 1 \end{pmatrix} e^{-r^2}, \quad (\boldsymbol{x} \in \mathbb{R}^3) \tag{2.2.14}$$

Figures 2.2.2 (a)-(c) respectively show the three vector fields '$\boldsymbol{u}(\boldsymbol{x})$', 'rot $\boldsymbol{u}(\boldsymbol{x})$', and 'rotrot$\boldsymbol{u}(\boldsymbol{x})$' in the $x_1 - x_2$ plane. The vectors for $\boldsymbol{u}$, shown in Fig. 2.2.2(a), are orthogonal to the $x_1 - x_2$ plane. They point upward as a result of Eq. (2.2.14). The vectors

[2]See §1.5.2.

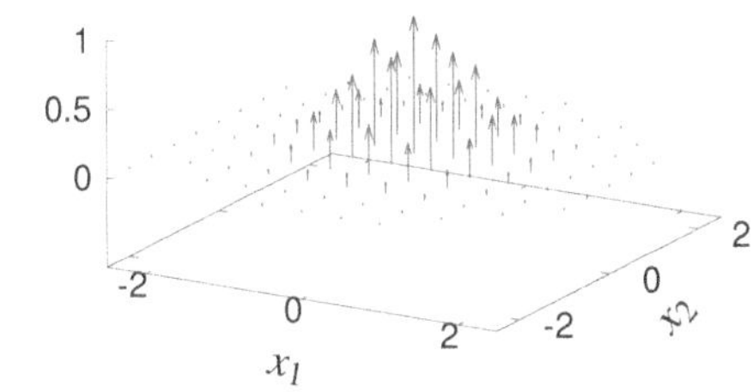

(a) Vector field $\boldsymbol{u}(\boldsymbol{x})$ on $x_1 - x_2$ plane.

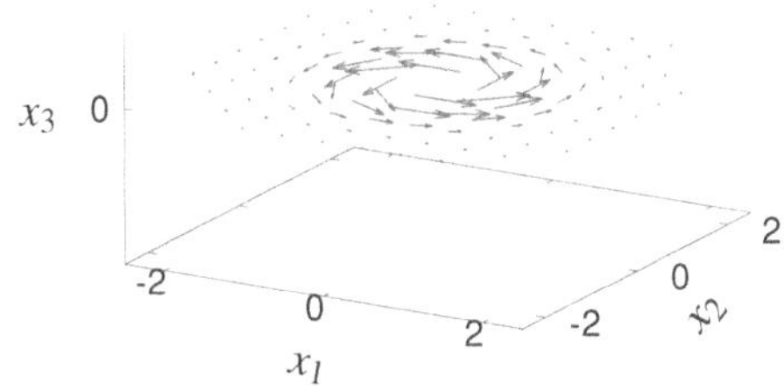

(b) Vector field 'rot $\boldsymbol{u}(\boldsymbol{x})$' on $x_1 - x_2$ plane.

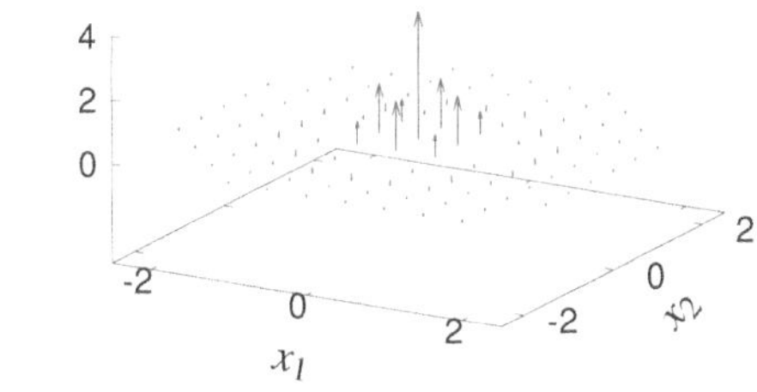

(c) Vector field 'rot rot $\boldsymbol{u}(\boldsymbol{x})$' on $x_1 - x_2$ plane.

Figure 2.2.2 Examples of effects of "rot" operation on vector fields. (a) Vector field $\boldsymbol{u}(\boldsymbol{x})$ on $x_1 - x_2$ plane. The vectors point upward from the plane and those with a high amplitude are concentrated around the x_3 axis. (b) Vector field 'rot $\boldsymbol{u}(\boldsymbol{x})$' on $x_1 - x_2$ plane. The rotation is in the anti-clockwise direction due to the right-handed screw direction of the vectors shown in (a). (c) Vector field 'rot rot $\boldsymbol{u}(\boldsymbol{x})$' on $x_1 - x_2$ plane. The vectors point upwards due to right-handed screw direction of the rotational field of 'rot $\boldsymbol{u}$'.

for rot $\boldsymbol{u}$, shown in Fig. 2.2.2(b), form a rotational field, which can be explained by Eq. (2.2.13). Note that the rotation is anti-clockwise as a result of the right-handed screw direction of $\boldsymbol{u}$. The vectors for 'rot rot $\boldsymbol{u}$' shown in Fig. 2.2.2(c), form a vector field that is orthogonal to the $x_1 - x_2$ plane. They point upwards, which is the right-handed screw direction of 'rot $\boldsymbol{u}$'. Figure 2.2.2 thus shows an intuitive explanation of the effects of the "rot" operation on vector fields.

2.2.2 DECOMPOSITION OF VECTOR FIELDS INTO IRROTATIONAL AND DIVERGENCE-FREE VECTOR FIELDS

We say that the vector field $\boldsymbol{u}(\boldsymbol{x})$ is irrotational in V if it satisfies

$$\mathrm{rot}\,\boldsymbol{u}(\boldsymbol{x})\big(=\nabla\times\boldsymbol{u}(\boldsymbol{x})\big)=\boldsymbol{0},\quad(\boldsymbol{x}\in V) \tag{2.2.15}$$

For example, the vector field $\nabla\varphi(\boldsymbol{x})$, $(\boldsymbol{x}\in V)$ is irrotational since

$$\nabla\times\nabla\varphi(\boldsymbol{x})=\boldsymbol{0},\quad(\boldsymbol{x}\in V) \tag{2.2.16}$$

where $\varphi(\boldsymbol{x})$ is an arbitrary scalar field twice differentiable in V. We also say that the vector field $\boldsymbol{u}(\boldsymbol{x})$ is divergence-free in V if it satisfies

$$\mathrm{div}\;\boldsymbol{u}(\boldsymbol{x})=\partial_i u_i(\boldsymbol{x})=0,\quad(\boldsymbol{x}\in V) \tag{2.2.17}$$

For example, the vector field $\nabla\times\boldsymbol{B}(\boldsymbol{x})$, $(\boldsymbol{x}\in V\subset\mathbb{R}^3)$ is divergence-free since

$$\mathrm{div}\,\nabla\times\boldsymbol{B}(\boldsymbol{x})=0,\quad(\boldsymbol{x}\in V) \tag{2.2.18}$$

where $\boldsymbol{B}(\boldsymbol{x})$ is an arbitrary twice-differentiable vector field.

The Helmholtz theorem states that any arbitrary vector field can be decomposed into two components: an irrotational vector field and a divergence-free vector field. This decomposition is expressed as follows:

$$\boldsymbol{u}(\boldsymbol{x})=\nabla\phi(\boldsymbol{x})+\nabla\times\boldsymbol{B}(\boldsymbol{x}),\quad(\boldsymbol{x}\in V\subset\mathbb{R}^3) \tag{2.2.19}$$

We refer to ϕ and $\boldsymbol{B}$ the scalar and vector potentials for a given vector field $\boldsymbol{u}$, respectively.

To show that this decomposition is possible, let $\boldsymbol{U}(\boldsymbol{x})$ be a solution for the Poisson equation

$$\nabla^2\boldsymbol{U}(\boldsymbol{x})=\boldsymbol{u}(\boldsymbol{x})\quad(\boldsymbol{x}\in V) \tag{2.2.20}$$

The construction of the solution for Eq. (2.2.20) in a bounded domain of V is shown later. We achieve the Helmholtz decomposition based on the function given by Eq. (2.2.20) and the properties of the Laplace operator for a vector field expressed as[3]

$$\nabla^2\boldsymbol{U}(\boldsymbol{x})\quad=\quad\nabla\nabla\cdot\boldsymbol{U}(\boldsymbol{x})-\nabla\times\nabla\times\boldsymbol{U}(\boldsymbol{x}) \tag{2.2.21}$$

where "$\nabla\cdot$" operation stands for "div ".

[3] See problems presented in §2.6.4 at the end of this chapter.

If we define ϕ and $\boldsymbol{B}$ as

$$\begin{aligned}\phi(\boldsymbol{x}) &= \nabla\cdot\boldsymbol{U}(\boldsymbol{x})\\ \boldsymbol{B}(\boldsymbol{x}) &= -\nabla\times\boldsymbol{U}(\boldsymbol{x})\end{aligned} \tag{2.2.22}$$

then we obtain Eq. (2.2.19) from Eq. (2.2.22), which verifies the Helmholtz decomposition.

At this point, let us construct a solution for Eq. (2.2.20) in a bounded domain of V. In general, a solution for Eq. (2.2.20) is formulated as follows:

$$\boldsymbol{U}(\boldsymbol{x}) = -\int_V g(\boldsymbol{x},\boldsymbol{y})\boldsymbol{u}(\boldsymbol{y})d\boldsymbol{y}, \quad (\boldsymbol{x}\in V) \tag{2.2.23}$$

for a case V is a bounded region. Here, $g(\boldsymbol{x},\boldsymbol{y})$ is defined by

$$g(\boldsymbol{x},\boldsymbol{y}) = \frac{1}{4\pi|\boldsymbol{x}-\boldsymbol{y}|} \tag{2.2.24}$$

To establish that Equation (2.2.23) is indeed a solution for Eq. (2.2.20), we introduce $M_\varepsilon(\boldsymbol{y})$ and $S_\varepsilon(\boldsymbol{y})$ $M_\varepsilon(\boldsymbol{y})$ and $S_\varepsilon(\boldsymbol{y})$ as follows:

$$\begin{aligned}M_\varepsilon(\boldsymbol{y}) &= \{\boldsymbol{x}\in\mathbb{R}^3 \mid |\boldsymbol{x}-\boldsymbol{y}|<\varepsilon\}\\ S_\varepsilon(\boldsymbol{y}) &= \{\boldsymbol{x}\in\mathbb{R}^3 \mid |\boldsymbol{x}-\boldsymbol{y}|=\varepsilon\}\end{aligned} \tag{2.2.25}$$

for $\varepsilon > 0$. Consequently, we have

$$\nabla^2 g(\boldsymbol{x},\boldsymbol{y}) = 0, \qquad (\boldsymbol{x}\notin M_\varepsilon(\boldsymbol{y})) \tag{2.2.26}$$

Furthermore, in accordance with the Gauss divergence theorem, as expressed by Eq. (1.3.51), we also derive:

$$\begin{aligned}\int_{M_\varepsilon(\boldsymbol{y})} \nabla^2 g(\boldsymbol{x},\boldsymbol{y})d\boldsymbol{x} &= \int_{S_\varepsilon(\boldsymbol{y})} n_i\frac{\partial}{\partial x_i}g(\boldsymbol{x},\boldsymbol{y})dS\\ &= \int_{S_\varepsilon(\boldsymbol{y})} \frac{\partial g(\boldsymbol{x},\boldsymbol{y})}{\partial n}dS = -1\end{aligned} \tag{2.2.27}$$

Therefore, based on Eqs. (2.2.26) and (2.2.27), it follows that

$$\nabla^2 g(\boldsymbol{x},\boldsymbol{y}) = -\delta(\boldsymbol{x}-\boldsymbol{y}) \tag{2.2.28}$$

where $\delta(\cdot)$ is the Dirac delta function (Appendix B) and $g(\boldsymbol{x},\boldsymbol{y})$ defined by Eq. (2.2.24) serves as Green's function for the Laplace operator. Thus, we arrive at

$$\begin{aligned}-\nabla^2\int_V g(\boldsymbol{x},\boldsymbol{y})\boldsymbol{u}(\boldsymbol{y})d\boldsymbol{y} &= \int_V -\nabla^2 g(\boldsymbol{x},\boldsymbol{y})\boldsymbol{u}(\boldsymbol{y})d\boldsymbol{y}\\ &= \int_V \delta(\boldsymbol{x}-\boldsymbol{y})\boldsymbol{u}(\boldsymbol{y})d\boldsymbol{y}\\ &= \boldsymbol{u}(\boldsymbol{x}), \quad (\boldsymbol{x}\in V)\end{aligned} \tag{2.2.29}$$

This confirms the validity of Eq. (2.2.23). According to the representation of the solution for the Poisson equation shown in Eq. (2.2.23), a region V for the vector field has to be bounded in order for the integral to converge. Note that the Helmholtz decomposition of a vector field also holds for an infinite domain for the case where the vector field satisfies

$$|\boldsymbol{u}(\boldsymbol{x})| = O(|\boldsymbol{x}|^{-2}), \quad (|\boldsymbol{x}| \to \infty) \tag{2.2.30}$$

(Achenbach, 1975).

Equation (2.2.30) indicates that the applicability of the Helmholtz decomposition to a vector field in an infinite domain may have certain restrictions. However, in the subsequent section, our discussion is centered on solutions for the elastic wave equation of irrotational and divergence-free vector fields.

2.3 P AND S WAVE SOLUTIONS DERIVED FROM ELASTIC WAVE EQUATION

2.3.1 SCALAR POTENTIALS FOR ELASTIC WAVE EQUATION

Now, let us return to the elastic wave equation presented in Eq. (1.4.14) in Chapter 1. We apply the discussion in the previous two sections of the present chapter to the equation to find solutions. We seek solutions for Eq. (1.4.14) defined in the whole 3D space under the assumption that $f_i = 0$.

In our search for the solutions, we cannot expect the solutions to satisfy Eq. (2.2.30), which ensures the application of the Helmholtz decomposition of vector fields. For example, the plane wave shown in Eq. (2.1.13) is a case for which Eq. (2.2.30) does not hold, since we have

$$u(\boldsymbol{x},t) \neq O(|\boldsymbol{x}|^{-2}), \quad (|\boldsymbol{x}| \to \infty) \tag{2.3.1}$$

for plane waves.

Nevertheless, we seek solutions that are irrotational, divergence-free, or both. The superposition of such solutions is also a solution for the elastic wave equation. We assume that a solution for Eq. (1.4.14) can be expressed in the form

$$\boldsymbol{u}(\boldsymbol{x},t) = \nabla\phi(\boldsymbol{x},t) + \nabla \times \psi_k(\boldsymbol{x},t)\boldsymbol{e}_k \tag{2.3.2}$$

where ϕ and $\psi_k (k = 1,2,3)$ are scalar potentials. Equation (2.3.2) can also be written as

$$u_i(\boldsymbol{x},t) = \partial_i\phi(\boldsymbol{x},t) + e_{ijk}\partial_j\psi_k(\boldsymbol{x},t) \tag{2.3.3}$$

in terms of the component of the coordinate system. ϕ and ψ_k are expected to yield irrotational and divergence-free solutions, respectively.

The substitution of Eq. (2.3.3) into the left-hand side of Eq. (1.4.14) yields

$$\begin{aligned} &L_{ij}(\partial_1,\partial_2,\partial_3)\left(\partial_j\phi(\boldsymbol{x},t) + e_{jkl}\partial_k\psi_l(\boldsymbol{x},t)\right) \\ &= (\lambda + 2\mu)\,\partial_i\nabla^2\phi(\boldsymbol{x},t) + \mu\, e_{ijk}\partial_j\nabla^2\psi_k(\boldsymbol{x},t) \end{aligned} \tag{2.3.4}$$

As a result, the left-hand side Eq. (1.4.14) written in terms of the scalar potentials becomes

$$\begin{aligned}(\lambda+2\mu)\,\partial_i\nabla^2\phi(\boldsymbol{x},t)&+\mu\, e_{ijk}\partial_j\nabla^2\psi_k(\boldsymbol{x},t)\\ &=\rho\,\partial_i\frac{\partial^2}{\partial t^2}\phi(\boldsymbol{x},t)+\rho e_{ijk}\partial_j\frac{\partial^2}{\partial t^2}\psi_k(\boldsymbol{x},t)\end{aligned}\tag{2.3.5}$$

Therefore, if ϕ and ψ_k solve the scalar wave equations

$$(\lambda+2\mu)\nabla^2\phi(\boldsymbol{x},t) = \rho\frac{\partial^2}{\partial t^2}\phi(\boldsymbol{x},t)\tag{2.3.6}$$

$$\mu\nabla^2\psi_k(\boldsymbol{x},t) = \rho\frac{\partial^2}{\partial t^2}\psi_k(\boldsymbol{x},t)\tag{2.3.7}$$

then Eq. (2.3.3) is a homogeneous solution for Eq. (1.4.14). Equations (2.3.3), (2.3.6), and (2.3.7) show that solutions for the elastic wave equation can be found using the two kinds of the scalar wave equation. In the next section, we construct solutions for the elastic wave equation using the scalar potentials.

2.3.2 PLANE WAVE SOLUTIONS FOR ELASTIC WAVE EQUATION

According to the discussion on solutions for the scalar wave equation given in §2.1, plane wave solutions for Eqs. (2.3.6) and (2.3.7) can be expressed as

$$\phi(\boldsymbol{x},t) = f_0\left(t-\frac{\boldsymbol{p}\cdot\boldsymbol{x}}{c_L}\right)+g_0\left(t+\frac{\boldsymbol{p}\cdot\boldsymbol{x}}{c_L}\right)\tag{2.3.8}$$

$$\psi_k(\boldsymbol{x},t) = f_k\left(t-\frac{\boldsymbol{p}\cdot\boldsymbol{x}}{c_T}\right)+g_k\left(t+\frac{\boldsymbol{p}\cdot\boldsymbol{x}}{c_T}\right),\tag{2.3.9}$$

where f_k and g_k, $(k=0,1,2,3)$ are arbitrary functions, $\boldsymbol{p}\in\mathbb{R}^3$ whose components are

$$\boldsymbol{p}=\begin{pmatrix} p_1 & p_2 & p_3\end{pmatrix},\ (|\boldsymbol{p}|=1)\tag{2.3.10}$$

is a unit vector that describes the direction of wave propagation, and c_L and c_T are wave velocities, respectively defined as

$$c_L = \sqrt{\frac{\lambda+2\mu}{\rho}}\tag{2.3.11}$$

$$c_T = \sqrt{\frac{\mu}{\rho}}\tag{2.3.12}$$

Later, we will see that c_L and c_T are P and S wave velocities, respectively.

Let the wavefield due to the potential ϕ be denoted as $u_i^{(L)}$, which is expressed as

$$\begin{aligned}u_i^{(L)}(\boldsymbol{x},t) &= \partial_i\phi(\boldsymbol{x},t)\\ &= \frac{p_i}{c_L}\left(-f_0'\left(t-\frac{\boldsymbol{p}\cdot\boldsymbol{x}}{c_L}\right)+g_0'\left(t+\frac{\boldsymbol{p}\cdot\boldsymbol{x}}{c_L}\right)\right)\end{aligned}\tag{2.3.13}$$

Equation (2.3.13) shows that the polarization of the wave agrees with the direction of wave propagation, and thus the wave can be characterized as a *longitudinal wave*. Therefore, we can see that $u_i^{(L)}$ is the P wavefield and c_L is the P wave velocity, where the subscript or superscript L stands for "longitudinal". The vector field $u_i^{(L)}$ is irrotational.

Likewise, let the wavefield due to the potential ψ_k be denoted as $u_i^{(T)}$, which is expressed as

$$\begin{aligned} u_i^{(T)}(\boldsymbol{x},t) &= e_{ijk}\partial_j\psi_k(\boldsymbol{x},t) \\ &= \frac{e_{ijk}p_j}{c_T}\Big(-f_k'\big(t-\frac{\boldsymbol{p}\cdot\boldsymbol{x}}{c_T}\big)+g_k'\big(t+\frac{\boldsymbol{p}\cdot\boldsymbol{x}}{c_T}\big)\Big) \end{aligned} \tag{2.3.14}$$

The polarization of the wave $u_i^{(T)}$ has the properties

$$\begin{aligned} p_i u_i^{(T)}(\boldsymbol{x},t) &= \frac{e_{ijk}p_ip_j}{c_T}\Big(-f_k'\big(t-\frac{\boldsymbol{p}\cdot\boldsymbol{x}}{c_T}\big)+g_k'\big(t+\frac{\boldsymbol{p}\cdot\boldsymbol{x}}{c_T}\big)\Big) \\ &= 0 \end{aligned} \tag{2.3.15}$$

This shows that the polarization of the wave is orthogonal to the direction of wave propagation, and thus the wave is characterized as a *transverse wave*. We can see that $u_i^{(T)}$ is the S wavefield and c_T is the S wave velocity, where the superscript or subscript T stands for "transverse". The vector field $u_i^{(T)}$ is divergence-free.

Together with the wave velocities c_L and c_T, we later use the wavenumbers of the P and S waves represented by k_L and k_T, respectively, which are defined as

$$\begin{aligned} k_L &= \frac{\omega}{c_L} \\ k_T &= \frac{\omega}{c_T} \end{aligned} \tag{2.3.16}$$

At this point, we have to be aware that we have used four scalar potentials shown in Eqs. (2.3.8) and (2.3.9) for a solution of the 3D elastic wave equation. In general, it is not necessary to use four scalar potentials for 3D elastic wavefield. We introduce two scalar potentials ψ and χ [4] for the S wave potentials and use totally three scalar potentials to span the 3D displacement field. In terms of ψ and χ, let us define three scalar potentials ψ_k, $(k=1,2,3)$ for the S wave by

$$\begin{aligned} \psi_1(\boldsymbol{x},t) &= \partial_2\psi(\boldsymbol{x},t) \\ \psi_2(\boldsymbol{x},t) &= -\partial_1\psi(\boldsymbol{x},t) \\ \psi_3(\boldsymbol{x},t) &= \chi(\boldsymbol{x},t) \end{aligned} \tag{2.3.17}$$

As a result of Eq. (2.3.17) , Eq.(2.3.2) can be written as

$$\boldsymbol{u}(\boldsymbol{x},t) = \nabla\phi(\boldsymbol{x},t) + \nabla\times\nabla\times\psi(\boldsymbol{x},t)\boldsymbol{e}_3 + \nabla\times\chi(\boldsymbol{x},t)\boldsymbol{e}_3 \tag{2.3.18}$$

[4] χ is different from what was used in Eq. (1.3.5) in Chapter 1.

Note that the potential χ yields the rotational motions around the x_3 axis, while ψ yields those around x_1 and x_2 axes.

Let $u_i^{(T_V)}$ and $u_i^{(T_H)}$ be the displacement field due to the potentials ψ and χ, respectively. For the case that the scalar potentials ψ and χ have the following forms:

$$\begin{aligned}\psi(\boldsymbol{x},t) &= f_1\left(t-\frac{\boldsymbol{p}\cdot\boldsymbol{x}}{c_T}\right)+g_1\left(t+\frac{\boldsymbol{p}\cdot\boldsymbol{x}}{c_T}\right)\\ \chi(\boldsymbol{x},t) &= f_2\left(t-\frac{\boldsymbol{p}\cdot\boldsymbol{x}}{c_T}\right)+g_2\left(t+\frac{\boldsymbol{p}\cdot\boldsymbol{x}}{c_T}\right)\end{aligned} \tag{2.3.19}$$

$u_i^{(T_V)}$ and $u_i^{(T_H)}$ are expressed by

$$\begin{aligned}u_1^{(T_V)}(\boldsymbol{x},t) &= \frac{p_1p_3}{c_T^2}\left(f_1''\left(t-\frac{\boldsymbol{p}\cdot\boldsymbol{x}}{c_T}\right)+g_1''\left(t+\frac{\boldsymbol{p}\cdot\boldsymbol{x}}{c_T}\right)\right)\\ u_2^{(T_V)}(\boldsymbol{x},t) &= \frac{p_2p_3}{c_T^2}\left(f_1''\left(t-\frac{\boldsymbol{p}\cdot\boldsymbol{x}}{c_T}\right)+g_1''\left(t+\frac{\boldsymbol{p}\cdot\boldsymbol{x}}{c_T}\right)\right)\\ u_3^{(T_V)}(\boldsymbol{x},t) &= -\frac{p_1^2+p_2^2}{c_T^2}\left(f_1''\left(t-\frac{\boldsymbol{p}\cdot\boldsymbol{x}}{c_T}\right)+g_1''\left(t+\frac{\boldsymbol{p}\cdot\boldsymbol{x}}{c_T}\right)\right)\end{aligned} \tag{2.3.20}$$

$$\begin{aligned}u_1^{(T_H)}(\boldsymbol{x},t) &= \frac{p_2}{c_T}\left(-f_2'\left(t-\frac{\boldsymbol{p}\cdot\boldsymbol{x}}{c_T}\right)+g_2'\left(t+\frac{\boldsymbol{p}\cdot\boldsymbol{x}}{c_T}\right)\right)\\ u_2^{(T_H)}(\boldsymbol{x},t) &= -\frac{p_1}{c_T}\left(-f_2'\left(t-\frac{\boldsymbol{p}\cdot\boldsymbol{x}}{c_T}\right)+g_2'\left(t+\frac{\boldsymbol{p}\cdot\boldsymbol{x}}{c_T}\right)\right)\\ u_3^{(T_H)}(\boldsymbol{x},t) &= 0\end{aligned} \tag{2.3.21}$$

which yields $(\nabla\times\nabla\times\psi(\boldsymbol{x},t)\boldsymbol{e}_3)\perp(\nabla\times\chi(\boldsymbol{x},t)\boldsymbol{e}_3)$. Namely, the orthogonal decomposition of the S wavefield is established.

Now, according to Eqs. (2.3.13), (2.3.20) and (2.3.21), the P and S wavefields are also found to be orthogonal each other. Therefore, we have the following orthogonal decomposition of 3D displacement field:

$$(\nabla\phi(\boldsymbol{x},t))\perp(\nabla\times\nabla\times\psi(\boldsymbol{x},t)\boldsymbol{e}_3)\perp(\nabla\times\chi(\boldsymbol{x},t)\boldsymbol{e}_3) \tag{2.3.22}$$

by the three scalar potentials, when the three scalar potentials have the identical wave propagation direction.

Figure 2.3.1 shows the propagation of plane waves and the polarization of the P and S waves for this case. The unit vector $\boldsymbol{p}$ that describes the direction of wave propagation is a normal vector to the plane defined as

$$M(s)=\{\boldsymbol{x}\in\mathbb{R}^3 \mid s=\boldsymbol{p}\cdot\boldsymbol{x}\} \tag{2.3.23}$$

On the plane $M(s)$, $f_k(t-s/c)$ and $g_k(t+s/c)$ take constant values, where $k=0,1,2$ for Eq. (2.3.19) and c is c_L or c_T. The polarizations of the P and S waves are indicated by "P wave motion" and "S wave motion", respectively, in Fig. 2.3.1. The polarization of the S wave motions are decomposed into two independent orthogonal vectors on the plane of $M(s)$. One of the decomposed vector is parallel to the x_1-x_2 plane, which is due to the potential χ.

2.4 GREEN'S FUNCTION FOR ELASTIC WAVE EQUATION

2.4.1 DEFINITION OF GREEN'S FUNCTION FOR ELASTIC WAVE EQUATION

2.4.1.1 Definition of Green's function in the time domain

We present the following problem to introduce the concept of Green's function.

Problem 1 *Given that a unit impulse has been applied to a certain space-time point* $(\boldsymbol{y}, \tau) \in \mathbb{R}^3 \times \mathbb{R}$ *in an elastic wavefield, describe the effects of the applied impulse at any point* $(\boldsymbol{x}, t) \in \mathbb{R}^3 \times \mathbb{R} \setminus \{(\boldsymbol{y}, \tau)\}$.

For a solution to this problem, we define the function $G_{ij}(\boldsymbol{x}, t : \boldsymbol{y}, \tau)$. The function describes the displacement in the i-th direction at $(\boldsymbol{x}, t)$ due to the impulse in the j-th direction at $(\boldsymbol{y}, \tau)$. We refer to G_{ij} as *Green's function*. In addition, we refer to $(\boldsymbol{x}, t)$ and $(\boldsymbol{y}, \tau)$ as the field and source points, respectively. An important property of Green's function, called the time translation property, is

$$G_{ij}(\boldsymbol{x}, t : \boldsymbol{y}, \tau) = G_{ij}(\boldsymbol{x}, t + a : \boldsymbol{y}, \tau + a), \ (a \in \forall \mathbb{R}) \tag{2.4.1}$$

If we set $a = -\tau$ in Eq. (2.4.1), we have

$$G_{ij}(\boldsymbol{x}, t : \boldsymbol{y}, \tau) = G_{ij}(\boldsymbol{x}, t - \tau : \boldsymbol{y}, 0) \tag{2.4.2}$$

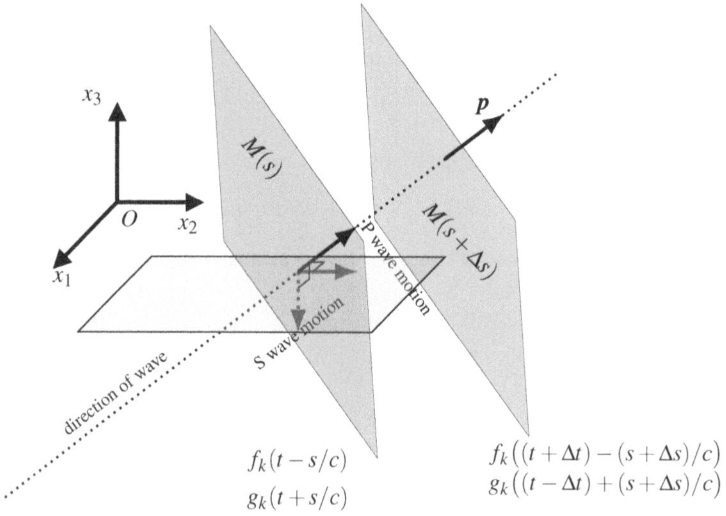

Figure 2.3.1 Plane wave propagation and polarization of P and S waves. The polarization of the P wave is orthogonal to the plane of $M(s)$, while that of the S wave is on the plane of $M(s)$. An important fact for the S wave motion is that it can be decomposed into two linearly independent orthogonal vectors on the plane of $M(s)$. One of the decomposed vector in this figure is parallel to the $x_1 - x_2$ plane, which is due to the potential χ.

Green's function can thus be expressed by $\boldsymbol{x}$, $\boldsymbol{y}$, and $t-\tau$. Therefore, we express Green's function as $G_{ij}(\boldsymbol{x},\boldsymbol{y},t-\tau)$ in the discussion below.

We use the Dirac delta function (Appendix B) for expressing the unit impulse. According to Eq. (1.4.14), the equation for the definition of Green's function becomes

$$\begin{aligned}&\left[L_{ij}(\partial_1,\partial_2,\partial_3)-\delta_{ij}\rho\frac{\partial^2}{\partial t^2}\right]G_{jk}(\boldsymbol{x},\boldsymbol{y},t-\tau)\\ =\ &-\delta_{ik}\delta(\boldsymbol{x}-\boldsymbol{y})\delta(t-\tau),\ (\boldsymbol{x},\boldsymbol{y}\in\mathbb{R}^3,\ t,\tau\in\mathbb{R})\end{aligned}\tag{2.4.3}$$

where $\delta(\cdot)$ is the Dirac delta function.

The solution for the inhomogeneous elastic wave equation

$$\left(L_{ij}(\partial_1,\partial_2,\partial_3)-\delta_{ij}\rho\frac{\partial^2}{\partial t^2}\right)u_j(\boldsymbol{x},t)=-f_i(\boldsymbol{x},t),\ (\boldsymbol{x}\in\mathbb{R}^3,t\in\mathbb{R})\tag{2.4.4}$$

can be expressed by the use of Green's function as

$$u_i(\boldsymbol{x},t)=\int_{-\infty}^{+\infty}\int_{\mathbb{R}^3}G_{ij}(\boldsymbol{x},\boldsymbol{y},t-\tau)f_j(\boldsymbol{y},\tau)d\boldsymbol{y}d\tau\tag{2.4.5}$$

where f_i in Eq. (2.4.4) is an arbitrary force density applied to the elastic wavefield. The derivation of Eq. (2.4.5) is almost clear according to the following process:

$$\begin{aligned}&\left[L_{ij}(\partial_1,\partial_2,\partial_3)-\delta_{ij}\rho\frac{\partial^2}{\partial t^2}\right]\int_{-\infty}^{+\infty}\int_{\mathbb{R}^3}G_{jk}(\boldsymbol{x},\boldsymbol{y},t-\tau)f_k(\boldsymbol{y},\tau)d\boldsymbol{y}\,d\tau\\ =\ &\int_{-\infty}^{+\infty}\int_{\mathbb{R}^3}\left[L_{ij}(\partial_1,\partial_2,\partial_3)-\delta_{ij}\rho\frac{\partial^2}{\partial t^2}\right]G_{jk}(\boldsymbol{x},\boldsymbol{y},t-\tau)f_k(\boldsymbol{y},\tau)d\boldsymbol{y}\,d\tau\\ =\ &-\int_{-\infty}^{+\infty}\int_{\mathbb{R}^3}\delta_{ik}\delta(\boldsymbol{x}-\boldsymbol{y})\,\delta(t-\tau)f_k(\boldsymbol{y},\tau)\,d\boldsymbol{y}\,d\tau=-f_i(\boldsymbol{x},t)\end{aligned}\tag{2.4.6}$$

2.4.1.2 Definition of Green's function in the frequency domain

We also need to deal with the application of a time harmonic point force to an elastic wavefield, which yields the following problem.

Problem 2 *Given that a time harmonic point force of unit amplitude has been applied to a certain spatial point* $\boldsymbol{y}\in\mathbb{R}^3$*, describe the effects of the applied time harmonic force at any point* $\boldsymbol{x}\in\mathbb{R}^3\setminus\{\boldsymbol{y}\}$ *for the case where the time factor of the point force is* $e^{i\omega t}$.

For a solution to this problem, we define the function $\widehat{G}_{ij}(\boldsymbol{x},\boldsymbol{y},\omega)$, which describes the displacement in the i-th direction at $\boldsymbol{x}$ due to the time harmonic point force in the j-th direction at $\boldsymbol{y}$. We also refer to $\widehat{G}_{ij}(\boldsymbol{x},\boldsymbol{y},\omega)$ as Green's function, where $\boldsymbol{x}$ and $\boldsymbol{y}$ are the field and source points, respectively. When we need to distinguish G and $\widehat{G}$, we refer to G as Green's function in the time domain and to $\widehat{G}$ as Green's function in the frequency domain. We also sometimes refer to G as Green's function in the space-time domain and to $\widehat{G}$ as Green's function in the space-frequency domain.

The equation for $\widehat{G}$ is

$$\left[L_{ij}(\partial_1,\partial_2,\partial_3)+\delta_{ij}\rho\omega^2\right]\widehat{G}_{jk}(\boldsymbol{x},\boldsymbol{y},\omega)=-\delta_{ik}\delta(\boldsymbol{x}-\boldsymbol{y}),\ \ (\boldsymbol{x},\boldsymbol{y}\in\mathbb{R}^3) \tag{2.4.7}$$

The solution for the inhomogeneous elastic wave equation

$$\left(L_{ij}(\partial_1,\partial_2,\partial_3)+\delta_{ij}\rho\omega^2\right)\hat{u}_j(\boldsymbol{x})=-\hat{f}_i(\boldsymbol{x}) \tag{2.4.8}$$

is expressed as the integral of the product of Green's function and the inhomogeneous term, given as

$$\hat{u}_i(\boldsymbol{x})=\int_{\mathbb{R}^3}\widehat{G}_{ij}(\boldsymbol{x},\boldsymbol{y},\omega)\hat{f}_j(\boldsymbol{y})d\boldsymbol{y} \tag{2.4.9}$$

The derivation of Eq. (2.4.9) from Eq. (2.4.8) can be carried out using almost the same procedure as that used for Eq. (2.4.6).

Equations (2.4.5) as well as (2.4.9) show the importance of Green's function. Once we obtain Green's function, the solution for the inhomogeneous elastic wave equation can be expressed as the integral of the product of Green's function and the inhomogeneous term, which is the benefit of using Green's function. This importance is not only for theoretical research but also for engineering analysis. It is worth while to derive Green's function, even if the derivation of Green's function is very complicated.

This section shows the derivation process of Green's functions in the time and frequency domains for an elastic wavefield. For the derivation, we obtain Green's function in the frequency domain first. Then, Green's function in the time domain is derived using a Fourier integral transform, since Green's functions in the frequency and time domains are connected by the Fourier integral transforms

$$\begin{aligned}\widehat{G}_{ij}(\boldsymbol{x},\boldsymbol{y},\omega) &= \int_{-\infty}^{+\infty} G_{ij}(\boldsymbol{x},\boldsymbol{y},t-\tau)\exp(-i\omega(t-\tau))dt\\ G_{ij}(\boldsymbol{x},\boldsymbol{y},t-\tau) &= \frac{1}{2\pi}\int_{-\infty}^{+\infty}\widehat{G}_{ij}(\boldsymbol{x},\boldsymbol{y},\omega)\exp(i\omega(t-\tau))d\omega\end{aligned} \tag{2.4.10}$$

The derivation of Green's function for the elastic wave equation uses the same process as that for the scalar wave equation. Therefore, we start with the derivation of Green's function for the scalar wave equation, which provides an important and useful mathematical technique related to the Dirac delta function and the Fourier transform.

2.4.2 GREEN'S FUNCTION FOR SCALAR WAVE EQUATION

2.4.2.1 Green's function in the frequency domain

Now, let us derive Green's function for the scalar wave equation. First, we obtain Green's function in the frequency domain (i.e., Green's function for the Helmholtz equation) defined as

$$\left(\nabla^2 + k^2\right)\widehat{G}(\boldsymbol{x},\boldsymbol{y},\omega) = -\delta(\boldsymbol{x}-\boldsymbol{y}),\ (\boldsymbol{x},\boldsymbol{y}\in\mathbb{R}^3) \tag{2.4.11}$$

We solve Eq. (2.4.11) using the Fourier transform, which connects the space-frequency domain and the wavenumber-frequency domain. The Fourier transform is given by

$$\begin{aligned}\overline{\widehat{G}}(\boldsymbol{\xi},\boldsymbol{y},\omega) &= \int_{\mathbb{R}^3}\widehat{G}(\boldsymbol{x},\boldsymbol{y},\omega)\exp(-i\boldsymbol{x}\cdot\boldsymbol{\xi})d\boldsymbol{x}\\ \widehat{G}(\boldsymbol{x},\boldsymbol{y},\omega) &= \frac{1}{8\pi^3}\int_{\mathbb{R}^3}\overline{\widehat{G}}(\boldsymbol{\xi},\boldsymbol{y},\omega)\exp(i\boldsymbol{x}\cdot\boldsymbol{\xi})d\boldsymbol{\xi}\end{aligned} \tag{2.4.12}$$

where $\boldsymbol{\xi} = (\xi_1\ \xi_2\ \xi_3) \in \mathbb{R}^3$ is a point in the wavenumber space. The result of the application of the Fourier transform to Eq. (2.4.11) is

$$\begin{aligned}(\xi^2-k^2)\overline{\widehat{G}}(\boldsymbol{\xi},\boldsymbol{y},\omega) &= \exp(-i\boldsymbol{\xi}\cdot\boldsymbol{y})\\ &\quad +C\,(\xi^2-k^2)\delta(\xi^2-k^2)\end{aligned} \tag{2.4.13}$$

where

$$\xi^2 = |\boldsymbol{\xi}|^2 = \xi_1^2+\xi_2^2+\xi_3^2 \tag{2.4.14}$$

and C is an undetermined coefficient. In Eq. (2.4.13), we use

$$(\xi^2-k^2)\delta(\xi^2-k^2) = 0 \tag{2.4.15}$$

from Appendix B. The second term on the right-hand side of Eq. (2.4.13) is used for the treatment of the singularity of $\overline{\widehat{G}}$ at the point $\xi^2 = k^2$. Due to the presence of this second term, we have the following representation of $\overline{\widehat{G}}$.

$$\overline{\widehat{G}}(\boldsymbol{\xi},\boldsymbol{y},\omega) = \text{P.V.}\frac{\exp(-i\boldsymbol{\xi}\cdot\boldsymbol{y})}{\xi^2-k^2} + C\delta(\xi^2-k^2) \tag{2.4.16}$$

where P.V. denotes the Cauchy principal value (Appendix B). For the case where $C = \pm i\pi$, $\overline{\widehat{G}}$ has the form

$$\overline{\widehat{G}}^{(\pm)}(\boldsymbol{\xi},\boldsymbol{y},\omega) = \frac{\exp(-i\boldsymbol{\xi}\cdot\boldsymbol{y})}{\xi^2-k^2\mp i\varepsilon},\quad \text{(double-sign corresponds)} \tag{2.4.17}$$

where ε is an infinitesimally small positive number. Equation (2.4.17) shows how to avoid the singularity of $\overline{\widehat{G}}^{(\pm)}$ at $\xi^2 = k^2$ in the complex wavenumber plane.

The inverse Fourier transform in Eq. (2.4.17) is carried out based on the spherical coordinate system. According to Note 2.1, the inverse Fourier transform in Eq. (2.4.17) becomes

$$\begin{aligned}\widehat{G}^{(\pm)}(\boldsymbol{x},\boldsymbol{y},\omega) &= \frac{1}{8\pi^3}\int_{\mathbb{R}^3}\frac{\exp\left(i\boldsymbol{\xi}\cdot(\boldsymbol{x}-\boldsymbol{y})\right)}{\xi^2-k^2\mp i\varepsilon}d\boldsymbol{\xi}\\ &= \frac{1}{8\pi^2 ir}\int_{-\infty}^{\infty}\frac{\xi\left(e^{i\xi r}-e^{-i\xi r}\right)}{\xi^2-k^2\mp i\varepsilon}d\xi,\\ &\qquad \text{(double-sign corresponds)}\end{aligned} \tag{2.4.18}$$

which is to be evaluated in the complex ξ plane.

Based on Jordan's lemma presented in Note 2.2 and the residue theorem, we have the following formulas for evaluating Eq. (2.4.18).

$$\begin{aligned}&\int_{-\infty}^{\infty} \frac{\xi e^{i\xi r}}{\xi^2 - k^2 - i\varepsilon} d\xi \\ &= \lim_{R\to\infty} \int_{[-R,R]+M_{R_+}} \frac{\xi e^{i\xi r}}{\xi^2 - k^2 - i\varepsilon} d\xi = \pi i e^{ikr}, \\ &\quad \left(\text{see Fig. 2.4.1 (a)}\right)\end{aligned} \tag{2.4.19}$$

$$\begin{aligned}&\int_{-\infty}^{\infty} \frac{\xi e^{-i\xi r}}{\xi^2 - k^2 - i\varepsilon} d\xi \\ &= \lim_{R\to\infty} \int_{[-R,R]+M_{R_-}} \frac{\xi e^{-i\xi r}}{\xi^2 - k^2 - i\varepsilon} d\xi = -\pi i e^{ikr}, \\ &\quad \left(\text{see Fig. 2.4.1 (b)}\right)\end{aligned} \tag{2.4.20}$$

$$\begin{aligned}&\int_{-\infty}^{\infty} \frac{\xi e^{i\xi r}}{\xi^2 - k^2 + i\varepsilon} d\xi \\ &= \lim_{R\to\infty} \int_{[-R,R]+M_{R_+}} \frac{\xi e^{i\xi r}}{\xi^2 - k^2 + i\varepsilon} d\xi = \pi i e^{-ikr}, \\ &\quad \left(\text{see Fig. 2.4.1 (c)}\right)\end{aligned} \tag{2.4.21}$$

$$\begin{aligned}&\int_{-\infty}^{\infty} \frac{\xi e^{-i\xi r}}{\xi^2 - k^2 + i\varepsilon} d\xi \\ &= \lim_{R\to\infty} \int_{[-R,R]+M_{R_-}} \frac{\xi e^{-i\xi r}}{\xi^2 - k^2 + i\varepsilon} d\xi = -\pi i e^{-ikr}, \\ &\quad \left(\text{see Fig. 2.4.1 (d)}\right)\end{aligned} \tag{2.4.22}$$

where the paths of the integral M_{R+} and M_{R-} are shown in Fig.2.4.1(a)-(d).

From Eqs. (2.4.19) to (2.4.22), we have the following result for Green's function in the frequency domain:

$$\widehat{G}^{(\pm)}(\boldsymbol{x},\boldsymbol{y},\omega) = \frac{1}{4\pi|\boldsymbol{x}-\boldsymbol{y}|} \exp\left(\pm i\omega|\boldsymbol{x}-\boldsymbol{y}|/c\right), \quad \text{(double-sign corresponds)} \tag{2.4.23}$$

This equation gives the incoming and outgoing spherical waves due to the time factor $e^{i\omega t}$. We presented the solution for spherical waves for the scalar Helmholtz equation in Eq. (2.1.16). These spherical waves are related to Green's function.

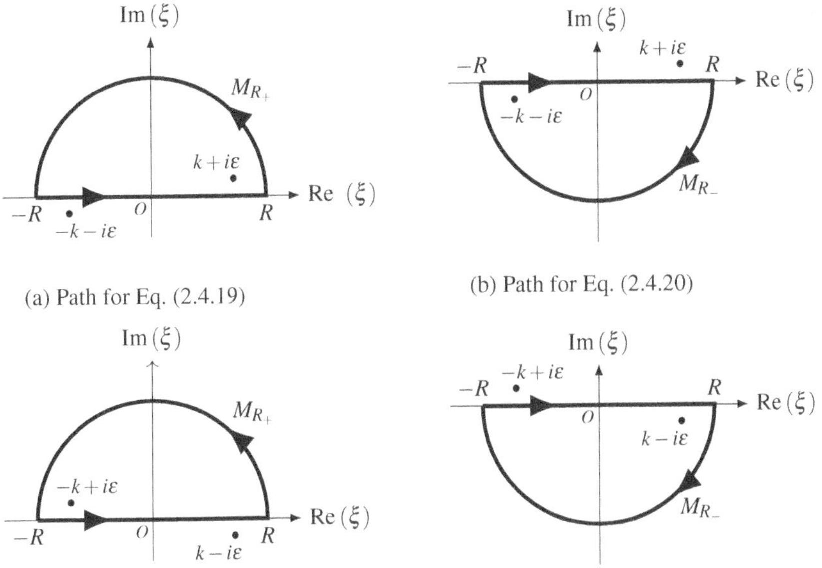

(a) Path for Eq. (2.4.19)

(b) Path for Eq. (2.4.20)

(c) Path for Eq. (2.4.21)

(d) Path for Eq. (2.4.22)

Figure 2.4.1 Paths for integral in complex wavenumber plane used to obtain Eqs. (2.4.19) to (2.4.21).

2.4.2.2 Green's function in the time domain

Green's function for the scalar wave equation in the time domain is defined as

$$\begin{aligned}\left(\nabla^2 - \frac{1}{c^2}\frac{\partial^2}{\partial t^2}\right)G(\boldsymbol{x},\boldsymbol{y},t-\tau) &= -\delta(\boldsymbol{x}-\boldsymbol{y})\delta(t-\tau) \\ &\quad (\boldsymbol{x},\boldsymbol{y}\in\mathbb{R}^3, t-\tau\in\mathbb{R})\end{aligned} \tag{2.4.24}$$

which can be derived from the inverse Fourier transform of Green's function in the frequency domain in Eq. (2.4.23). The result is

$$\begin{aligned}G^{(\pm)}(\boldsymbol{x},\boldsymbol{y},t-\tau) &= \frac{1}{2\pi}\int_{-\infty}^{+\infty}\widehat{G}^{(\pm)}(\boldsymbol{x},\boldsymbol{y},\omega)\exp(i\omega(t-\tau))d\omega \\ &= \frac{1}{4\pi|\boldsymbol{x}-\boldsymbol{y}|}\delta\left(t-\tau\pm|\boldsymbol{x}-\boldsymbol{y}|/c\right)\end{aligned} \tag{2.4.25}$$

According to Eq. (2.4.25), the difference between $G^{(+)}$ and $G^{(-)}$ is $\pm|\boldsymbol{x}-\boldsymbol{y}|/c$. For later discussion, we define

$$t_D(\boldsymbol{x},\boldsymbol{y}) = \frac{|\boldsymbol{x}-\boldsymbol{y}|}{c} \tag{2.4.26}$$

which describes the time needed for wave propagation over a distance of $|\boldsymbol{x}-\boldsymbol{y}|$ with a wave velocity c. In order to have a physical understanding of the two kinds of Green's function, let us construct the solutions of the inhomogeneous scalar wave equation

$$\left(\nabla^2 - \rho\frac{\partial^2}{\partial t^2}\right)u(\boldsymbol{x},t) = -\delta(\boldsymbol{x}-\boldsymbol{y}_0)f(t) \tag{2.4.27}$$

where $\boldsymbol{y}_0$ and $f(t)$ are the position and the time signal of the point force applied to the wavefield, respectively. Using the two kinds of Green's function, we have the following expressions for the solutions:

$$\begin{aligned} u^{(\pm)}(\boldsymbol{x},t) &= \int_{-\infty}^{\infty}\int_{\mathbb{R}^3} G^{(\pm)}(\boldsymbol{x},\boldsymbol{y},t-\tau)\delta(\boldsymbol{y}-\boldsymbol{y}_0)f(\tau)d\boldsymbol{y}\,d\tau \\ &= \int_{-\infty}^{+\infty}\frac{1}{4\pi|\boldsymbol{x}-\boldsymbol{y}_0|}\delta(t-\tau\pm|\boldsymbol{x}-\boldsymbol{y}_0|/c)f(\tau)d\tau \\ &= \frac{1}{4\pi|\boldsymbol{x}-\boldsymbol{y}_0|}f(t\pm t_D(\boldsymbol{x},\boldsymbol{y}_0)), \text{ (double-sign corresponds)} \end{aligned} \tag{2.4.28}$$

Equation (2.4.28) shows how the solution of the wave equation keeps its waveform as it propagates, although the waveform's amplitude decays in inverse proportion to the distance $|\boldsymbol{x}-\boldsymbol{y}_0|$ as $|\boldsymbol{x}|\to\infty$.

Note that t for $u^{(\pm)}$ in Eq. (2.4.28) is the time required for the observation of the signal and $u^{(+)}(\boldsymbol{x},t)$ observes the signal of the time at $t+t_D(\boldsymbol{x},\boldsymbol{y}_0)$, which is ahead of the current time t. This time advance decreases as the field point $\boldsymbol{x}$ approaches the source point $\boldsymbol{y}_0$, which shows that $u^{(+)}(\boldsymbol{x},t)$ is the incoming spherical wave from infinity to the source point. Similarly, $u^{(-)}(\boldsymbol{x},t)$ describes the outgoing spherical wave from the source point to infinity. That is, it observes the signal of the time at $t-t_D(\boldsymbol{x},\boldsymbol{y}_0)$, which is behind the current time t. This time retardation increases as the field point $\boldsymbol{x}$ moves away from the source point $\boldsymbol{y}_0$. $G^{(+)}$ is referred to as the advanced Green's function and $G^{(-)}$ is referred to as the retarded Green's function. Notably, the retarded Green's function satisfies causality and therefore, in general, $G^{(-)}$ is used for engineering problems such as scattering analysis.

Note 2.1 Inverse Fourier transform of Eq. (2.4.17)

The inverse Fourier transform of Eq. (2.4.17) is carried out using the spherical coordinate system. Figure N2.1 shows the relationships between the components of the Cartesian and spherical coordinate systems in wavenumber space. According to Fig. N2.1:

$$\begin{aligned} \xi_1 &= \xi \sin\theta \cos\varphi \\ \xi_2 &= \xi \sin\theta \sin\varphi \\ \xi_3 &= \xi \cos\theta \end{aligned} \tag{N2.1.1}$$

In addition, we set the direction of ξ_3 in wavenumber space to agree with the direction of $\boldsymbol{x}-\boldsymbol{y}$, which yields

$$\boldsymbol{\xi}\cdot(\boldsymbol{x}-\boldsymbol{y}) = \xi r \cos\theta \tag{N2.1.2}$$

where $r=|\boldsymbol{x}-\boldsymbol{y}|$. Therefore, the inverse Fourier transform of Eq. (2.4.17) is given by

$$\begin{aligned} \widehat{G}^{(\pm)}(\boldsymbol{x},\boldsymbol{y},\omega) &= \frac{1}{8\pi^3}\int_{\mathbb{R}^3}\frac{\exp(i(\boldsymbol{x}-\boldsymbol{y})\cdot\boldsymbol{\xi})}{\xi^2-k^2\mp i\varepsilon}d\boldsymbol{\xi} \\ &= \frac{1}{8\pi^3}\int_0^\infty \frac{\xi^2\exp(i\xi r\cos\theta)}{\xi^2-k^2\mp i\varepsilon}d\xi\int_0^\pi \sin\theta\, d\theta \int_0^{2\pi} d\varphi \end{aligned} \tag{N2.1.3}$$

For Eq. (N2.1.3), we use

$$d\xi_1 d\xi_2 d\xi_3 = |J|\, d\xi d\theta d\varphi \tag{N2.1.4}$$

where J is a Jacobian, given as

$$J = \begin{vmatrix} \frac{\partial\xi_1}{\partial\xi} & \frac{\partial\xi_1}{\partial\theta} & \frac{\partial\xi_1}{\partial\varphi} \\ \frac{\partial\xi_2}{\partial\xi} & \frac{\partial\xi_2}{\partial\theta} & \frac{\partial\xi_2}{\partial\varphi} \\ \frac{\partial\xi_3}{\partial\xi} & \frac{\partial\xi_3}{\partial\theta} & \frac{\partial\xi_3}{\partial\varphi} \end{vmatrix} = \xi^2\sin\theta \tag{N2.1.5}$$

The integration of Eq. (N2.1.3) with respect to the components (θ,φ) yields

$$\widehat{G}^{(\pm)}(\boldsymbol{x},\boldsymbol{y},\omega) = \frac{1}{8\pi^2 ir}\int_{-\infty}^{\infty}\frac{\xi(e^{i\xi r}-e^{-i\xi r})}{\xi^2-k^2\mp i\varepsilon}d\xi \tag{N2.1.6}$$

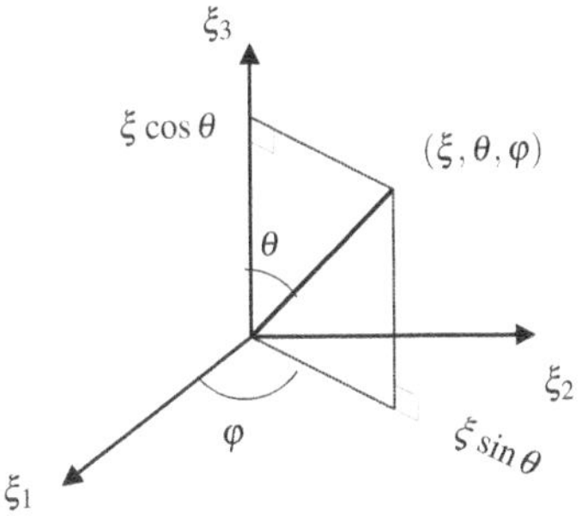

Figure: N2.1 Relationships between components of Cartesian and spherical coordinate systems in wavenumber space.

Note 2.2 Jordan's lemma

Lemma N2.2 *Let $F(\xi)$ be a regular function except for a finite number of poles $\alpha_1, \alpha_2, \cdots, \alpha_m$ in the complex plane. For the case where $F(\xi)$ satisfies*

$$\lim_{|\xi|\to\infty} F(\xi) = 0$$

we have

$$\begin{aligned}\lim_{R\to\infty}\int_{M_{R+}} e^{i\xi r}F(\xi)d\xi &= 0\\ \lim_{R\to\infty}\int_{M_{R-}} e^{-i\xi r}F(\xi)d\xi &= 0\end{aligned} \qquad \text{(N2.2.1)}$$

where $r>0$ is a positive real number and M_{R+} and M_{R-} are semi-circular contours of radius R lying in the upper and lower half-plane of the complex ξ plane, respectively.

[Proof] First, we prove

$$\lim_{R\to\infty}\int_{M_{R+}} e^{i\xi r}F(\xi)d\xi = 0 \qquad \text{(N2.2.2)}$$

Let $\xi \in M_{R+}$ be denoted as

$$\xi = Re^{i\chi}, \quad (0\le\chi\le\pi) \qquad \text{(N2.2.3)}$$

Then, we have the following inequalities for the above integral:

$$\begin{aligned}\left|\int_{M_{R+}} e^{i\xi r}F(\xi)d\xi\right| &\le \int_0^{\pi}\left|F(\xi)\,e^{ir(R\cos\chi+iR\sin\chi)}\,iRe^{i\chi}\right|d\chi\\ &\le R\sup_{\xi\in M_{R+}}|F(\xi)|\int_0^{\pi}e^{-rR\sin\chi}d\chi\\ &= 2R\sup_{\xi\in M_{R+}}|F(\xi)|\int_0^{\pi/2}e^{-rR\sin\chi}d\chi\\ &\le 2R\sup_{\xi\in M_{R+}}|F(\xi)|\int_0^{\pi/2}e^{-2rR\chi/\pi}d\chi\\ &= \frac{\pi}{r}\sup_{\xi\in M_{R+}}|F(\xi)|(1-e^{-rR})\to 0,\ (R\to\infty)\end{aligned} \qquad \text{(N2.2.4)}$$

For the derivation of Eq. (N2.2.4), we use

$$\sin\chi \ge \frac{2}{\pi}\chi,\ (0\le\chi\le\frac{\pi}{2}) \qquad \text{(N2.2.5)}$$

Likewise, we can also verify

$$\lim_{R\to\infty}\int_{M_{R-}} e^{-i\xi r}F(\xi)d\xi = 0 \qquad \text{(N2.2.6)}$$

□

2.4.3 DERIVATION OF GREEN'S FUNCTION FOR ELASTIC WAVE EQUATION

2.4.3.1 Green's function in the frequency domain

We will now derive Green's function for the elastic wave equation in the frequency domain. Taking the Fourier transform of Eq. (2.4.7) from the space-frequency domain to the wavenumber-frequency domain yields:

$$\left[L_{ij}(i\xi_1,i\xi_2,i\xi_3)+\rho\omega^2\delta_{ij}\right]\overline{\widehat{G}}_{jk}(\boldsymbol{\xi},\boldsymbol{y},\omega) = -\delta_{ik}\exp(-i\boldsymbol{\xi}\cdot\boldsymbol{y}) \quad (2.4.29)$$

where

$$L_{ij}(i\xi_1,i\xi_2,i\xi_3) = -(\lambda+\mu)\xi_i\xi_j - \mu\xi^2\delta_{ij} \quad (2.4.30)$$

It is not very difficult to see that

$$\begin{aligned}&\left[L_{ij}(i\xi_1,i\xi_2,i\xi_3)+\rho\omega^2\delta_{ij}\right]^{-1}\\ &= \frac{1}{2\mu(1-\nu)\overline{\widehat{\square}}_T\overline{\widehat{\square}}_L}\left[2(1-\nu)\overline{\widehat{\square}}_L\delta_{ij}-(i\xi_i)(i\xi_j)\right] \quad (2.4.31)\end{aligned}$$

where

$$\begin{aligned}\overline{\widehat{\square}}_L &= -\xi^2+k_L^2\\ \overline{\widehat{\square}}_T &= -\xi^2+k_T^2 \quad (2.4.32)\end{aligned}$$

and ν is the Poisson ratio. Note that Eq. (2.4.31) holds true when $\xi^2 \neq k_L^2$ and $\xi^2 \neq k_T^2$. Remember that k_L and k_T are the wavenumbers for the P and S waves, respectively, as defined by Eq. (2.3.16). By multiplying both sides of Eq. (2.4.29) with the expression

$$(1-\nu)\overline{\widehat{\square}}_T\overline{\widehat{\square}}_L\left[L_{li}(i\xi_1,i\xi_2,i\xi_3)+\rho\omega^2\delta_{li}\right]^{-1}$$

we arrive at the following result:

$$\begin{aligned}&(1-\nu)\overline{\widehat{\square}}_L\overline{\widehat{\square}}_T\overline{\widehat{G}}_{ij}(\boldsymbol{\xi},\boldsymbol{y},\omega)\\ &= -\frac{1}{2\mu}\left[2(1-\nu)\overline{\widehat{\square}}_L\delta_{ij}-(i\xi_i)(i\xi_j)\right]\exp(-i\boldsymbol{\xi}\cdot\boldsymbol{y}) \quad (2.4.33)\end{aligned}$$

Therefore, if we have $\overline{\widehat{g}}(\boldsymbol{\xi},\omega)$ that fulfills the equation

$$\overline{\widehat{\square}}_L\overline{\widehat{\square}}_T\overline{\widehat{g}}(\boldsymbol{\xi},\omega) = -\frac{1}{1-\nu}\left[1+C\overline{\widehat{\square}}_L\overline{\widehat{\square}}_T\,\delta\left(\overline{\widehat{\square}}_L\overline{\widehat{\square}}_T\right)\right] \quad (2.4.34)$$

where C is an undetermined coefficient, then the solution for Eq. (2.4.7) can be expressed as

$$\begin{aligned}&2\mu\widehat{G}_{ij}(\boldsymbol{x},\boldsymbol{y},\omega)\\ &= \frac{1}{8\pi^3}\int_{\mathbb{R}^3}\left[2(1-\nu)\overline{\widehat{\square}}_L\delta_{ij}-(i\xi_i)(i\xi_j)\right]\\ &\quad\times\overline{\widehat{g}}(\boldsymbol{\xi},\omega)\exp(i(\boldsymbol{x}-\boldsymbol{y})\cdot\boldsymbol{\xi})d\boldsymbol{\xi}\\ &= \left[2(1-\nu)\widehat{\square}_L\delta_{ij}-\frac{\partial}{\partial x_i}\frac{\partial}{\partial x_j}\right]\widehat{g}(\boldsymbol{x}-\boldsymbol{y},\omega) \quad (2.4.35)\end{aligned}$$

where $\widehat{\square}_L$ and $\hat{g}(\boldsymbol{x}-\boldsymbol{y},\omega)$ are respectively defined as

$$\widehat{\square}_L = \nabla^2 + k_L^2 \tag{2.4.36}$$

$$\hat{g}(\boldsymbol{x}-\boldsymbol{y},\omega) = \frac{1}{8\pi^3}\int_{\mathbb{R}^3} \bar{g}(\boldsymbol{\xi},\omega)\exp\big(i(\boldsymbol{x}-\boldsymbol{y})\cdot\boldsymbol{\xi}\big)d\boldsymbol{\xi} \tag{2.4.37}$$

Equation (2.4.35) shows that construction of Green's function in the frequency domain defined by Eq. (2.4.7) is possible via $\widehat{g}(\boldsymbol{x}-\boldsymbol{y},\omega)$. Therefore, let us investigate Eq. (2.4.37) to obtain $\widehat{g}(\boldsymbol{x}-\boldsymbol{y},\omega)$ from $\overline{\widehat{g}}(\boldsymbol{\xi},\omega)$. According to Eq. (2.4.34), we have the following expression for $\overline{\widehat{g}}(\boldsymbol{\xi},\omega)$:

$$\begin{aligned}\overline{\widehat{g}}(\boldsymbol{\xi},\omega) &= \frac{1}{1-\nu}\Big[-\text{P.V.}\frac{1}{(\xi^2-k_T^2)(\xi^2-k_L^2)} \\ &\quad -C\delta\Big((\xi^2-k_T^2)(\xi^2-k_L^2)\Big)\Big]\end{aligned} \tag{2.4.38}$$

For the case where $C=\pm i\pi$, Eq. (2.4.38) becomes

$$\overline{\widehat{g}}^{(\pm)}(\boldsymbol{\xi},\omega) = -\frac{1}{1-\nu}\frac{1}{(\xi^2-k_T^2\mp i\varepsilon)(\xi^2-k_L^2\mp i\varepsilon)},$$

$$\text{(double-sign corresponds)} \tag{2.4.39}$$

The inverse Fourier transform in terms of the spherical coordinate system for $\overline{\widehat{g}}^{(\pm)}(\boldsymbol{\xi},\omega)$ is

$$\begin{aligned}\widehat{g}^{(\pm)}(\boldsymbol{x}-\boldsymbol{y},\omega) &= -\frac{1}{8\pi^3(1-\nu)}\int_0^\infty \frac{\xi^2 d\xi}{(\xi^2-k_T^2\mp i\varepsilon)(\xi^2-k_L^2\mp i\varepsilon)} \\ &\quad \times\int_0^\pi \sin\theta d\theta \int_0^{2\pi} d\varphi \exp(i\xi r\cos\theta) \\ &= -\frac{1}{8\pi^2(1-\nu)ir}\int_{-\infty}^{+\infty}\frac{\xi(e^{i\xi r}-e^{-i\xi r})}{(\xi^2-k_T^2\mp i\varepsilon)(\xi^2-k_L^2\mp i\varepsilon)}d\xi \\ &= -\frac{1}{4\pi(1-\nu)r}\frac{e^{\pm ik_T r}-e^{\pm ik_L r}}{k_T^2-k_L^2},\end{aligned}$$

$$\text{(double-sign corresponds)} \tag{2.4.40}$$

where $r=|\boldsymbol{x}-\boldsymbol{y}|$. Computing $\widehat{G}_{ij}$ from $\widehat{g}$ using Eq. (2.4.37) is relatively simple. We use

$$2(1-\nu)\widehat{\square}_L\widehat{g}^{\pm} = \frac{1}{2\pi r}e^{\pm ik_T r}, \quad \text{(double-sign corresponds)} \tag{2.4.41}$$

and

$$(1-\nu)(k_T^2-k_L^2) = \frac{k_T^2}{2} \tag{2.4.42}$$

As a result, we have

$$\widehat{G}_{ij}^{(\pm)}(\boldsymbol{x},\boldsymbol{y},\omega) = \frac{1}{4\pi\mu}\Big[\widehat{A}^{(\pm)}(r,\omega)\delta_{ij}+\widehat{B}^{(\pm)}(r,\omega)\frac{x_i-y_i}{r}\frac{x_j-y_j}{r}\Big],$$

$$\text{(double-sign corresponds)} \tag{2.4.43}$$

where

$$
\begin{aligned}
\widehat{A}^{(\pm)}(r,\omega) &= \left[1-\frac{1}{k_T^2r^2}\pm\frac{i}{k_Tr}\right]\frac{e^{\pm ik_Tr}}{r} \\
&\quad -(c_T/c_L)^2\left[-\frac{1}{k_L^2r^2}\pm\frac{i}{k_Lr}\right]\frac{e^{\pm ik_Lr}}{r}, \\
\widehat{B}^{(\pm)}(r,\omega) &= -\left[1-\frac{3}{k_T^2r^2}\pm\frac{3i}{k_Tr}\right]\frac{e^{\pm ik_Tr}}{r} \\
&\quad +(c_T/c_L)^2\left[1-\frac{3}{k_L^2r^2}\pm\frac{3i}{k_Lr}\right]\frac{e^{\pm ik_Lr}}{r}, \\
&\qquad \text{(double-sign corresponds)} \qquad (2.4.44)
\end{aligned}
$$

Equations (2.4.43) and (2.4.44) show that Green's function for the elastic wave equation consists of spherical P and S waves. For the case where $r\to\infty$, Green's function can be expressed as:

$$
\widehat{G}_{ij}^{(\pm)}(\boldsymbol{x},\boldsymbol{y},\omega) = \widehat{G}_{ij}^{(\pm)T\infty}(r,\omega)+\widehat{G}_{ij}^{(\pm)L\infty}(r,\omega)+O(r^{-2}) \qquad (2.4.45)
$$

where

$$
\begin{aligned}
\widehat{G}_{ij}^{(\pm)T\infty}(r,\omega) &= \frac{1}{4\pi\mu}\left[\delta_{ij}-\frac{x_i-y_i}{r}\frac{x_j-y_j}{r}\right]\frac{e^{\pm k_Tr}}{r} \\
\widehat{G}_{ij}^{(\pm)L\infty}(r,\omega) &= \frac{1}{4\pi(\lambda+2\mu)}\frac{x_i-y_i}{r}\frac{x_j-y_j}{r}\frac{e^{\pm k_Lr}}{r} \\
&\qquad \text{(double-sign corresponds)} \qquad (2.4.46)
\end{aligned}
$$

These expressions result in the following properties of $\widehat{G}_{ij}^{(\pm)T\infty}$ and $\widehat{G}_{ij}^{(\pm)T\infty}$:

$$
\begin{aligned}
\widehat{G}_{ij}^{(\pm)T\infty}(r,\omega) &\perp (x_i-y_i)/r \\
\widehat{G}_{ij}^{(\pm)L\infty}(r,\omega) &\parallel (x_i-y_i)/r
\end{aligned} \qquad (2.4.47)
$$

This shows that $\widehat{G}_{ij}^{(\pm)T\infty}$ and $\widehat{G}_{ij}^{(\pm)L\infty}$ respectively exhibit their proper polarizations as the S and P waves in spherical wave propagation.

2.4.3.2 Green's function in the time domain

At the end of this section, we carry out the inverse Fourier transform of Eq. (2.4.43) with respect to frequency to obtain Green's function in the time domain. To accomplish the inverse Fourier transform, we employ the following formulas based on the

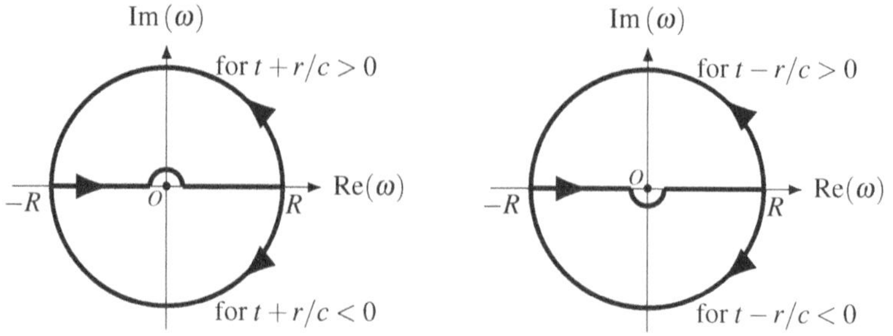

(a) Paths for Eqs. (2.4.48) and (2.4.49) (b) Paths for Eqs. (2.4.50) and (2.4.51)

Figure 2.4.2 Paths of integrals used to obtain space-time domain solution in complex ω plane.

paths of integrals in the complex ω plane as shown in Fig. 2.4.2:

$$\frac{1}{2\pi}\int_{-\infty}^{\infty}\frac{i}{kr}e^{ikr+i\omega t}d\omega = \frac{c}{r}H(-(t+r/c)) \tag{2.4.48}$$

$$\frac{1}{2\pi}\int_{-\infty}^{\infty}\frac{1}{k^2r^2}e^{ikr+i\omega t}d\omega = \frac{c^2}{r^2}(t+r/c)H(-(t+r/c)) \tag{2.4.49}$$

$$\frac{1}{2\pi}\int_{-\infty}^{\infty}-\frac{i}{kr}e^{-ikr+i\omega t}d\omega = \frac{c}{r}H(t-r/c) \tag{2.4.50}$$

$$\frac{1}{2\pi}\int_{-\infty}^{\infty}\frac{1}{k^2r^2}e^{-ikr+i\omega t}d\omega = -\frac{c^2}{r^2}(t-r/c)H(t-r/c) \tag{2.4.51}$$

where $k = \omega/c$ and $H(\cdot)$ is the Heaviside unit step function expressed as

$$H(t) = \begin{cases} 1 & (t>0) \\ 0 & (t<0) \end{cases} \tag{2.4.52}$$

As a result, the inverse Fourier transform of $\hat{A}^{(\pm)}$ and $\hat{B}^{(\pm)}$ becomes

$$\begin{aligned} A^{(\pm)}(r,t) &= \frac{1}{2\pi}\int_{-\infty}^{\infty}\hat{A}^{(\pm)}(r,\omega)\exp(i\omega t)d\omega \\ &= \frac{\delta(t\pm r/c_T)}{r} \pm \frac{c_T^2 t}{r^3}\Big(H(\mp t - r/c_L) - H(\mp t - r/c_T)\Big) \\ B^{(\pm)}(r,t) &= \frac{1}{2\pi}\int_{-\infty}^{\infty}\hat{B}^{(\pm)}(r,\omega)\exp(i\omega t)d\omega \\ &= -\frac{\delta(t\pm r/c_T)}{r} + (c_T/c_L)^2\frac{\delta(t\pm r/c_L)}{r} \\ &\quad \mp\frac{3c_T^2 t}{r^3}\Big(H(\mp t - r/c_L) - H(\mp t - r/c_T)\Big) \\ &\qquad \text{(double-sign corresponds)} \end{aligned} \tag{2.4.53}$$

Therefore, we get a representation of Green's function defined in Eq. (2.4.3) in the form

$$\begin{aligned} G_{ij}^{(\pm)}(\boldsymbol{x},\boldsymbol{y},t-\tau) &= \frac{1}{4\pi\mu}\Big[A^{(\pm)}(r,t-\tau)\,\delta_{ij} \\ &\quad +B^{(\pm)}(r,t-\tau)\frac{x_i-y_i}{r}\frac{x_i-y_i}{r}\Big] \\ &\quad \text{(double-sign corresponds)} \end{aligned} \tag{2.4.54}$$

Equations (2.4.53) and (2.4.54) show that how the S and P wavefunctions contribute to Green's function. For the case where $r \to \infty$, Green's function in the time domain is expressed as:

$$\widehat{G}_{ij}^{(\pm)}(\boldsymbol{x},\boldsymbol{y},t) = G_{ij}^{(\pm)T\infty}(r,t) + G_{ij}^{(\pm)L\infty}(r,t) + O(r^{-3}) \tag{2.4.55}$$

where

$$\begin{aligned} G_{ij}^{(\pm)T\infty}(r,t) &= \frac{1}{4\pi\mu}\left[\delta_{ij} - \frac{x_i-y_i}{r}\frac{x_j-y_j}{r}\right]\frac{\delta(t\pm r/c_T)}{r} \\ G_{ij}^{(\pm)L\infty}(r,t) &= \frac{1}{4\pi(\lambda+2\mu)}\frac{x_i-y_i}{r}\frac{x_j-y_j}{r}\frac{\delta(t\pm r/c_L)}{r} \\ &\quad \text{(double-sign corresponds)} \end{aligned} \tag{2.4.56}$$

Equation (2.4.56) clearly shows the proper polarizations as the S and P wavefunctions, which was also discussed for Green's function in the frequency domain.

2.5 REPRESENTATION THEOREM FOR SOLUTIONS FOR ELASTIC WAVEFIELD

2.5.1 OVERVIEW OF CONSIDERED PROBLEM

In the previous section, we proposed representations of solutions for the elastic wave equation in terms of Green's functions, as shown in Eqs. (2.4.5) and (2.4.9). For these representations, we did not take into account the initial-boundary conditions or the boundary conditions since we assumed that the inhomogeneous terms f_j and $\hat{f}_j$ were defined on $\mathbb{R}^3 \times \mathbb{R}$ and $\mathbb{R}^3$.

In the present section, we propose a representation of a solution for the elastic wave equation for initial-boundary value problems or boundary value problems. For this representation, we employ Green's function that expresses the outgoing wave with a time factor of $\exp(i\omega t)$ in the frequency domain or Green's function that satisfies causality in the time domain. That is, we employ $\widehat{G}^-$ and G^- from Eqs. (2.4.43) and (2.4.54), respectively. In the following, we omit the superscript $-$ for Green's function (i.e., $\widehat{G}$ and G are used).

According to the explicit form of Green's function in Eqs. (2.4.43) and (2.4.54), we can see that Green's function has the following properties:

$$\widehat{G}_{ij}(\boldsymbol{x},\boldsymbol{y},\omega) = \widehat{G}_{ji}(\boldsymbol{y},\boldsymbol{x},\omega) \tag{2.5.1}$$

$$G_{ij}(\boldsymbol{x},\boldsymbol{y},t-\tau) = G_{ji}(\boldsymbol{y},\boldsymbol{x},\tau'-t') \tag{2.5.2}$$

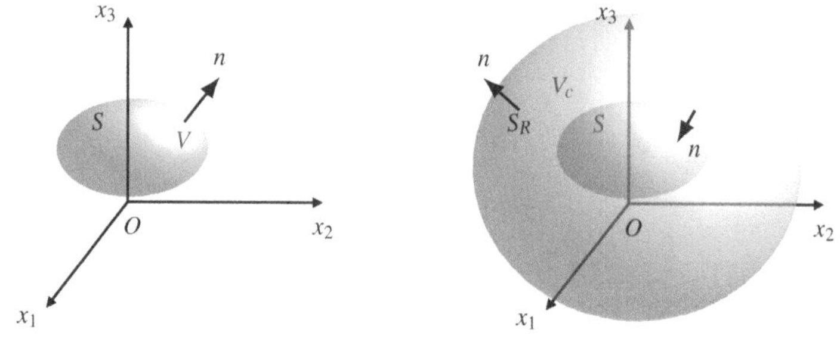

(a) Interior region V (b) Exterior region V_c

Figure 2.5.1 Interior and exterior regions for representation theorem.

where $\tau' = -\tau$ and $t' = -t$. We call these properties the *reciprocity of Green's function.* The reciprocity of Green's function and that of the elastic wavefield derived at the end of the previous chapter, shown in Eqs. (1.4.27) and (1.4.31), have important roles in the derivation of the representation of the solution.

Figures 2.5.1 (a) and (b) show regions of V and V_C in $\mathbb{R}^3$, respectively, to which the reciprocity of the elastic wavefield is applied. Region V is bounded by S and region V_c (outside the domain V) is surrounded by the boundaries S and S_R. Note that the boundary S_R is the surface of a sphere of radius R whose center is the origin of the coordinate system. Later, we take the limit of $R \to \infty$ to derive the representation of the solution for the exterior region of V. For the discussion below, we call the representation of the solution inside the region V the *interior problem* and that outside the region V the *exterior problem.* The discussion in this section begins with the application of the reciprocity of the elastic wavefield in the frequency domain to the regions shown in Figs. 2.5.1(a) and (b).

2.5.2 REPRESENTATION OF SOLUTIONS FOR ELASTIC WAVEFIELD IN THE FREQUENCY DOMAIN

2.5.2.1 Interior problem

We apply the reciprocity of the elastic wavefield in the frequency domain shown in Eq. (1.4.27) to region V and its boundary S shown in Fig. 2.5.1 (a). The result is

$$\begin{aligned}\int_S \left[\hat{u}_j^\star(\mathbf{y})\hat{p}_j(\mathbf{y}) - \hat{u}_i(\mathbf{y})\hat{p}_j^\star(\mathbf{y})\right] dS(\mathbf{y}) \\ = \int_V \Big[\hat{u}_i^\star(\mathbf{y})\left(L_{ij}(\partial_1,\partial_2,\partial_3) + \rho\omega^2\delta_{ij}\right)\hat{u}_j(\mathbf{y}) \\ -\hat{u}_i(\mathbf{y})\left(L_{ij}(\partial_1,\partial_2,\partial_3) + \rho\omega^2\delta_{ij}\right)\hat{u}_j^\star(\mathbf{y})\Big] d\mathbf{y}\end{aligned} \tag{2.5.3}$$

We define the following two states of the wavefield for Eq. (2.5.3):

$$\left(L_{ij}(\partial_{y_1},\partial_{y_2},\partial_{y_3})+\delta_{ij}\rho\omega^2\right)\hat{u}_j(\boldsymbol{y}) = -\hat{f}_i(\boldsymbol{y}) \tag{2.5.4}$$

$$\left(L_{ij}(\partial_{y_1},\partial_{y_2},\partial_{y_3})+\delta_{ij}\rho\omega^2\right)\hat{u}^\star_j(\boldsymbol{y}) = -\delta_{ik}\delta(\boldsymbol{y}-\boldsymbol{x}) \tag{2.5.5}$$

where $\boldsymbol{x},\boldsymbol{y}\in V$. According to Eqs. (2.5.5) and (2.5.1), we have

$$\hat{u}^\star_j(\boldsymbol{y}) = \widehat{G}_{jk}(\boldsymbol{y},\boldsymbol{x},\omega) = \widehat{G}_{kj}(\boldsymbol{x},\boldsymbol{y},\omega) \tag{2.5.6}$$

An important point for the formulation is to use Green's function for one of the two states of the wavefields for Eq. (2.5.3). The substitution of Eqs. (2.5.4) and (2.5.6) into the right-hand side of Eq. (2.5.3) yields

$$\begin{aligned}
&\int_V \Big[\hat{u}^\star_i(\boldsymbol{y})\left(L_{ij}(\partial_{y_1},\partial_{y_2},\partial_{y_3})+\rho\omega^2\delta_{ij}\right)\hat{u}_j(\boldsymbol{y}) \\
&\qquad -\hat{u}_i(\boldsymbol{y})\left(L_{ij}(\partial_{y_1},\partial_{y_2},\partial_{y_3})+\rho\omega^2\delta_{ij}\right)\hat{u}^\star_j(\boldsymbol{y})\Big]dy \\
&= -\int_V \left(\widehat{G}_{ki}(\boldsymbol{x},\boldsymbol{y},\omega)\hat{f}_i(\boldsymbol{y}) - \hat{u}_i(\boldsymbol{y})\delta_{ik}\delta(\boldsymbol{x}-\boldsymbol{y})\right)dy \\
&= -\int_V \widehat{G}_{ki}(\boldsymbol{x},\boldsymbol{y},\omega)\hat{f}_i(\boldsymbol{y})dy + \hat{u}_k(\boldsymbol{x})
\end{aligned} \tag{2.5.7}$$

Therefore, Eq. (2.5.3) becomes

$$\begin{aligned}
\hat{u}_k(\boldsymbol{x}) &= \int_V \widehat{G}_{kj}(\boldsymbol{x},\boldsymbol{y},\omega)\hat{f}_j(\boldsymbol{y})dy \\
&\quad + \int_S \left[\widehat{G}_{kj}(\boldsymbol{x},\boldsymbol{y},\omega)\hat{p}_j(\boldsymbol{y}) - \widehat{P}_{kj}(\boldsymbol{x},\boldsymbol{y},\omega)\hat{u}_j(\boldsymbol{y})\right] dS(\boldsymbol{y}), \quad (\boldsymbol{x}\in V)
\end{aligned} \tag{2.5.8}$$

as a result of the substitution of Eqs. (2.5.4) and (2.5.6) into Eq. (2.5.3). Note that $\widehat{P}_{ki}$ in Eq. (2.5.8), which is called *Green's function for traction*, is defined as

$$\begin{aligned}
\widehat{P}_{ki}(\boldsymbol{x},\boldsymbol{y},\omega) &= \Big[\lambda\delta_{ij}\frac{\partial\widehat{G}_{km}(\boldsymbol{x},\boldsymbol{y},\omega)}{\partial y_m} \\
&\quad +\mu\frac{\partial\widehat{G}_{kj}(\boldsymbol{x},\boldsymbol{y},\omega)}{\partial y_i} + \mu\frac{\partial\widehat{G}_{ki}(\boldsymbol{x},\boldsymbol{y},\omega)}{\partial y_j}\Big]n_j(\boldsymbol{y})
\end{aligned} \tag{2.5.9}$$

based on Eqs. (1.4.27) and (1.4.29). Equation (2.5.8) is the representation of the solution for the elastic wave equation in V in the frequency domain in terms of Green's function. Equation (2.5.8) is referred to as the representation theorem of the elastic wave equation for the interior problem in the frequency domain. It is found from Eq. (2.5.8) that the solution is determined by the boundary values of the displacement and traction and the body force acting inside V.

2.5.2.2 Exterior problem

Next, we apply the representation of the solution for Eq. (2.5.8) to the region V_c surrounded by the boundaries S and S_R shown in Fig. 2.5.1 (b). Then, the representation of the solution becomes

$$\begin{aligned}\hat{u}_k(\boldsymbol{x}) &= \int_S \left[\widehat{G}_{ki}(\boldsymbol{x},\boldsymbol{y},\omega)\hat{p}_i(\boldsymbol{y}) - \widehat{P}_{ki}(\boldsymbol{x},\boldsymbol{y},\omega)\hat{u}_i(\boldsymbol{y})\right]dS(\boldsymbol{y}) \\ &+ \int_{S_R} \left[\widehat{G}_{ki}(\boldsymbol{x},\boldsymbol{y},\omega)\hat{p}_i(\boldsymbol{y}) - \widehat{P}_{ki}(\boldsymbol{x},\boldsymbol{y},\omega)\hat{u}_i(\boldsymbol{y})\right]dS_R(\boldsymbol{y}) \\ &+ \int_{V_C} \widehat{G}_{kj}(\boldsymbol{x},\boldsymbol{y},\omega)\hat{f}_j(\boldsymbol{y})d\boldsymbol{y}, \quad (\boldsymbol{x}\in V_c) \qquad (2.5.10)\end{aligned}$$

We will now examine the conditions under which the effects of the boundary integral for S_R on the wavefield become negligible, when S_R is defined by

$$S_R = \left\{\boldsymbol{y}\in\mathbb{R}^3 \mid |\boldsymbol{y}| = R\right\} \qquad (2.5.11)$$

as $R \to \infty$. To perform this examination, we need to clarify the properties of Green's function for the case where $R \to \infty$ while keeping $\boldsymbol{x}$ fixed. Let $r = |\boldsymbol{y}-\boldsymbol{x}|$ and $R = |\boldsymbol{y}| \to \infty$. According to Eqs. (2.4.43) and (2.4.44), Green's function $\widehat{G}_{ij}$ can be represented as

$$\widehat{G}_{ij}(\boldsymbol{x},\boldsymbol{y},\omega) = \widehat{G}^{(L)}_{ij}(\boldsymbol{x},\boldsymbol{y},\omega) + \widehat{G}^{(T)}_{ij}(\boldsymbol{x},\boldsymbol{y},\omega) \qquad (2.5.12)$$

where $\widehat{G}^{(L)}_{ij}$ and $\widehat{G}^{(T)}_{ij}$ are given by

$$\begin{aligned}\widehat{G}^{(L)}_{ij}(\boldsymbol{x},\boldsymbol{y},\omega) &= \frac{1}{4\pi(\lambda+2\mu)R}\frac{y_i}{R}\frac{y_j}{R}\beta_L\exp(-ik_LR) + O(R^{-2}) \\ \widehat{G}^{(T)}_{ij}(\boldsymbol{x},\boldsymbol{y},\omega) &= \frac{1}{4\pi\mu R}\left[\delta_{ij} - \frac{y_i}{R}\frac{y_j}{R}\right]\beta_T\exp(-ik_TR) + O(R^{-2}) \qquad (2.5.13)\end{aligned}$$

Note that β_L and β_T in Eq. (2.5.13) are expressed as:

$$\begin{aligned}\beta_L &= \exp\left(ik_L\boldsymbol{x}\cdot\boldsymbol{y}/R\right) \\ \beta_T &= \exp\left(ik_T\boldsymbol{x}\cdot\boldsymbol{y}/R\right) \qquad (2.5.14)\end{aligned}$$

For the derivation of Eq. (2.5.13), we utilized the following processes:

$$\begin{aligned}
\frac{R}{r} &= \frac{R}{|\boldsymbol{y}|\sqrt{1-2\frac{\boldsymbol{y}\cdot\boldsymbol{x}}{|\boldsymbol{y}|^2}+\frac{|\boldsymbol{x}|^2}{|\boldsymbol{y}|^2}}} \\
&= 1+O(R^{-1}) \\
r-R &= R(r/R-1) \\
&= R\left(1-\frac{\boldsymbol{x}\cdot\boldsymbol{y}}{R^2}+O(R^{-2})-1\right) \\
&= -\frac{\boldsymbol{x}\cdot\boldsymbol{y}}{R}+O(R^{-1}) \\
\frac{e^{-ikr}}{r} &= \frac{e^{-ikR}}{R}\frac{R}{r}e^{-ik(r-R)} \\
\frac{y_i-x_i}{r} &= \left[\frac{y_i}{R}-\frac{x_i}{R}\right]\frac{R}{r}=\frac{y_i}{R}+O(R^{-1})
\end{aligned} \tag{2.5.15}$$

Green's function for the traction defined in Eq. (2.5.9) can also be decomposed into the form

$$\widehat{P}_{ij}(\boldsymbol{x},\boldsymbol{y},\omega)=\widehat{P}_{ij}^{(T)}(\boldsymbol{x},\boldsymbol{y},\omega)+\widehat{P}_{ij}^{(L)}(\boldsymbol{x},\boldsymbol{y},\omega) \tag{2.5.16}$$

where

$$\begin{aligned}
\widehat{P}_{ij}^{(L)}(\boldsymbol{x},\boldsymbol{y},\omega) &= \frac{-ik_L}{4\pi R}\frac{y_i}{R}\frac{y_j}{R}\beta_L\exp(-ik_LR)+O(R^{-2}) \\
\widehat{P}_{ij}^{(T)}(\boldsymbol{x},\boldsymbol{y},\omega) &= \frac{-ik_T}{4\pi R}\left[\delta_{ij}-\frac{y_i}{R}\frac{y_j}{R}\right]\beta_T\exp(-ik_TR)+O(R^{-2})
\end{aligned} \tag{2.5.17}$$

For constructing Eq. (2.5.17), the normal vector at the boundary S_R is expressed as:

$$n_j(\boldsymbol{y})=\frac{y_j}{R},\ (\boldsymbol{y}\in S_R) \tag{2.5.18}$$

Based on Eqs. (2.5.13) and (2.5.17), Green's function is found to satisfy the subsequent conditions:

$$\begin{aligned}
\widehat{P}_{ij}^{(L)}(\boldsymbol{x},\boldsymbol{y},\omega)+ik_L(\lambda+2\mu)\widehat{G}_{ij}^{(L)}(\boldsymbol{x},\boldsymbol{y},\omega) &= O(R^{-2}) \\
\frac{y_i^{\perp}}{|\boldsymbol{y}^{\perp}|}\widehat{G}_{ij}^{(L)}(\boldsymbol{x},\boldsymbol{y},\omega) &= O(R^{-2}) \\
\widehat{G}_{ij}^{(L)}(\boldsymbol{x},\boldsymbol{y},\omega) &= O(R^{-1}) \\
\widehat{P}_{ij}^{(T)}(\boldsymbol{x},\boldsymbol{y},\omega)+ik_T\mu\widehat{G}_{ij}^{(T)}(\boldsymbol{x},\boldsymbol{y},\omega) &= O(R^{-2}) \\
\frac{y_i}{|\boldsymbol{y}|}\widehat{G}_{ij}^{(T)}(\boldsymbol{x},\boldsymbol{y},\omega) &= O(R^{-2}) \\
\widehat{G}_{ij}^{(T)}(\boldsymbol{x},\boldsymbol{y},\omega) &= O(R^{-1})
\end{aligned} \tag{2.5.19}$$

These conditions are commonly referred to as the *Sommerfeld-Kupradze radiation conditions*. In Eq. (2.5.19), note that $y_i^{\perp}$ is the component of the vector defined as

$$y_i^{\perp}y_i=0 \tag{2.5.20}$$

Now, we discuss the effects of the boundary integral on S_R on the wavefield in Eq. (2.5.10). We impose the Sommerfeld-Kupradze radiation conditions on the wavefield u_i:

$$\begin{aligned}
\hat{p}_i^{(L)}(\mathbf{y}) + i(\lambda+2\mu)k_L\hat{u}_i^{(L)}(\mathbf{y}) &= O(R^{-2}) \\
\frac{y_i^{\perp}}{|\mathbf{y}^{\perp}|}\hat{u}_i^{(L)}(\mathbf{y}) &= O(R^{-2}) \\
\widehat{u}_i^{(L)}(\mathbf{y}) &= O(R^{-1}) \\
\hat{p}_i^{(T)}(\mathbf{y}) + i\mu k_T\hat{u}_i^{(T)}(\mathbf{y}) &= O(R^{-2}) \\
\frac{y_i}{|\mathbf{y}|}\hat{u}_i^{(T)}(\mathbf{y}) &= O(R^{-2}) \\
\widehat{u}_i^{(T)}(\mathbf{y}) &= O(R^{-1})
\end{aligned} \tag{2.5.21}$$

where $\hat{u}_i^{(L)}$ and $\hat{u}_i^{(T)}$ are the P and S wave components of the wavefield, respectively, obtained from

$$\begin{aligned}
\hat{u}_i^{(L)}(\mathbf{y}) &= \frac{1}{k_T^2 - k_L^2}\left(\nabla^2 + k_T^2\right)\hat{u}_i(\mathbf{y}) \\
\hat{u}_i^{(T)}(\mathbf{y}) &= \frac{1}{k_L^2 - k_T^2}\left(\nabla^2 + k_L^2\right)\hat{u}_i(\mathbf{y})
\end{aligned} \tag{2.5.22}$$

and $\hat{p}_i^{(L)}$ and $\hat{p}_i^{(T)}$ correspond to the traction of the P and S wave components, respectively. According to the decomposition of Green's function and the wavefield, the boundary integral on S_R for Eq. (2.5.10) becomes

$$\begin{aligned}
&\int_{S_R}\left[\widehat{G}_{ki}(\mathbf{x},\mathbf{y},\omega)\hat{p}_i(\mathbf{y}) - \widehat{P}_{ki}(\mathbf{x},\mathbf{y},\omega)\hat{u}_i(\mathbf{y})\right]dS_R(\mathbf{y}) \\
&= \int_{S_R}\left[\widehat{G}_{ki}^{(T)}(\mathbf{x},\mathbf{y},\omega)\hat{p}_i^{(T)}(\mathbf{y}) - \widehat{P}_{ki}^{(T)}(\mathbf{x},\mathbf{y},\omega)\hat{u}_i^{(T)}(\mathbf{y})\right]dS_R(\mathbf{y}) \\
&\quad + \int_{S_R}\left[\widehat{G}_{ki}^{(L)}(\mathbf{x},\mathbf{y},\omega)\hat{p}_i^{(L)}(\mathbf{y}) - \widehat{P}_{ki}^{(L)}(\mathbf{x},\mathbf{y},\omega)\hat{u}_i^{(L)}(\mathbf{y})\right]dS_R(\mathbf{y}) \\
&\quad + \int_{S_R}\left[\widehat{G}_{ki}^{(L)}(\mathbf{x},\mathbf{y},\omega)\hat{p}_i^{(T)}(\mathbf{y}) - \widehat{P}_{ki}^{(L)}(\mathbf{x},\mathbf{y},\omega)\hat{u}_i^{(T)}(\mathbf{y})\right]dS_R(\mathbf{y}) \\
&\quad + \int_{S_R}\left[\widehat{G}_{ki}^{(T)}(\mathbf{x},\mathbf{y},\omega)\hat{p}_i^{(L)}(\mathbf{y}) - \widehat{P}_{ki}^{(T)}(\mathbf{x},\mathbf{y},\omega)\hat{u}_i^{(L)}(\mathbf{y})\right]dS_R(\mathbf{y})
\end{aligned} \tag{2.5.23}$$

The evaluation of each term in the integrals on the right-hand side of Eq. (2.5.23) using the Sommerfeld-Kupradze radiation conditions is not difficult. For example, the first term becomes

$$\begin{aligned}
&\int_{S_R}\left[\widehat{G}_{ki}^{(T)}(\boldsymbol{x},\boldsymbol{y},\omega)\hat{p}_i^{(T)}(\boldsymbol{y})-\widehat{P}_{ki}^{(T)}(\boldsymbol{x},\boldsymbol{y},\omega)\hat{u}_i^{(T)}(\boldsymbol{y})\right]dS_R(\boldsymbol{y})\\
&=\int_{S_R}\left[\widehat{G}_{ki}^{(T)}(\boldsymbol{x},\boldsymbol{y},\omega)\hat{p}_i^{(T)}(\boldsymbol{y})-\left(-ik_T\mu\widehat{G}_{ik}^{(T)}(\boldsymbol{x},\boldsymbol{y},\omega)+O(R^{-2})\right)\hat{u}_i^{(T)}(\boldsymbol{y})\right]dS_R(\boldsymbol{y})\\
&=\int_{S_R}\left[\widehat{G}_{ki}^{(T)}(\boldsymbol{x},\boldsymbol{y},\omega)\left(\hat{p}_i^{(T)}(\boldsymbol{y})+i\mu k_T\hat{u}_i^{(T)}(\boldsymbol{y})\right)+O(R^{-2})\hat{u}_l^{(T)}(\boldsymbol{y})\right]dS_R(\boldsymbol{y})\\
&=\int_{S_R}O(R^{-3})dS_R=O(R^{-1})\to 0,\quad (R\to\infty)
\end{aligned}\tag{2.5.24}$$

In addition, the third term becomes

$$\begin{aligned}
&\int_{S_R}\left[\widehat{G}_{ki}^{(L)}(\boldsymbol{x},\boldsymbol{y},\omega)\hat{p}_i^{(T)}(\boldsymbol{y})-\widehat{P}_{ki}^{(L)}(\boldsymbol{x},\boldsymbol{y},\omega)\hat{u}_i^{(T)}(\boldsymbol{y})\right]dS_R(\boldsymbol{y})\\
&=\int_{S_R}\left[\widehat{G}_{ki}^{(L)}(\boldsymbol{x},\boldsymbol{y},\omega)\hat{p}_i^{(T)}(\boldsymbol{y})-\left(-ik_L(\lambda+2\mu)\widehat{G}_{ik}^{(L)}(\boldsymbol{x},\boldsymbol{y},\omega)+O(R^{-2})\right)\hat{u}_i^{(T)}(\boldsymbol{y})\right]dS_R(\boldsymbol{y})\\
&=\int_{S_R}\left[\widehat{G}_{ki}^{(L)}(\boldsymbol{x},\boldsymbol{y})\left(\hat{p}_i^{(T)}(\boldsymbol{y},\omega)+i(\lambda+2\mu)k_L\hat{u}_i^{(T)}(\boldsymbol{y})\right)+O(R^{-2})\hat{u}_i^{(T)}(\boldsymbol{y})\right]dS_R(\boldsymbol{y})\\
&=\int_{S_R}\left[\widehat{G}_{ki}^{(L)}(\boldsymbol{x},\boldsymbol{y},\omega)\left(-ik_T\mu+ik_L(\lambda+2\mu)\right)\hat{u}_i^{(T)}(\boldsymbol{y})\right.\\
&\qquad\left.+\widehat{G}_{ki}^{(L)}(\boldsymbol{x},\boldsymbol{y},\omega)O(R^{-2})+O(R^{-2})\hat{u}_i^{(T)}(\boldsymbol{y})\right]dS_R(\boldsymbol{y})\\
&=\int_{S_R}O(R^{-3})dS_R=O(R^{-1})\to 0,\quad (R\to\infty)
\end{aligned}\tag{2.5.25}$$

In the derivation of Eq. (2.5.25), we use the following evaluation:

$$\begin{aligned}
&\widehat{G}_{ki}^{(L)}(\boldsymbol{x},\boldsymbol{y},\omega)\hat{u}_i^{(T)}(\boldsymbol{y})\\
&=\left(\frac{\beta_L}{4\pi(\lambda+2\mu)}\frac{e^{-ik_LR}}{R}\frac{y_k}{R}\frac{y_i}{R}+O(R^{-2})\right)\hat{u}_i^{(T)}(\boldsymbol{y})=O(R^{-3})
\end{aligned}\tag{2.5.26}$$

based on Eqs. (2.5.17) and (2.5.21). Each term on the right-hand side of Eq. (2.5.23) becomes zero by the similar way of the evaluation when $R\to\infty$. Therefore, for the case where the wavefield u_i satisfies the Sommerfeld-Kupradze radiation conditions, the representation of the solution outside V becomes

$$\begin{aligned}
\hat{u}_k(\boldsymbol{x}) &= \int_S\left[\widehat{G}_{ki}(\boldsymbol{x},\boldsymbol{y},\omega)\hat{p}_i(\boldsymbol{y})-\widehat{P}_{ki}(\boldsymbol{x},\boldsymbol{y},\omega)\hat{u}_i(\boldsymbol{y})\right]dS(\boldsymbol{y})\\
&\quad+\int_{\mathbb{R}^3\setminus(V\cup S)}\widehat{G}_{kj}(\boldsymbol{x},\boldsymbol{y},\omega)\hat{f}_j(\boldsymbol{y})d\boldsymbol{y},\quad \left(\boldsymbol{x}\in\mathbb{R}^3\setminus(V\cup S)\right)
\end{aligned}\tag{2.5.27}$$

Equation (2.5.27) is the representation theorem of the elastic wave equation for the exterior problem in the frequency domain.

2.5.3 REPRESENTATION OF SOLUTION FOR ELASTIC WAVEFIELD IN THE TIME DOMAIN

Now, let us return to Fig. 2.5.1 (a) to apply the reciprocity of the elastic wavefield in the time domain shown in Eq. (1.4.31). Recall that the reciprocity of the elastic wavefield in the time domain is expressed as:

$$\begin{aligned}
\int_0^\infty d\tau \int_V \Big[u_i^\star(\mathbf{y},\tau)\Big(L_{ij}(\partial_{y_1},\partial_{y_2},\partial_{y_3}) - \delta_{ij}\rho\frac{\partial^2}{\partial\tau^2}\Big)u_j(\mathbf{y},\tau) \\
- u_i(\mathbf{y},\tau)\Big(L_{ij}(\partial_{y_1},\partial_{y_2},\partial_{y_3}) - \delta_{ij}\rho\frac{\partial^2}{\partial\tau^2}\Big)u_j^\star(\mathbf{y},\tau)\Big] d\mathbf{y} \\
= \int_0^\infty d\tau \int_S \Big[u_i^\star(\mathbf{y},\tau)p_i(\mathbf{y},\tau) - u_i(\mathbf{y},\tau)p_i^\star(\mathbf{y},\tau)\Big] dS(\mathbf{y}) \\
- \int_V \rho \Big[u_i^\star(\mathbf{y},\tau)\frac{\partial}{\partial\tau}u_i(\mathbf{y},\tau) - u_i(\mathbf{y},\tau)\frac{\partial}{\partial\tau}u_i^\star(\mathbf{y},\tau)\Big]_0^\infty d\mathbf{y}
\end{aligned} \tag{2.5.28}$$

For the expression of the reciprocity of Eq. (2.5.28), we consider the time interval $0 \le \tau < \infty$.

Let the two states of the wavefield for the reciprocity be defined as

$$\begin{aligned}
\Big(L_{ij}(\partial_{y_1},\partial_{y_2},\partial_{y_3}) - \delta_{ij}\rho\frac{\partial^2}{\partial\tau^2}\Big)u_j(\mathbf{y},\tau) &= -f_i(\mathbf{y},\tau) \\
\Big(L_{ij}(\partial_{y_1},\partial_{y_2},\partial_{y_3}) - \delta_{ij}\rho\frac{\partial^2}{\partial\tau^2}\Big)u_j^\star(\mathbf{y},\tau) &= -\delta_{ik}\delta(\mathbf{y}-\mathbf{x})\delta(t-\tau)
\end{aligned} \tag{2.5.29}$$

According to the definition of $u_j^\star$ shown in Eq. (2.5.29), we see that

$$u_j^\star(\mathbf{y},\tau) = G_{jk}(\mathbf{y},\mathbf{x},t-\tau) = G_{kj}(\mathbf{x},\mathbf{y},t-\tau) \tag{2.5.30}$$

Therefore, substitution of Eq. (2.5.29) into Eq. (2.5.28) yields

$$\begin{aligned}
u_k(\mathbf{x},t) \\
= \int_0^t d\tau \int_V G_{ij}(\mathbf{x},\mathbf{y},t-\tau)f_j(\mathbf{y},\tau)d\mathbf{y} \\
+ \int_0^t d\tau \int_S (G_{ij}(\mathbf{x},\mathbf{y},t-\tau)p_j(\mathbf{y},\tau) - P_{ij}(\mathbf{x},\mathbf{y},t-\tau)u_j(\mathbf{y},\tau))\,dS(\mathbf{y}) \\
+ \int_V \rho\left(G_{ki}(\mathbf{x},\mathbf{y},t)\dot{u}_i(\mathbf{y},0) - \dot{G}_{ki}(\mathbf{x},\mathbf{y},t)u_i(\mathbf{y},0)\right)d\mathbf{y}, \quad (\mathbf{x}\in V)
\end{aligned} \tag{2.5.31}$$

which is the representation of the solution for initial boundary value problems inside region V. We have considered the causality of Green's function, as shown in the following expressions:

$$\begin{aligned}
G_{ij}(\mathbf{x},\mathbf{y},t-\tau) &= 0 \\
P_{ij}(\mathbf{x},\mathbf{y},t-\tau) &= 0, \quad (\text{when } t<\tau)
\end{aligned} \tag{2.5.32}$$

to the first and second terms of right-hand side of Eq. (2.5.31). In addition, to the third term of right-hand side of Eq. (2.5.31), we have also considered

$$G_{ij}(\boldsymbol{x},\boldsymbol{y},0) = \dot{G}_{ij}(\boldsymbol{x},\boldsymbol{y},0) = 0, \ \ (\text{when } |\boldsymbol{x}-\boldsymbol{y}| \neq 0) \tag{2.5.33}$$

We have arrived at Eq. (2.5.31 as the representation theorem of the elastic wave equation for the interior problem in the time domain.

Next, we apply the representation of the solution in the time domain in Eq. (2.5.31) to the region V_c shown in Fig. 2.5.1 (b) under the assumption that

$$\dot{u}_i(\boldsymbol{y},0) = u_i(\boldsymbol{y},0) = 0, \ \ (\boldsymbol{y} \in V_c) \tag{2.5.34}$$

The result is

$$\begin{aligned}
u_k(\boldsymbol{x},t) &= \int_0^t d\tau \int_{V_c} G_{ij}(\boldsymbol{x},\boldsymbol{y},t-\tau) f_j(\boldsymbol{y},\tau) d\boldsymbol{y} \\
&+ \int_0^t d\tau \int_S (G_{ij}(\boldsymbol{x},\boldsymbol{y},t-\tau)p_j(\boldsymbol{y},\tau) - P_{ij}(\boldsymbol{x},\boldsymbol{y},t-\tau)u_j(\boldsymbol{y},\tau))\, dS(\boldsymbol{y}) \\
&+ \int_0^t d\tau \int_{S_R} (G_{ij}(\boldsymbol{x},\boldsymbol{y},t-\tau)p_j(\boldsymbol{y},\tau) - P_{ij}(\boldsymbol{x},\boldsymbol{y},t-\tau)u_j(\boldsymbol{y},\tau))\, dS(\boldsymbol{y}) \\
&(\boldsymbol{x} \in V_c)
\end{aligned} \tag{2.5.35}$$

We discuss the effect of the boundary integral on S_R, which appears in the third term on the right-hand side of Eq. (2.5.35). According to Eqs. (2.4.53), (2.4.54), and (2.5.17), we see that

$$\begin{aligned}
G_{ij}^{(L)}(\boldsymbol{x},\boldsymbol{y},t) &= \frac{1}{4\pi(\lambda+2\mu)R}\frac{y_i}{R}\frac{y_j}{R}\delta(t - R/c_L + \boldsymbol{x}\cdot\boldsymbol{y}/(Rc_L)) + O(R^{-2}) \\
G_{ij}^{(T)}(\boldsymbol{x},\boldsymbol{y},t) &= \frac{1}{4\pi\mu R}\left[\delta_{ij} - \frac{y_i}{R}\frac{y_j}{R}\right]\delta(t - R/c_T + \boldsymbol{x}\cdot\boldsymbol{y}/(Rc_T)) + O(R^{-2}) \\
P_{ij}^{(L)}(\boldsymbol{x},\boldsymbol{y},t) &= -\rho c_L \frac{\partial}{\partial t} G_{ij}^{(L)}(x,y,t) + O(R^{-2}) \\
P_{ij}^{(T)}(\boldsymbol{x},\boldsymbol{y},t) &= -\rho c_T \frac{\partial}{\partial t} G_{ij}^{(T)}(x,y,t) + O(R^{-2})
\end{aligned} \tag{2.5.36}$$

where $G_{ij}^{(L)}$ and $G_{ij}^{(T)}$ are the P and S wave components, respectively, of Green's function in the space-time domain. Likewise, $P_{ij}^{(L)}$ and $P_{ij}^{(T)}$ are the P and S wave components, respectively, of Green's function for the traction in the space-time domain. According to Eq. (2.5.36), we see that Green's function in the space-time domain satisfies the following Sommerfeld-Kupradze radiation condition :

$$\begin{aligned}
P_{ij}^{(L)}(\boldsymbol{x},\boldsymbol{y},t)+\rho c_L\frac{\partial}{\partial t}G_{ij}^{(L)}(\boldsymbol{x},\boldsymbol{y},t) &= O(R^{-2})\\
\frac{y_i^{\perp}}{|\boldsymbol{y}^{\perp}|}G_{ij}^{(L)}(\boldsymbol{x},\boldsymbol{y},t) &= O(R^{-2})\\
G_{ij}^{(L)}(\boldsymbol{x},\boldsymbol{y},t) &= O(R^{-1})\\
P_{ij}^{(T)}(\boldsymbol{x},\boldsymbol{y},t)+\rho c_T\frac{\partial}{\partial t}G_{ij}^{(T)}(\boldsymbol{x},\boldsymbol{y},t) &= O(R^{-2})\\
\frac{y_i}{|\boldsymbol{y}|}G_{ij}^{(T)}(\boldsymbol{x},\boldsymbol{y},t) &= O(R^{-2})\\
G_{ij}^{(T)}(\boldsymbol{x},\boldsymbol{y},t) &= O(R^{-1})
\end{aligned} \tag{2.5.37}$$

We also impose the Sommerfeld-Kupradze radiation conditions on u_i and p_i for Eq. (2.5.35) in the form

$$\begin{aligned}
p_j^{(L)}(\boldsymbol{y},\tau)+\rho c_L\frac{\partial}{\partial \tau}u_j^{(L)}(\boldsymbol{y},\tau) &= O(R^{-2})\\
\frac{y_i^{\perp}}{|\boldsymbol{y}^{\perp}|}u_j^{(L)}(\boldsymbol{y},\tau) &= O(R^{-2})\\
u_j^{(L)}(\boldsymbol{y},\tau) &= O(R^{-1})\\
p_j^{(T)}(\boldsymbol{y},\tau)+\rho c_T\frac{\partial}{\partial \tau}u_j^{(T)}(\boldsymbol{y},\tau) &= O(R^{-2})\\
\frac{y_i}{|\boldsymbol{y}|}u_j^{(T)}(\boldsymbol{y},\tau) &= O(R^{-2})\\
u_j^{(T)}(\boldsymbol{y},\tau) &= O(R^{-1})
\end{aligned} \tag{2.5.38}$$

From Eqs. (2.5.37) and (2.5.38), we find that the effects of the boundary integral on S_R on the wavefield vanish as $R \to \infty$. For example, we see that

$$\begin{aligned}
&\int_0^t d\tau \int_{S_R} \left(G_{ij}^{(T)}(\boldsymbol{x},\boldsymbol{y},t-\tau)p_j^{(T)}(\boldsymbol{y},\tau) - P_{ij}^{(T)}(\boldsymbol{x},\boldsymbol{y},t-\tau)u_j(\boldsymbol{y},\tau)\right) dS_R(\boldsymbol{y})\\
&= \int_0^t d\tau \int_{S_R} \Big(G_{ij}^{(T)}(\boldsymbol{x},\boldsymbol{y},t-\tau)p_j^{(T)}(\boldsymbol{y},\tau)\\
&\qquad -(\rho c_T\frac{\partial}{\partial \tau}G_{ij}(\boldsymbol{x},\boldsymbol{y},t-\tau)+O(R^{-2}))u_j(\boldsymbol{y},\tau)\Big] dS_R(\boldsymbol{y})\\
&= \int_0^t d\tau \int_{S_R} \Big(G_{ij}^{(T)}(\boldsymbol{x},\boldsymbol{y},t-\tau)(p_j^{(T)}(\boldsymbol{y},\tau)\\
&\qquad +\rho c_T\frac{\partial}{\partial \tau}u_j(\boldsymbol{y},\tau))+O(R^{-2})u_j(\boldsymbol{y})\Big) dS_R(\boldsymbol{y})\\
&= \int_0^t d\tau \int_{S_R} O(R^{-3}) dS_R(\boldsymbol{y}) \to 0, \ \ (R\to\infty)
\end{aligned} \tag{2.5.39}$$

Therefore, the representation theorem in the time domain for the exterior problem is expressed as:

$$\begin{aligned} u_k(\boldsymbol{x},t) &= \int_0^t d\tau \int_{\mathbb{R}^3\setminus(S\cup V)} G_{ij}(\boldsymbol{x},\boldsymbol{y},t-\tau) f_j(\boldsymbol{y},\tau) d\boldsymbol{y} \\ &+ \int_0^t d\tau \int_S \Big(G_{ij}(\boldsymbol{x},\boldsymbol{y},t-\tau) p_j(\boldsymbol{y},\tau) - P_{ij}(\boldsymbol{x},\boldsymbol{y},t-\tau) u_j(\boldsymbol{y},\tau) \Big) dS(\boldsymbol{y}) \\ & \big(\boldsymbol{x} \in \mathbb{R}^3 \setminus (V \cup S)\big) \end{aligned} \tag{2.5.40}$$

Note 2.3 Representation theorem for solution for Helmholtz equation

The representation theorem for the solution for the Helmholtz equation is also important. Based on the same procedure as that shown in Eqs. (2.5.4) and (2.5.5), we define two states of the wavefield such that

$$\left(\nabla_y^2 + k^2\right)\hat{u}(\boldsymbol{y}) = -\hat{f}(\boldsymbol{y}), \quad \left(\nabla_y^2 + k^2\right)\hat{u}^\star(\boldsymbol{y}) = -\delta(\boldsymbol{y}-\boldsymbol{x}) \tag{N2.3.1}$$

where $\boldsymbol{x}, \boldsymbol{y} \in V$ for the interior problem presented in Fig. 2.5.1 (a) and

$$\nabla_y^2 = \partial_{y_1}^2 + \partial_{y_2}^2 + \partial_{y_3}^2 \tag{N2.3.2}$$

According to Eqs. (2.4.11) and (2.4.23), we have

$$\hat{u}^\star(\boldsymbol{y}) = \hat{G}(\boldsymbol{x},\boldsymbol{y}) = \frac{\exp(-ikr)}{4\pi r} \tag{N2.3.3}$$

where $r = |\boldsymbol{x}-\boldsymbol{y}|$. Equations (N2.3.1) and (1.3.54) yield

$$\begin{aligned} & \int_V \left[\hat{u}^\star(\boldsymbol{y}) \left(\nabla^2 + k^2\right) \hat{u}(\boldsymbol{y}) - \hat{u}(\boldsymbol{y}) \left(\nabla^2 + k^2\right) \hat{u}^\star(\boldsymbol{y}) \right] d\boldsymbol{y} \\ &= -\int_V \hat{G}(\boldsymbol{x},\boldsymbol{y}) \hat{f}(\boldsymbol{y}) d\boldsymbol{y} + \hat{u}(\boldsymbol{x}) \\ &= \int_S \left[\hat{G}(\boldsymbol{x},\boldsymbol{y}) \hat{p}(\boldsymbol{y}) - \hat{P}(\boldsymbol{x},\boldsymbol{y}) \hat{u}(\boldsymbol{y}) \right] dS(\boldsymbol{y}) \end{aligned} \tag{N2.3.4}$$

where $\hat{P}$ and $\hat{p}$ are defined as

$$\begin{aligned} \hat{P}(\boldsymbol{x},\boldsymbol{y}) &= n_i(\boldsymbol{y}) \partial_{y_i} \hat{G}(\boldsymbol{x},\boldsymbol{y}) \\ \hat{p}(\boldsymbol{y}) &= n_i(\boldsymbol{y}) \partial_{y_i} \hat{u}(\boldsymbol{y}) \end{aligned} \tag{N2.3.5}$$

Equation (N2.3.4) yields the following representation theorem for the Helmholtz equation for the interior problem:

$$\hat{u}(\boldsymbol{x}) = \int_V \hat{G}(\boldsymbol{x},\boldsymbol{y}) \hat{f}(\boldsymbol{y}) d\boldsymbol{y} + \int_S \left[\hat{G}(\boldsymbol{x},\boldsymbol{y}) \hat{p}(\boldsymbol{y}) - \hat{P}(\boldsymbol{x},\boldsymbol{y}) \hat{u}(\boldsymbol{y}) \right] dS(\boldsymbol{y}) \tag{N2.3.6}$$

A representation of the solution for the exterior problem can also be obtained by imposing the *Sommerfeld radiation condition* on the solution for the Helmholtz equation:

$$\hat{p}(\boldsymbol{y}) + ik\hat{u}(\boldsymbol{y}) = O(|\boldsymbol{y}|^{-2}) \tag{N2.3.7}$$

which yields

$$\begin{aligned} \hat{u}(\boldsymbol{x}) &= \int_{\mathbb{R}^3\setminus S\cup V} \hat{G}(\boldsymbol{x},\boldsymbol{y}) \hat{f}(\boldsymbol{y}) d\boldsymbol{y} + \int_S \left[\hat{G}(\boldsymbol{x},\boldsymbol{y}) \hat{p}(\boldsymbol{y}) - \hat{P}(\boldsymbol{x},\boldsymbol{y}) \hat{u}(\boldsymbol{y}) \right] dS(\boldsymbol{y}) \\ & (\boldsymbol{x} \in \mathbb{R}^3 \setminus S \cup V) \end{aligned} \tag{N2.3.8}$$

Note that $\hat{G}$ also satisfies the Sommerfeld radiation condition:

$$\hat{P}(\boldsymbol{x},\boldsymbol{y}) + ik\hat{G}(\boldsymbol{x},\boldsymbol{y}) = O(|\boldsymbol{y}|^{-2}) \tag{N2.3.9}$$

Note 2.4 Representation theorem for solution for scalar wave equation

The representation of the solution for the scalar wave equation is expressed as

$$\begin{aligned} u(\boldsymbol{x},t) &= \int_0^t d\tau \int_V G(\boldsymbol{x},\boldsymbol{y},t-\tau)f(\tau)d\boldsymbol{y} \\ &+ \int_0^t d\tau \int_S \Big(G(\boldsymbol{x},\boldsymbol{y},t-\tau)p(\boldsymbol{y},\tau) - P(\boldsymbol{x},\boldsymbol{y},t-\tau)u(\boldsymbol{y},\tau)\Big)dS(\boldsymbol{y}) \\ &+ \int_V \frac{1}{c^2}\Big(G(\boldsymbol{x},\boldsymbol{y},t)\dot{u}(\boldsymbol{y},0) - \dot{G}(\boldsymbol{x},\boldsymbol{y},t)u(\boldsymbol{y},0)\Big)d\boldsymbol{y} \end{aligned} \tag{N2.4.1}$$

for the interior problem, where

$$\begin{aligned} G(\boldsymbol{x},\boldsymbol{y},t) &= \frac{1}{4\pi|\boldsymbol{x}-\boldsymbol{y}|}\delta\Big(t - \frac{|\boldsymbol{x}-\boldsymbol{y}|}{c}\Big) \\ P(\boldsymbol{x},\boldsymbol{y},t) &= n_i(\boldsymbol{y})\partial_{y_i}G(\boldsymbol{x},\boldsymbol{y},t) \\ p(\boldsymbol{y},t) &= n_i(\boldsymbol{y})\partial_{y_i}p(\boldsymbol{y},t) \end{aligned} \tag{N2.4.2}$$

For the exterior problem, we assume that

$$\begin{aligned} u(\boldsymbol{y},t) &= \dot{u}(\boldsymbol{y},t) = 0 \\ & t \le 0, \ \ \boldsymbol{y} \in \mathbb{R}^3 \setminus V \cup S \end{aligned} \tag{N2.4.3}$$

and impose the Sommerfeld radiation condition:

$$p(\boldsymbol{y},t) + \frac{1}{c}\dot{u}(\boldsymbol{y},t) = O(|\boldsymbol{y}|^{-2}) \tag{N2.4.4}$$

which yields

$$\begin{aligned} u(\boldsymbol{x},t) &= \int_0^t d\tau \int_{\mathbb{R}^3\setminus S\cup V} G(\boldsymbol{x},\boldsymbol{y},t-\tau)f(\boldsymbol{y},\tau)d\boldsymbol{y} \\ &+ \int_0^t d\tau \int_S \Big[G(\boldsymbol{x},\boldsymbol{y},t-\tau)p(\boldsymbol{y},\tau) - P(\boldsymbol{x},\boldsymbol{y},t-\tau)u(\boldsymbol{y},\tau)\Big]dS(\boldsymbol{y}) \\ & (\boldsymbol{x} \in \mathbb{R}^3 \setminus S \cup V) \end{aligned} \tag{N2.4.5}$$

2.6 PROBLEMS

1. Consider the following Helmholtz equation given by

$$(\nabla^2 + k^2)u(\boldsymbol{x}) = 0, \ \ (\boldsymbol{x} \in \mathbb{R}^3) \tag{2.6.1}$$

   a. Solve the Helmholtz equation by using the method of separation of variables in a Cartesian coordinate system. Assume that the solution is expressed as:

$$u(\boldsymbol{x}) = U_1(x_1)U_2(x_2)U_3(x_3) \tag{2.6.2}$$

   Substitute this expression into the Helmholtz equation, and derive the ordinary differential equations to solve U_1, U_2 and U_3.

   b. Verify that the Laplace operator for the scalar field in terms of the cylindrical coordinate system is expressed as

$$\nabla^2\phi(\boldsymbol{x}) = \Big(\frac{\partial^2}{\partial r^2} + \frac{1}{r}\frac{\partial}{\partial r} + \frac{1}{r^2}\frac{\partial^2}{\partial \varphi^2} + \frac{\partial^2}{\partial z}\Big)\phi(\boldsymbol{x}) \tag{2.6.3}$$

   where (r,φ,z) denotes the cylindrical coordinate system.

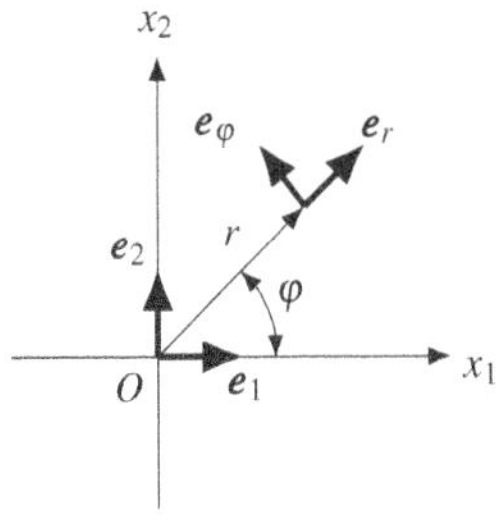

Figure 2.6.1 Relationship of the base vectors between a Cartesian and a cylindrical coordinate system in $x_1 - x_2$ plane. Note that $\boldsymbol{e}_z = \boldsymbol{e}_3$.

c. Solve the Helmholtz equation by the method of separation of variables in a cylindrical coordinate system. Assume that the solution is expressed as

$$u(\boldsymbol{x}) = U_r(r)U_\varphi(\varphi)U_z(z) \tag{2.6.4}$$

2. Solve the following 1D wave equation:

$$\frac{\partial^2 u(x,t)}{\partial x^2} = \frac{1}{c^2}\frac{\partial^2 u(x,t)}{\partial t^2} \tag{2.6.5}$$

by the transformation of the variables:

$$\begin{aligned} \xi &= x - ct \\ \eta &= x + ct \end{aligned} \tag{2.6.6}$$

3. Verify Eqs. (2.1.15) and (2.1.18).
4. Let $\phi(\boldsymbol{x})$ and $\boldsymbol{u}(\boldsymbol{x})$, $\boldsymbol{x} \in \mathbb{R}^3$ be a scalar and a vector fields, respectively.
   a. Let $\boldsymbol{e}_r$, $\boldsymbol{e}_\varphi$ and $\boldsymbol{e}_z$ are the base vectors for a cylindrical coordinate system. The definition of the base vectors $\boldsymbol{e}_r$, $\boldsymbol{e}_\varphi$ is given in Fig. 2.6.1. We express a vector field in terms of the cylindrical coordinate system by

$$\boldsymbol{u}(\boldsymbol{x}) = u_r(\boldsymbol{x})\boldsymbol{e}_r + u_\varphi(\boldsymbol{x})\boldsymbol{e}_\varphi + u_z(\boldsymbol{x})\boldsymbol{e}_z \tag{2.6.7}$$

   Then, prove the following identities:

$$\nabla\phi(\boldsymbol{x}) = \frac{\partial\phi(\boldsymbol{x})}{\partial r}\boldsymbol{e}_r + \frac{1}{r}\frac{\partial\phi(\boldsymbol{x})}{\partial\varphi}\boldsymbol{e}_\varphi + \frac{\partial\phi(\boldsymbol{x})}{\partial z}\boldsymbol{e}_z \tag{2.6.8}$$

$$\operatorname{div}\boldsymbol{u}(\boldsymbol{x}) = \frac{\partial u_r(\boldsymbol{x})}{\partial r} + \frac{u_r(\boldsymbol{x})}{r} + \frac{1}{r}\frac{\partial u_\varphi(\boldsymbol{x})}{\partial\varphi} + \frac{\partial u_z(\boldsymbol{x})}{\partial z} \tag{2.6.9}$$

$$\begin{aligned} \operatorname{rot}\boldsymbol{u}(\boldsymbol{x}) &= \left(\frac{1}{r}\frac{\partial u_z}{\partial\varphi} - \frac{\partial u_\varphi}{\partial z}\right)\boldsymbol{e}_r + \left(\frac{\partial u_r}{\partial z} - \frac{\partial u_z}{\partial r}\right)\boldsymbol{e}_\varphi \\ &+ \left(\frac{\partial u_\varphi}{\partial r} - \frac{1}{r}\frac{\partial u_r}{\partial\varphi} + \frac{u_\varphi}{r}\right)\boldsymbol{e}_z \end{aligned} \tag{2.6.10}$$

   b. Let $\boldsymbol{e}_1$, $\boldsymbol{e}_2$ and $\boldsymbol{e}_3$ are the base vectors for a Cartesian coordinate system. Verify the following identity:

$$\begin{aligned} &\left(\nabla\nabla\cdot - \nabla\times\nabla\times\right)(u_1\boldsymbol{e}_1 + u_2\boldsymbol{e}_2 + u_3\boldsymbol{e}_3) \\ &= \left(\frac{\partial^2 u_1}{\partial x_1^2} + \frac{\partial^2 u_1}{\partial x_2^2} + \frac{\partial^2 u_1}{\partial x_3^2}\right)\boldsymbol{e}_1 \\ &+ \left(\frac{\partial^2 u_2}{\partial x_1^2} + \frac{\partial^2 u_2}{\partial x_2^2} + \frac{\partial^2 u_2}{\partial x_3^2}\right)\boldsymbol{e}_2 \\ &+ \left(\frac{\partial^2 u_3}{\partial x_1^2} + \frac{\partial^2 u_3}{\partial x_2^2} + \frac{\partial^2 u_3}{\partial x_3^2}\right)\boldsymbol{e}_3 \end{aligned} \tag{2.6.11}$$

c. In the cylindrical coordinate system, unfortunately, we can not establish the formula as can be seen in Eq. (2.6.11). Make sure the following properties of the Laplace operator:

$$\begin{aligned}&\left(\nabla\nabla\cdot-\nabla\times\nabla\times\right)\left(u_r\boldsymbol{e}_r+u_\varphi\boldsymbol{e}_\varphi+u_z\boldsymbol{e}_z\right)\\ \neq\;&\left(\frac{\partial^2 u_r}{\partial r^2}+\frac{1}{r}\frac{\partial u_r}{\partial r}+\frac{1}{r^2}\frac{\partial^2 u_r}{\partial\varphi^2}+\frac{\partial^2 u_r}{\partial z^2}\right)\boldsymbol{e}_r\\ &+\left(\frac{\partial^2 u_\varphi}{\partial r^2}+\frac{1}{r}\frac{\partial u_\varphi}{\partial r}+\frac{1}{r^2}\frac{\partial^2 u_\varphi}{\partial\varphi^2}+\frac{\partial^2 u_\varphi}{\partial z^2}\right)\boldsymbol{e}_\varphi\\ &+\left(\frac{\partial^2 u_z}{\partial r^2}+\frac{1}{r}\frac{\partial u_z}{\partial r}+\frac{1}{r^2}\frac{\partial^2 u_z}{\partial\varphi^2}+\frac{\partial^2 u_z}{\partial z^2}\right)\boldsymbol{e}_z\end{aligned}\tag{2.6.12}$$

d. Verify the following identity in stead of Eq. (2.6.12):

$$\begin{aligned}&\left(\nabla\nabla\cdot-\nabla\times\nabla\times\right)\left(u_r\boldsymbol{e}_r+u_\varphi\boldsymbol{e}_\varphi+u_z\boldsymbol{e}_z\right)\\ =\;&\left(\frac{\partial^2}{\partial r^2}+\frac{1}{r}\frac{\partial}{\partial r}+\frac{1}{r^2}\frac{\partial^2}{\partial\varphi^2}+\frac{\partial^2}{\partial z^2}\right)\left(u_r\boldsymbol{e}_r+u_\varphi\boldsymbol{e}_\varphi+u_z\boldsymbol{e}_z\right)\end{aligned}\tag{2.6.13}$$

by taking into account the spatial derivative of the base vectors in the cylindrical coordinate system for the right-hand side of Eq. (2.6.13).

It is of course that Eq. (2.6.11) can also be expressed by

$$\begin{aligned}&\left(\nabla\nabla\cdot-\nabla\times\nabla\times\right)\left(u_1\boldsymbol{e}_1+u_2\boldsymbol{e}_2+u_3\boldsymbol{e}_3\right)\\ =\;&\left(\frac{\partial^2}{\partial x_1^2}+\frac{\partial^2}{\partial x_2^2}+\frac{\partial^2}{\partial x_3^2}\right)\left(u_1\boldsymbol{e}_1+u_2\boldsymbol{e}_2+u_3\boldsymbol{e}_3\right)\end{aligned}\tag{2.6.14}$$

Note that in general, by taking into account the spatial derivative of the base vectors, we can establish the following identity

$$\nabla^2=\nabla\nabla\cdot-\nabla\times\nabla\times\tag{2.6.15}$$

for the Laplace operator.

5. Prove Eq. (N2.3.8).
6. Prove Eq. (N2.4.5).

3 Elastic wave propagation in 3D elastic half-space

3.1 PLANE WAVE PROPAGATION

3.1.1 COORDINATE SYSTEM, GOVERNING EQUATION, FREE BOUNDARY CONDITIONS, AND SCALAR POTENTIALS

This chapter deals with several topics on the propagation of elastic waves in a 3D half-space. A wide range of solutions, describing complex wave phenomena, are obtained for the governing equation. The first topic discussed is the fundamental example of wave propagation in an elastic half-space, namely plane wave propagation. This example introduces several important and interesting topics related to wave propagation problems.

The coordinate system for the elastic half-space is shown in Fig. 3.1.1, where x_1 and x_2 are horizontal coordinates and x_3 is the vertical coordinate, for which the positive direction is downward and $x_3 = 0$ denotes the free surface. A position vector in the half-space is expressed as:

$$\boldsymbol{x} = \begin{pmatrix} x_1 & x_2 & x_3 \end{pmatrix} \in \mathbb{R}^2 \times \mathbb{R}_+ = \mathbb{R}^3_+ \tag{3.1.1}$$

According to Eqs. (1.4.12) and (1.4.14), the governing equation and boundary conditions for wave propagation in the elastic half-space are respectively given as

$$\left[L_{ij}(\partial_1, \partial_2, \partial_3) - \rho\delta_{ij}\frac{\partial^2}{\partial t^2}\right]u_j(\boldsymbol{x},t) = 0 \tag{3.1.2}$$

and

$$\sigma_{3j} = \lambda\delta_{3j}\partial_i u_i + \mu(\partial_3 u_j + \partial_j u_3) = 0, \quad (\text{at } x_3 = 0) \tag{3.1.3}$$

Drawing from the discussion for Eq. (2.3.18) in Chapter 2, we adopt the expression for a displacement field of the elastic half-space in the following form:

$$\boldsymbol{u}(\boldsymbol{x},t) = \nabla\phi(\boldsymbol{x},t) + \nabla \times \left(\chi(\boldsymbol{x},t)\boldsymbol{e}_3\right) + \nabla \times \nabla \times \left(\psi(\boldsymbol{x},t)\boldsymbol{e}_3\right) \tag{3.1.4}$$

where ϕ , χ, and ψ are the scalar potentials for the P wave, the S wave of rotational motion around the x_3 axis, and the S wave of rotational motion around the x_1 and/or x_2 axes, respectively.

The S wave due to the potential χ is called the SH wave, since the wave motion is in a horizontal plane, and that due to ψ is called the SV wave, since the motion is

DOI: 10.1201/9781003251729-3

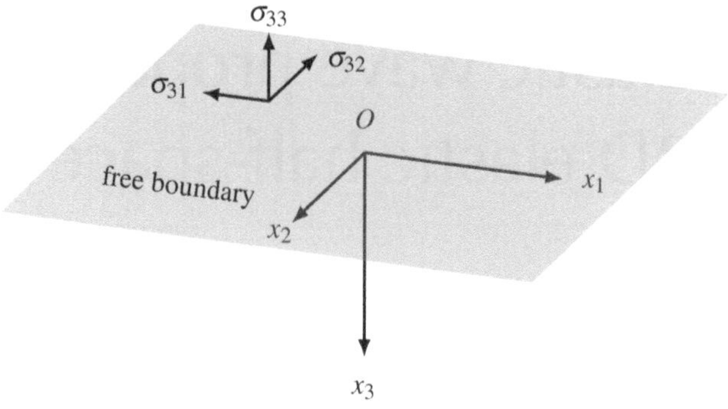

Figure 3.1.1 Coordinate system for elastic half-space, where $x_3 = 0$ denotes free surface. The free boundary conditions at $x_3 = 0$ are given by $\sigma_{31} = \sigma_{32} = \sigma_{33} = 0$, where σ_{ij} represents the components of the stress tensor.

in a vertical plane. The three scalar potentials satisfy the following equations:

$$\begin{aligned}
\left[(\lambda + 2\mu)\nabla^2 - \rho\frac{\partial^2}{\partial t^2}\right]\phi(\boldsymbol{x},t) &= 0 \\
\left[\mu\nabla^2 - \rho\frac{\partial^2}{\partial t^2}\right]\chi(\boldsymbol{x},t) &= 0 \\
\left[\mu\nabla^2 - \rho\frac{\partial^2}{\partial t^2}\right]\psi(\boldsymbol{x},t) &= 0
\end{aligned} \tag{3.1.5}$$

The boundary conditions at $x_3 = 0$ shown in Eq. (3.1.3) can be written in terms of the above scalar potentials as

$$\begin{aligned}
\sigma_{31} &= \mu\left(\partial_1 u_3 + \partial_3 u_1\right) \\
&= 2\mu\partial_1\partial_3\phi + \mu\partial_1(-\nabla^2 + 2\partial_3^2)\psi + \mu\partial_2\partial_3\chi = 0 \\
\sigma_{32} &= \mu\left(\partial_2 u_3 + \partial_3 u_2\right) \\
&= 2\mu\partial_2\partial_3\phi + \mu\partial_2(-\nabla^2 + 2\partial_3^2)\psi - \mu\partial_1\partial_3\chi = 0 \\
\sigma_{33} &= \lambda\partial_k u_k + 2\mu\partial_3 u_3 \\
&= \left(\lambda\nabla^2 + 2\mu\partial_3^2\right)\phi - 2\mu\partial_3\left(\partial_1^2 + \partial_2^2\right)\psi = 0
\end{aligned} \tag{3.1.6}$$

Because of the presence of the free boundary, the scalar potentials have to be constituted by the up- and down-going waves due to the reflection of waves at the free boundary, except for the Rayleigh wave which will be discussed at the end of this section. The expression of the superposition of the up- and down-going waves to ensure the free boundary conditions in Eq. (3.1.6) are in the following form:

$$\begin{aligned}
\phi(\boldsymbol{x},t) &= A^{\downarrow}\exp\left(i\boldsymbol{d}_L^{(+)}\cdot\boldsymbol{x} - i\omega t\right) + A^{\uparrow}\exp\left(i\boldsymbol{d}_L^{(-)}\cdot\boldsymbol{x} - i\omega t\right) \\
\psi(\boldsymbol{x},t) &= B^{\downarrow}\exp\left(i\boldsymbol{d}_T^{(+)}\cdot\boldsymbol{x} - i\omega t\right) + B^{\uparrow}\exp\left(i\boldsymbol{d}_T^{(-)}\cdot\boldsymbol{x} - i\omega t\right) \\
\chi(\boldsymbol{x},t) &= C^{\downarrow}\exp\left(i\boldsymbol{d}_T^{(+)}\cdot\boldsymbol{x} - i\omega t\right) + C^{\uparrow}\exp\left(i\boldsymbol{d}_T^{(-)}\cdot\boldsymbol{x} - i\omega t\right)
\end{aligned} \tag{3.1.7}$$

where $\boldsymbol{d}_L^{(\pm)}$ and $\boldsymbol{d}_T^{(\pm)}$ are respectively wavenumber vectors defined by

$$\begin{aligned}\boldsymbol{d}_L^{(\pm)} &= (\ \xi_1,\ \ \xi_2,\ \ \pm\bar{\gamma}\ ) \\ \boldsymbol{d}_T^{(\pm)} &= (\ \xi_1,\ \ \xi_2,\ \ \pm\bar{\nu}\ ) \quad \text{(double-sign corresponds)}\end{aligned} \tag{3.1.8}$$

$A^{\downarrow}$, $B^{\downarrow}$, and $C^{\downarrow}$ are the coefficients of the potentials for the amplitudes of down-going waves and $A^{\uparrow}$, $B^{\uparrow}$, and $C^{\uparrow}$ are those of up-going waves. Note that we employ the expression for scalar potentials for plane waves.

In Eq. (3.1.8), $\bar{\gamma}$ and $\bar{\nu}$ are respectively defined by

$$\begin{aligned}\bar{\gamma} &= \sqrt{k_L^2-\xi_1^2-\xi_2^2} = \sqrt{k_L^2-\xi_r^2} \\ \bar{\nu} &= \sqrt{k_T^2-\xi_1^2-\xi_2^2} = \sqrt{k_T^2-\xi_r^2}\end{aligned} \tag{3.1.9}$$

where

$$\xi_r^2 = \xi_1^2+\xi_2^2 \tag{3.1.10}$$

Recall that k_T and k_L are the wavenumbers for the S and P waves, respectively, defined by Eq. (2.3.16).

The substitution of Eq. (3.1.7) into Eq. (3.1.6) yields the relationship equation of the coefficients $A^{\uparrow}, A^{\downarrow}, \ldots$, and $C^{\downarrow}$, which enables us to discuss the reflection of plane waves at the free surface. This relationship is given by

$$\begin{aligned}&\begin{pmatrix} 2\xi_1\bar{\gamma} & i\xi_1(2\xi_r^2-k_T^2) & \xi_2\bar{\nu} \\ 2\xi_2\bar{\gamma} & i\xi_2(2\xi_r^2-k_T^2) & -\xi_1\bar{\nu} \\ (2\xi_r^2-k_T^2) & -2i\xi_r^2\bar{\nu} & 0 \end{pmatrix}\begin{pmatrix} A^{\uparrow} \\ B^{\uparrow} \\ C^{\uparrow}\end{pmatrix} \\ &= \begin{pmatrix} 2\xi_1\bar{\gamma} & -i\xi_1(2\xi_r^2-k_T^2) & \xi_2\bar{\nu} \\ 2\xi_2\bar{\gamma} & -i\xi_2(2\xi_r^2-k_T^2) & -\xi_1\bar{\nu} \\ -(2\xi_r^2-k_T^2) & -2i\xi_r^2\bar{\nu} & 0 \end{pmatrix}\begin{pmatrix} A^{\downarrow} \\ B^{\downarrow} \\ C^{\downarrow}\end{pmatrix}\end{aligned} \tag{3.1.11}$$

Later, we use Eq. (3.1.11) to determine the coefficients $A^{\downarrow}$, $B^{\downarrow}$, and $C^{\downarrow}$ by specifying $A^{\uparrow}$, $B^{\uparrow}$, and/or $C^{\uparrow}$.

Note that Eq. (3.1.11) is derived by eliminating the factor $\exp(i\xi_1x_1+i\xi_2x_2)$ from the boundary condition in Eq. (3.1.6). We can thus determine the coefficients of the potentials using the parameters ξ_1, ξ_2, k_T, and k_L. The elimination of the factor $\exp(i\xi_1x_1+i\xi_2x_2)$ is possible since the potentials defined by Eq. (3.1.7) have the following properties:

$$\left.\begin{aligned}\phi(\boldsymbol{x},t) \\ \psi(\boldsymbol{x},t) \\ \chi(\boldsymbol{x},t)\end{aligned}\right\} \propto \exp\big(i(\xi_1x_1+\xi_2x_2)-i\omega t\big), \qquad (\text{at } x_3=0) \tag{3.1.12}$$

As shown later, Eq. (3.1.12) yields interesting wave phenomena in the elastic half-space.

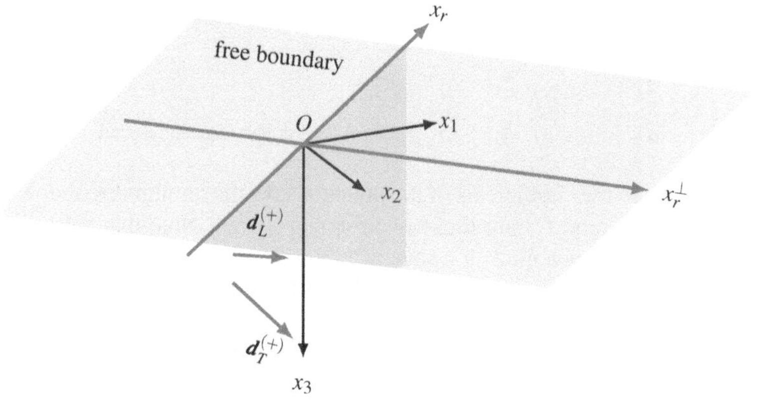

Figure 3.1.2 Plane that embeds wavenumber vectors defined in Eq. (3.1.8). The plane is spanned by the x_3 and x_r axes, where the direction of the x_r axis is $(\xi_1, \xi_2, 0)$. For later discussion, we also define $x_{r\perp}$, whose direction is $(-\xi_2, \xi_1, 0)$.

3.1.2 PROPERTIES OF SCALAR POTENTIALS DEFINED BY EQ. (3.1.7)

We examine Eq. (3.1.7) before discussing the reflection of plane waves at the free surface. An important conclusion derived from Eq. (3.1.7) is that the wavenumber spaces for the potentials, namely $\boldsymbol{d}_L^{(\pm)}$ and $\boldsymbol{d}_T^{(\pm)}$, are orthogonal to the vector

$$\boldsymbol{n} = \begin{pmatrix} \frac{\xi_2}{\xi_r} & -\frac{\xi_1}{\xi_r} & 0 \end{pmatrix} \tag{3.1.13}$$

This indicates that the wavenumber vectors $\boldsymbol{d}_L^{(\pm)}$ and $\boldsymbol{d}_T^{(\pm)}$ are embedded in a plane whose normal vector is $\boldsymbol{n}$. Figure 3.1.2 shows the plane that embeds the wavenumber vectors $\boldsymbol{d}_L^{(\pm)}$ and $\boldsymbol{d}_T^{(\pm)}$. It also shows the x_r and $x_{r\perp}$ axes. Note that the x_r and x_3 axes span the plane that embeds $\boldsymbol{d}_L^{(\pm)}$ and $\boldsymbol{d}_T^{(\pm)}$, and that the direction of the $x_{r\perp}$ axis is the vector $\boldsymbol{n}$. The direction of the x_r axis is found to be

$$\boldsymbol{s} = \begin{pmatrix} \frac{\xi_1}{\xi_r} & \frac{\xi_2}{\xi_r} & 0 \end{pmatrix} \tag{3.1.14}$$

It follows that $\boldsymbol{s} \perp \boldsymbol{n}$.

Based on the definitions of x_r and $x_{r\perp}$, the relationship between the $(x_r, x_{r\perp})$ and (x_1, x_2) coordinates is as follows:

$$\begin{pmatrix} x_1 \\ x_2 \end{pmatrix} = \begin{pmatrix} \mathrm{c} & -\mathrm{s} \\ \mathrm{s} & \mathrm{c} \end{pmatrix} \begin{pmatrix} x_r \\ x_{r\perp} \end{pmatrix} \Longleftrightarrow \begin{pmatrix} x_r \\ x_{r\perp} \end{pmatrix} = \begin{pmatrix} \mathrm{c} & \mathrm{s} \\ -\mathrm{s} & \mathrm{c} \end{pmatrix} \begin{pmatrix} x_1 \\ x_2 \end{pmatrix} \tag{3.1.15}$$

where

$$\begin{aligned} \mathrm{c} &= \xi_1/\xi_r \\ \mathrm{s} &= \xi_2/\xi_r \end{aligned} \tag{3.1.16}$$

We can define the displacement vector on the free boundary along x_r and $x_{r\perp}$ as $(u_r, u_{r\perp})$. Similar to Eq. (3.1.15), the relationship of the displacement between the $(u_r, u_{r\perp})$ and (u_1, u_2) coordinates is as follows:

$$\begin{pmatrix} u_1 \\ u_2 \end{pmatrix} = \begin{pmatrix} \mathrm{c} & -\mathrm{s} \\ \mathrm{s} & \mathrm{c} \end{pmatrix} \begin{pmatrix} u_r \\ u_{r\perp} \end{pmatrix} \Longleftrightarrow \begin{pmatrix} u_r \\ u_{r\perp} \end{pmatrix} = \begin{pmatrix} \mathrm{c} & \mathrm{s} \\ -\mathrm{s} & \mathrm{c} \end{pmatrix} \begin{pmatrix} u_1 \\ u_2 \end{pmatrix} \tag{3.1.17}$$

Now, by means of Eq. (3.1.15), the scalar potentials defined by Eq. (3.1.7) can be rewritten in the $x_r - x_3$ coordinate system in the following form:

$$\begin{aligned} \phi(\boldsymbol{x},t) &= A^{\uparrow}\exp\big(i(\xi_r x_r - \bar{\gamma}x_3) - i\omega t\big) + A^{\downarrow}\exp\big(i(\xi_r x_r + \bar{\gamma}x_3) - i\omega t\big) \\ \psi(\boldsymbol{x},t) &= B^{\uparrow}\exp\big(i(\xi_r x_r - \bar{\nu}x_3) - i\omega t\big) + B^{\downarrow}\exp\big(i(\xi_r x_r + \bar{\nu}x_3) - i\omega t\big) \\ \chi(\boldsymbol{x},t) &= C^{\uparrow}\exp\big(i(\xi_r x_r - \bar{\nu}x_3) - i\omega t\big) + C^{\downarrow}\exp\big(i(\xi_r x_r + \bar{\nu}x_3) - i\omega t\big) \end{aligned} \tag{3.1.18}$$

Equation (3.1.18) shows that the plane wave propagation in the 3D elastic half-space can be dealt with as a 2D problem in the $x_3 - x_r$ plane. We can define the incident and reflection angles of the plane waves in the 3D half-space in the $x_3 - x_r$ plane. Equation (3.1.9) yields

$$\xi_r^2 + \bar{\gamma}^2 = k_L^2 \tag{3.1.19}$$

$$\xi_r^2 + \bar{\nu}^2 = k_T^2 \tag{3.1.20}$$

This enables us to have two parameters θ_L and θ_T such that

$$\begin{aligned} \xi_r &= k_L \sin\theta_L, \quad \bar{\gamma} = k_L\cos\theta_L \\ \xi_r &= k_T \sin\theta_T, \quad \bar{\nu} = k_T\cos\theta_T \end{aligned} \tag{3.1.21}$$

The substitution of Eq. (3.1.21) into Eq. (3.1.18) yields

$$\begin{aligned} \phi(\boldsymbol{x},t) &= A^{\uparrow}\exp\big(ik_L(x_r\sin\theta_L - x_3\cos\theta_L) - i\omega t\big) \\ &\quad + A^{\downarrow}\exp\big(ik_L(x_r\sin\theta_L + x_3\cos\theta_L) - i\omega t\big) \\ \psi(\boldsymbol{x},t) &= B^{\uparrow}\exp\big(ik_T(x_r\sin\theta_T - x_3\cos\theta_T) - i\omega t\big) \\ &\quad + B^{\downarrow}\exp\big(ik_T(x_r\sin\theta_T + x_3\cos\theta_T) - i\omega t\big) \\ \chi(\boldsymbol{x},t) &= C^{\uparrow}\exp\big(ik_T(x_r\sin\theta_T - x_3\cos\theta_T) - i\omega t\big) \\ &\quad + C^{\downarrow}\exp\big(ik_T(x_r\sin\theta_T + x_3\cos\theta_T) - i\omega t\big) \end{aligned} \tag{3.1.22}$$

This clarifies the direction of wave propagation for each potential in the $x_r - x_3$ coordinate system. The directions of the up- and down-going P waves are

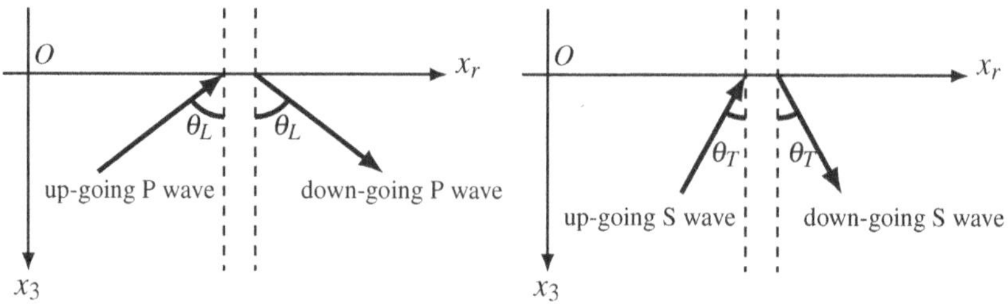

(a) Definition of θ_L (b) Definition of θ_T

Figure 3.1.3 Definition of angles θ_L and θ_T based on Eq. (3.1.21).

$(\sin\theta_L, \ -\cos\theta_L)$ and $(\sin\theta_L, \ \cos\theta_L)$, respectively, and those of the up- and down-going S waves are $(\sin\theta_T, -\cos\theta_T)$ and $(\sin\theta_T, \cos\theta_T)$, respectively. Figure 3.1.3 shows the geometry for the parameters θ_L and θ_T as the angles of incidence and reflection of the waves.

An important conclusion derived from Eq. (3.1.21) is that

$$\frac{\sin\theta_L}{c_L} = \frac{\sin\theta_T}{c_T} \tag{3.1.23}$$

This reminds us the Snell law. Since $c_L > c_T$, we also find that $\theta_L > \theta_T$. The propagation directions of the P and S waves are thus different when the free boundary conditions are imposed.

On the x_r axis, the expression of the potentials in Eq. (3.1.18) becomes

$$\begin{aligned} \phi(\boldsymbol{x},t) &= (A^{\uparrow}+A^{\downarrow})\exp\left(i\xi_r x_r - i\omega t\right) \\ \psi(\boldsymbol{x},t) &= (B^{\uparrow}+B^{\downarrow})\exp\left(i\xi_r x_r - i\omega t\right) \\ \chi(\boldsymbol{x},t) &= (C^{\uparrow}+C^{\downarrow})\exp\left(i\xi_r x_r - i\omega t\right) \end{aligned} \tag{3.1.24}$$

This shows that all potentials, both for the P and S waves, have the same wave velocity ω/ξ_r (and thus wavelength) along the x_r axis. This ensures that the set of potentials realize the free surface conditions. The wave velocity ω/ξ_r is called the *phase velocity* along the x_r axis.

Note that when using the potentials defined by Eq. (3.1.7), the polarizations of the P and SV waves are no longer orthogonal. However, the polarizations of the SH and SV waves, as well as those of the SH and P waves, remain orthogonal. This can be expressed as follows:

$$\nabla\times\left(\chi(\boldsymbol{x},t)\boldsymbol{e}_3\right)\perp\nabla\times\nabla\times\left(\psi(\boldsymbol{x},t)\boldsymbol{e}_3\right), \quad \nabla\times\left(\chi(\boldsymbol{x},t)\boldsymbol{e}_3\right)\perp\nabla\phi(\boldsymbol{x},t) \tag{3.1.25}$$

3.1.3 REFLECTION AND POLARIZATION OF PLANE WAVES AT FREE SURFACE

3.1.3.1 Plane incident P wave

We now investigate the reflection of plane waves at the free surface. The problem at this point is to determine the coefficients $A^{\downarrow}$, $B^{\downarrow}$, and $C^{\downarrow}$ from the up-going P wave. For this purpose, we employ the following three scalar potentials:

$$\begin{aligned}\phi(\boldsymbol{x},t) &= A^{\uparrow}\exp\big(i(\xi_r x_r - \bar{\gamma}x_3) - i\omega t\big) + A^{\downarrow}\exp\big(i(\xi_r x_r + \bar{\gamma}x_3) - i\omega t\big)\\ \psi(\boldsymbol{x},t) &= B^{\downarrow}\exp\big(i(\xi_r x_r + \bar{\nu}x_3) - i\omega t\big)\\ \chi(\boldsymbol{x},t) &= C^{\downarrow}\exp\big(i(\xi_r x_r + \bar{\nu}x_3) - i\omega t\big)\end{aligned} \tag{3.1.26}$$

where ξ_r and $A^{\uparrow}$ are given parameters. Recall that once the parameter ξ_r is known, the incident and reflection angles of the P wave is given by

$$\theta_L = \tan^{-1}\frac{\xi_r}{\bar{\gamma}} = \tan^{-1}\frac{\xi_r}{\sqrt{k_L^2 - \xi_r^2}} \tag{3.1.27}$$

and the reflection angle of the S wave is given by

$$\theta_T = \tan^{-1}\frac{\xi_r}{\bar{\nu}} = \tan^{-1}\frac{\xi_r}{\sqrt{k_T^2 - \xi_r^2}} \tag{3.1.28}$$

Note that $\theta_T < \theta_L$ based on Eqs. (3.1.27) and (3.1.28).

Figure 3.1.4 shows an incident P wave and its reflected P and SV waves in the $x_r - x_3$ plane. We restrict the range of ξ_r in order to keep $\bar{\gamma} = (k_L^2 - \xi_r^2)^{1/2} > 0$ for the discussion. Namely, the range of ξ_r is $0 \le \xi_r < k_L$. We discuss the case $\xi_r = 0$ separately from the case $0 < \xi_r < k_L$ since we cannot define the x_r axis.

Based on Eq. (3.1.11), the equation for determining $A^{\downarrow}$, $B^{\downarrow}$, and $C^{\downarrow}$ from $A^{\uparrow}$ for the case $0 < \xi_r < k_L$ is in the following form:

$$\begin{pmatrix} 2\xi_1\bar{\gamma} & -i\xi_1(2\xi_r^2 - k_T^2) & \xi_2\bar{\nu} \\ 2\xi_2\bar{\gamma} & -i\xi_2(2\xi_r^2 - k_T^2) & -\xi_1\bar{\nu} \\ -(2\xi_r^2 - k_T^2) & -2i\bar{\nu}\xi_r^2 & 0 \end{pmatrix}\begin{pmatrix} A^{\downarrow} \\ B^{\downarrow} \\ C^{\downarrow}\end{pmatrix} = \begin{pmatrix} 2\xi_1\bar{\gamma} \\ 2\xi_2\bar{\gamma} \\ (2\xi_r^2 - k_T^2)\end{pmatrix} A^{\uparrow} \tag{3.1.29}$$

The solution for Eq. (3.1.29) is

$$\begin{pmatrix} A^{\downarrow} \\ B^{\downarrow} \\ C^{\downarrow}\end{pmatrix} = \frac{1}{F(\xi_r)}\begin{pmatrix} -(2\xi_r^2 - k_T^2)^2 + 4\xi_r^2\bar{\gamma}\bar{\nu} \\ 4i\bar{\gamma}(2\xi_r^2 - k_T^2) \\ 0 \end{pmatrix} A^{\uparrow} \tag{3.1.30}$$

where

$$F(\xi_r) = (2\xi_r^2 - k_T^2)^2 + 4\xi_r^2\bar{\nu}\bar{\gamma} \tag{3.1.31}$$

Note that an important conclusion derived from Eq. (3.1.30) is that reflected P and SV waves, but not a reflected SH wave, are generated by the plane incident P wave.

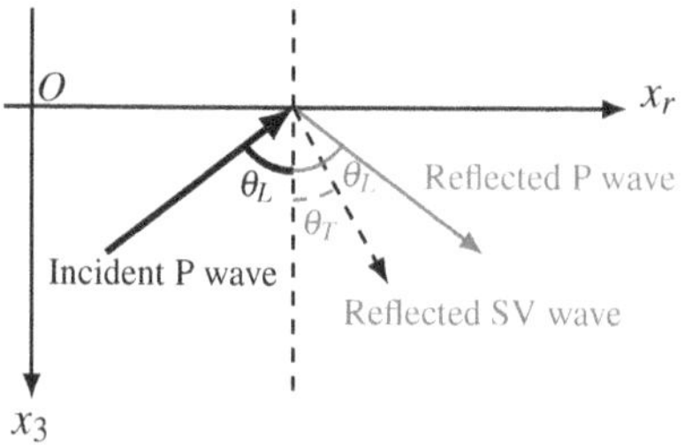

(a) Incident P wave with incident angle θ_L

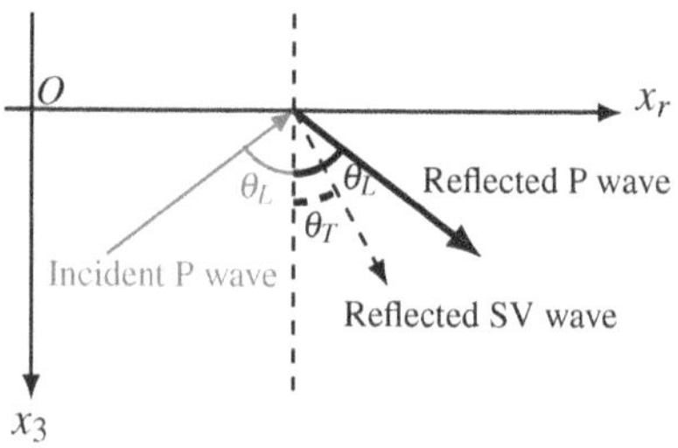

(b) Reflected P and SV waves

Figure 3.1.4 Incident P wave and its reflection. The incident angle θ_L is defined as $\tan^{-1}(\xi_r/\bar{\gamma})$. The reflection angle of the P wave is identical to θ_L and that of the S wave is $\theta_T = \tan^{-1}(\xi_r/\bar{\nu})$. Note that the reflected S resulting from the incident P wave is the SV wave, as shown in Eq. (3.1.30). There is no reflected SH wave present.

The substitution of Eq. (3.1.30) into Eq. (3.1.7) and the use of Eqs. (3.1.4) and (3.1.17) yield the surface response due to the plane incident P wave, expressed as

$$\begin{aligned} u_r(\boldsymbol{x},t) &= U_r^{(L)}(\xi_r)A^{\uparrow}\exp(i\xi_r x_r - i\omega t) \\ u_{r\perp}(\boldsymbol{x},t) &= 0 \\ u_3(\boldsymbol{x},t) &= U_3^{(L)}(\xi_r)A^{\uparrow}\exp(i\xi_r x_r - i\omega t) \end{aligned} \tag{3.1.32}$$

where $U_r^{(L)}(\xi_r)$ and $U_3^{(L)}(\xi_r)$ are given as

$$\begin{aligned} U_r^{(L)}(\xi_r) &= \frac{4i\xi_r\bar{\nu}\bar{\gamma}k_T^2}{F(\xi_r)} \\ U_3^{(L)}(\xi_r) &= \frac{2i\bar{\gamma}k_T^2(2\xi_r^2 - k_T^2)}{F(\xi_r)} \end{aligned} \tag{3.1.33}$$

For the case $\xi_r = 0$, the three scalar potentials shown in Eq. (3.1.26) become

$$\begin{aligned} \phi(\boldsymbol{x},t) &= A^{\uparrow}\exp(-ik_L x_3 - i\omega t) + A^{\downarrow}\exp(ik_L x_3 - i\omega t) \\ \psi(\boldsymbol{x},t) &= B^{\downarrow}\exp(ik_T x_3 - i\omega t) \\ \chi(\boldsymbol{x},t) &= C^{\downarrow}\exp(ik_T x_3 - i\omega t) \end{aligned} \tag{3.1.34}$$

The substitution of Eq. (3.1.34) into Eq. (3.1.6) yields

$$\begin{aligned} \sigma_{31} &= 0 \\ \sigma_{32} &= 0 \\ \sigma_{33} &= -k_L^2(\lambda + 2\mu)(A^{\uparrow} + A^{\downarrow}) = 0 \end{aligned} \tag{3.1.35}$$

Therefore, we only have

$$A^{\downarrow} = -A^{\uparrow} \tag{3.1.36}$$

Equation (3.1.36) shows that the reflection of an incident P wave propagating vertically generates a P wave propagating vertically with the same amplitude and a different sign of the coefficients $A^{\uparrow}$ and $A^{\downarrow}$. The surface response at the free surface due to the vertical incidence of the P wave is expressed as

$$\begin{aligned} u_r(\boldsymbol{x},t) &= 0 \\ u_r^{\perp}(\boldsymbol{x},t) &= 0 \\ u_3(\boldsymbol{x},t) &= U_3^{(L)}(0)A^{\uparrow}\exp(-i\omega t) \end{aligned} \tag{3.1.37}$$

where

$$U_3^{(L)}(0) = -2ik_L \tag{3.1.38}$$

Table 3.1 displays the reflection coefficients and responses at the free surface, specifically $U_3^{(L)}$ and $U_r^{(L)}$, for various incident angles. Figure 3.1.5 illustrates the polarization and amplitude of waves at the free surface, using the results from Table 3.1. The data presented in Table 3.1 and Fig. 3.1.5 was obtained under the conditions $c_T/c_L = 0.5$, $\omega = 1$ (rad/s), and $A^{\uparrow} = 1$

From Fig. 3.1.5 (a), the polarization of the wave due to the vertical incident P wave agrees with that of the incident wave, which is evident from Eq. (3.1.37). Figures 3.1.5 (b)-(c) show that the polarization and amplitude of waves at the free surface are nearly identical to those of the incident waves. For incident angles exceeding 60 degrees, shown in Figs. 3.1.5 (d) - (f), the amplitude become smaller than those of the incident waves, and the polarization (direction of the vibration) begins to deviate from the incident angle.

Table 3.1
Reflection coefficients and surface responses for incident P wave for $A^{\uparrow} = 1$.

$\theta(=\theta_L)$	θ_T	$A^{\downarrow}$	$B^{\downarrow}$	$U_3^{(L)}$	$U_r^{(L)}$
0°	—	-1.0	—	-1.0 i	0.0
20°	9.8°	-0.88	-1.9 i	-0.94 i	0.34i
40°	18.7°	-0.61	-1.6 i	-0.78 i	0.60 i
60°	25.7°	-0.40	-1.1 i	-0.56 i	0.70 i
70°	28.7°	-0.40	-0.85 i	-0.43 i	0.64 i
80°	29.5°	-0.57	-0.52 i	-0.26i	0.43 i

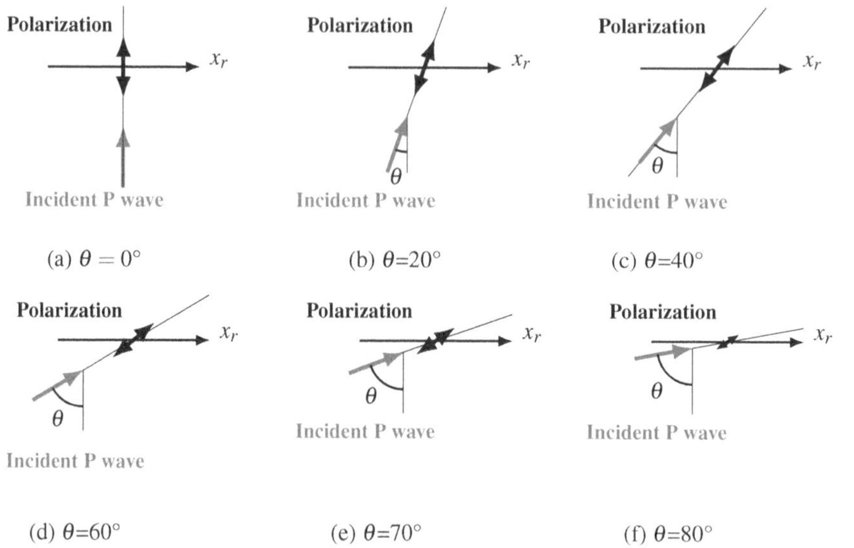

Figure 3.1.5 Polarization and amplitude of waves at free surface due to plane incident P waves with various incident angles.

3.1.3.2 Plane incident SV wave

Next, we analyze the reflection of an incident SV wave. For this problem, we use the following three scalar potentials:

$$\begin{aligned}\phi(\boldsymbol{x},t) &= A^{\downarrow}\exp\big(i(\xi_r x_r + \bar{\gamma}x_3) - i\omega t\big)\\ \psi(\boldsymbol{x},t) &= B^{\uparrow}\exp\big(i(\xi_r x_r - \bar{\nu}x_3) - i\omega t\big) + B^{\downarrow}\exp\big(i(\xi_r x_r + \bar{\nu}x_3) - i\omega t\big)\\ \chi(\boldsymbol{x},t) &= C^{\downarrow}\exp\big(i(\xi_r x_r + \bar{\nu}x_3) - i\omega t\big)\end{aligned} \tag{3.1.39}$$

where ξ_r and $B^{\uparrow}$ are given parameters that define the direction and amplitude of the incident SV wave. The unknown quantities that we have to obtain at present are the reflection coefficients $A^{\downarrow}$, $B^{\downarrow}$, and $C^{\downarrow}$. Recall that the incident angle of an SV wave θ_T can be determined from the parameter ξ_r using Eq. (3.1.28). In addition, $\xi_r = 0$ yields

$$\nabla \times \exp(i\boldsymbol{d}_T^{(\pm)} \cdot \boldsymbol{x} - i\omega t)\,\boldsymbol{e}_3 = 0 \tag{3.1.40}$$

Therefore, we exclude the discussion of an incident SV wave for the case $\xi_r = 0$ and set the range of ξ_r to $0 < \xi_r < k_T$.

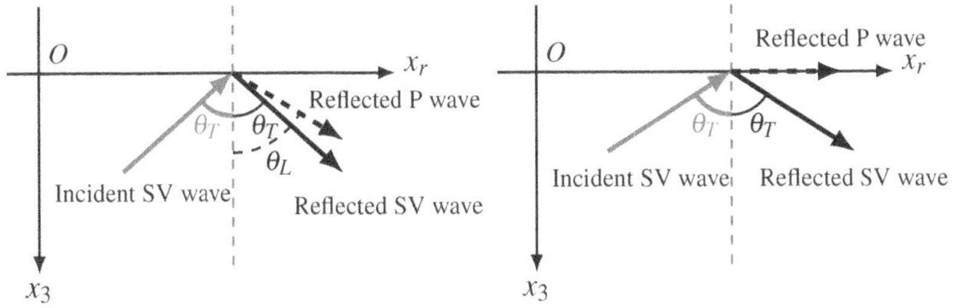

(a) Cause of reflected SV and P waves for $\theta_T < \theta_c$

(b) Cause of reflected SV and P waves for $\theta_T \geq \theta_c$.

Figure 3.1.6 Incident SV wave and its reflection. The incident angle θ_T is defined as $\tan^{-1}(\xi_r/\bar{\nu})$. The reflected waves consist of P and SV waves, whereas there is no reflected SH wave present. When the incident angle θ_T exceeds the critical angle θ_C, the reflected P wave propagates as an inhomogeneous wave.

Based on Eq. (3.1.11), the equation for determining the coefficients $A^{\downarrow}$, $B^{\downarrow}$, and $C^{\downarrow}$ is in the following form:

$$\begin{pmatrix} 2\xi_1\bar{\gamma} & -i\xi_1(2\xi_r^2-k_T^2) & \xi_2\bar{\nu} \\ 2\xi_2\bar{\gamma} & -i\xi_2(2\xi_r^2-k_T^2) & -\xi_1\bar{\nu} \\ -(2\xi_r^2-k_T^2) & -2i\bar{\nu}\xi_r^2 & 0 \end{pmatrix} \begin{pmatrix} A^{\downarrow} \\ B^{\downarrow} \\ C^{\downarrow} \end{pmatrix} = \begin{pmatrix} i\xi_1(2\xi_r^2-k_T^2) \\ i\xi_2(2\xi_r^2-k_T^2) \\ -2i\xi_r^2\bar{\nu} \end{pmatrix} B^{\uparrow} \tag{3.1.41}$$

which gives the coefficients

$$\begin{pmatrix} A^{\downarrow} \\ B^{\downarrow} \\ C^{\downarrow} \end{pmatrix} = \frac{1}{F(\xi_r)} \begin{pmatrix} 4i\bar{\nu}\xi_r^2(2\xi_r^2-k_T^2) \\ -(2\xi_r^2-k_T^2)^2+4\xi_r^2\bar{\nu}\bar{\gamma} \\ 0 \end{pmatrix} B^{\uparrow} \tag{3.1.42}$$

It is important to note that the reflected waves due to the plane incident SV wave consist of the P and SV waves, but not the SH wave. Therefore, based on the results of Eqs. (3.1.30) and (3.1.42), the P and SV waves are found to be coupled at the free surface, wheres the SH wave is not coupled with either the P or SV wave.

Figure 3.1.6 shows the incident SV wave and the reflected P and SV waves in the $x_r - x_3$ plane. The incident and reflection angles for the P and SV waves are denoted by θ_L and θ_T, respectively. As mentioned, the range of ξ_r is $0 < \xi_r < k_T$. Therefore, there is a case where ξ_r is in the range of $k_L < \xi_r < k_T$, where $\bar{\gamma}$ becomes an imaginary number. Recall that $\bar{\gamma}$ is defined by

$$\bar{\gamma} = \sqrt{k_L^2 - \xi_r^2}$$

in Eq. (3.1.9). Due to the requirement that $\phi(\boldsymbol{x},t)$ defined in Eq. (3.1.39) has to be bounded, $\bar{\gamma}$ has to be in the form

$$\bar{\gamma} = i\sqrt{\xi_r^2 - k_L^2} = i\gamma, \quad (\xi_r > k_L) \tag{3.1.43}$$

For this case, $\phi(\boldsymbol{x},t)$ is expressed as

$$\phi(\boldsymbol{x},t) = A^{\downarrow}\exp(i\xi_r x_r - \gamma x_3 - i\omega t) \tag{3.1.44}$$

A wavefunction decreasing exponentially along an axis as shown in Eq. (3.1.44) is referred to as an *inhomogeneous wave*. A reflected P wave generated by an incident plane SV wave becomes an inhomogeneous wave when the incident angle exceeds θ_c, which is defined by

$$\theta_c = \tan^{-1}\frac{k_L}{\sqrt{k_T^2 - k_L^2}} \tag{3.1.45}$$

The angle θ_c is called the *critical angle*.

The substitution of Eq. (3.1.42) into Eq. (3.1.7)) and the use of Eq. (3.1.4) yield the response at the free surface, which is expressed in the following form:

$$\begin{aligned} u_r(\boldsymbol{x},t) &= U_r^{(T_V)}(\xi_r)\,B^{\uparrow}\exp(i\xi_r x_r - i\omega t) \\ u_r^{\perp}(\boldsymbol{x},t) &= 0 \\ u_3(\boldsymbol{x},t) &= U_3^{(T_V)}(\xi_r)\,B^{\uparrow}\exp(i\xi_r x_r - i\omega t) \end{aligned} \tag{3.1.46}$$

where $U_r^{(T_V)}$ and $U_3^{(T_V)}$ are given as

$$\begin{aligned} U_r^{(T_V)}(\xi_r) &= -\frac{2k_T^2\xi_r\bar{\nu}(2\xi_r^2 - k_T^2)}{F(\xi_r)} \\ U_3^{(T_V)}(\xi_r) &= \frac{4k_T^2\bar{\gamma}\bar{\nu}\xi_r^2}{F(\xi_r)} \end{aligned} \tag{3.1.47}$$

Table 3.2 displays the reflection coefficients and the responses, specifically $U_r^{(T_V)}$ and $U_3^{(T_V)}$, for various incident angles. Figure 3.1.7 illustrates the polarization and amplitude of the waves at the free surface, using the results from Table 3.2. The data presented in Table 3.2 and Fig. 3.1.7 was obtained under the conditions $c_T/c_L = 0.5$, $\omega = 1.0$ (rad/s), and $B^{\uparrow} = 1$. The parameter $c_T/c_L = 0.5$ results in a critical angle of $\theta_c = 30°$ for the incident SV wave.

It is found from Fig. 3.1.7 that for the case $\theta \leq \theta_c$, the polarization of waves forms a plane, whereas for the case $\theta > \theta_c$, it forms an ellipse. For the latter case, the surface responses $U_r^{(T_V)}$ and $U_3^{(T_V)}$ take complex values, as shown in Table 3.2. In addition, from Eq. (3.1.47), we also have

$$\left|\mathrm{Arg}\left(U_3^{(T_V)}\right) - \mathrm{Arg}\left(U_r^{(T_V)}\right)\right| = \frac{\pi}{2} \tag{3.1.48}$$

Therefore, the axis of the ellipses of the polarization is orthogonal to the x_r axis. Note that the direction of the polarization shown in Fig. 3.1.7 (c) is different from that of the polarizations shown in Figs. 3.1.7 (d)-(f). This is due to the sign of $(2\,\xi_r^2 - k_T^2)$ in Eq. (3.1.47), which affects the phase of $U_r^{(T_V)}(\xi_r)$. For Fig. 3.1.7 (c), the sign of $(2\,\xi_r^2 - k_T^2)$ is negative, whereas for Figs. 3.1.7 (d)-(f), it is positive.

Table 3.2
Reflection coefficients for incident SV wave for $B^{\uparrow} = 1$.

$\theta(=\theta_T)$	θ_L	$A^{\downarrow}$	$B^{\downarrow}$	U_3^T	$U_r^{(T)}$
20°	43.2°	-0.45 i	-0.57	0.21	0.66
30°	90°	-1.7i	-1.0	0.0	1.7
40°	90°	-0.43- 0.03i	0.99+0.12 i	1.0+0.06i	0.02-0.33 i
60°	90°	0.67+0.16i	0.89+0.44 i	0.95+0.22i	-0.09+0.39i
70°	90°	0.70+0.42i	0.46+0.89i	0.73+0.44i	-0.23+0.37i
80°	90°	0.33+0.51i	-0.41+0.91i	0.30+0.45i	-0.26+0.17i

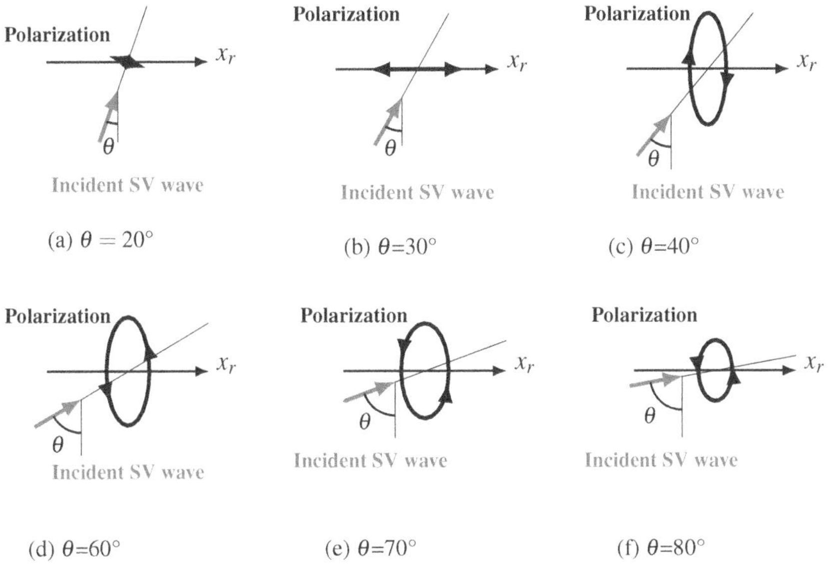

Figure 3.1.7 Polarization and amplitude of waves at the free surface due to plane incident SV waves with various incident angles.

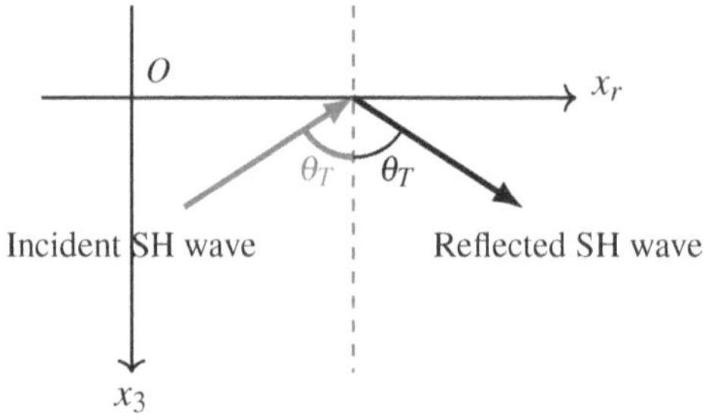

Figure 3.1.8 Reflection of incident SH wave at free surface. P and SV waves are not generated.

3.1.3.3 Incident SH wave

Finally, we investigate the reflection of an incident SH wave. For this problem, we employ the following three scalar potentials:

$$\begin{aligned}
\phi(\boldsymbol{x},t) &= A^{\downarrow}\exp\big(i(\xi_r x_r + \bar{\gamma}x_3) - i\omega t\big) \\
\psi(\boldsymbol{x},t) &= B^{\downarrow}\exp\big(i(\xi_r x_r + \bar{\nu}x_3) - i\omega t\big) \\
\chi(\boldsymbol{x},t) &= C^{\uparrow}\exp\big(i(\xi_r x_r - \bar{\nu}x_3 - i\omega t\big) + C^{\downarrow}\exp\big(i(\xi_r x_r + \bar{\nu}x_3 - i\omega t\big)
\end{aligned} \tag{3.1.49}$$

where $C^{\uparrow}$ and ξ_r are given parameters, from which we obtain $A^{\downarrow}$, $B^{\downarrow}$, and $C^{\downarrow}$. Equation (3.1.11) yields the following equation for obtaining the coefficients:

$$\begin{pmatrix} 2\xi_1\bar{\gamma} & -i\xi_1(2\xi_r^2 - k_T^2) & \xi_2\bar{\nu} \\ 2\xi_2\bar{\gamma} & -i\xi_2(2\xi_r^2 - k_T^2) & -\xi_1\bar{\nu} \\ -(2\xi_r^2 - k_T^2) & -2i\bar{\nu}\xi_r^2 & 0 \end{pmatrix} \begin{pmatrix} A^{\downarrow} \\ B^{\downarrow} \\ C^{\downarrow} \end{pmatrix} = \begin{pmatrix} \xi_2\bar{\nu} \\ -\xi_1\bar{\nu} \\ 0 \end{pmatrix} C^{\uparrow} \tag{3.1.50}$$

As a result, we have

$$\begin{pmatrix} A^{\downarrow} \\ B^{\downarrow} \\ C^{\downarrow} \end{pmatrix} = \begin{pmatrix} 0 \\ 0 \\ 1 \end{pmatrix} C^{\uparrow} \tag{3.1.51}$$

This indicates P and SV waves are not generated by an incident SH wave. Therefore, the P and SV waves are coupled but the SH wave is independent from the P and SV waves, as shown in Fig. 3.1.8, which shows the reflection of an incident SH wave. The substitution of Eq. (3.1.51) into Eq. (3.1.7) and the use of Eq. (3.1.7) yield the response at the free surface, expressed as

$$\begin{aligned}
u_r(\boldsymbol{x},t) &= 0 \\
u_{r\perp}(\boldsymbol{x},t) &= U_{r\perp}^{(T_H)}(\xi_r)C^{\uparrow}\exp(i\xi_r x_r - i\omega t) \\
u_3(\boldsymbol{x},t) &= 0
\end{aligned} \tag{3.1.52}$$

where $U_{r\perp}^{(T_H)}$ is expressed as

$$U_{r\perp}^{(T_H)}(\xi_r) = 2i\xi_r \tag{3.1.53}$$

We find that the motion of SH waves is horizontal, that is, orthogonal to the $x_r - x_3$ plane, and not coupled with the motions of P and SV waves. Such motion is called *anti-plane motion*. We also find that the motions of P and SV waves are coupled at the free surface and are in the $x_r - x_3$ plane. Such motions are called *in-plane motions*. The motions of P, SV, and SH waves, which constitute the 3D displacement field, are linearly independent so that three scalar potentials, namely ϕ, ψ, and χ, are sufficient for expressing the wave motions in a 3D elastic half-space.

Note that we used the superscripts (L), (T_V), and (T_H) for the surface responses caused by P, SV, and SH waves, respectively, in Eqs. (3.1.33), (3.1.37), (3.1.47), and (3.1.53). In later discussions, we sometimes use these notations to distinguish the wave motions of P, SV, and SH waves.

3.1.4 PROPAGATION OF THE RAYLEIGH WAVE

In Eqs. (3.1.30) and (3.1.42) for the expression of the expression of the reflection coefficients, we can see a function $F(\cdot)$ in the denominator. The function $F(\cdot)$ is defined by Eq. (3.1.31), which can also be expressed as

$$F(\xi_r) = (2\xi_r^2 - k_T^2)^2 - 4\xi_r^2 \nu\gamma \tag{3.1.54}$$

where

$$\gamma = \sqrt{\xi_r^2 - k_L^2}, \quad \nu = \sqrt{\xi_r^2 - k_T^2} \tag{3.1.55}$$

We call $F(\xi)$ the *Rayleigh function*. It is known that the Rayleigh function has a real root for the region of $\xi_r > k_T$ (Love, 1927). Therefore, the discussions for the plane wave propagation given above are not affected by the presence of the root of the Rayleigh function, since our discussions were always for $0 < \xi_r < k_T$. The root of the Rayleigh function, however, will play important roles for Green's function for an elastic half-space as shown later.

Now, let κ_R be the root for the Rayleigh function and investigate the wave function when $\xi_r = \kappa_R$. Since the Rayleigh function $F(\xi_r)$ is an even function, we have

$$F(\pm\kappa_R) = 0 \tag{3.1.56}$$

We define the three scalar potentials for the region of $\xi_r > k_T$ as[1]

$$\begin{aligned} \phi(\boldsymbol{x},t) &= A\exp(i\xi_1 x_1 + i\xi_2 x_2 - \gamma x_3 - i\omega t) \\ \phi(\boldsymbol{x},t) &= B\exp(i\xi_1 x_1 + i\xi_2 x_2 - \nu x_3 - i\omega t) \\ \chi(\boldsymbol{x},t) &= C\exp(i\xi_1 x_1 + i\xi_2 x_2 - \nu x_3 - i\omega t) \end{aligned} \tag{3.1.57}$$

[1] The scalar potentials describe inhomogeneous waves propagating horizontally. Therefore, we do not consider up- and down-going waves for the potentials as shown in Eq. (3.1.18).

By means of Eq. (3.1.6), the relationship of the coefficients A, B and C becomes as

$$\begin{pmatrix} -2\gamma i\xi_1 & i\xi_1(2\xi_r^2-k_T^2) & -i\xi_2\nu \\ -2\gamma i\xi_2 & i\xi_2(2\xi_r^2-k_T^2) & i\xi_1\nu \\ (2\xi_r^2-k_T^2) & -2\nu\xi_r^2 & 0 \end{pmatrix}\begin{pmatrix} A \\ B \\ C \end{pmatrix} = \begin{pmatrix} 0 \\ 0 \\ 0 \end{pmatrix} \quad (3.1.58)$$

Equation (3.1.58) can be decomposed into

$$\begin{pmatrix} 2\gamma\xi_r^2 & -\xi_r^2(2\xi_r^2-k_T^2) \\ (2\xi_r^2-k_T^2) & -2\nu\xi_r^2 \end{pmatrix}\begin{pmatrix} A \\ B \end{pmatrix} = \begin{pmatrix} 0 \\ 0 \end{pmatrix} \quad (3.1.59)$$

and

$$C = 0 \quad (3.1.60)$$

We find from Eq. (3.1.60) that the SH wave can not exist for the region of $\xi_r > k_T$. On the other hand, the P-SV motion is still possible for the region of $\xi_r > k_T$. The determinant of the matrix of the right-hand side of Eq. (3.1.59) is $\xi_r^2 F(\xi_r)$, and therefore, if $\xi_r = \kappa_R$, we have a non-trivial solution for A and B.

The above fact means that we can construct a non trivial solution for the elastic wave equation satisfying the free boundary conditions for an elastic half-space, if $\xi_r = \kappa_R$. In other words, for the following equation:

$$\begin{aligned} \left(L_{ij}(\partial_1,\partial_2,\partial_3) - \rho\frac{\partial^2}{\partial t^2}\right)u_j(\boldsymbol{x},t) &= 0 \\ P_{ij}(\partial_1,\partial_2,\partial_3)u_j(\boldsymbol{x},t) &= 0 \ (\text{at } x_3 = 0) \end{aligned} \quad (3.1.61)$$

we can construct a non-trivial solution by

$$\begin{aligned} u_i(\boldsymbol{x},t) &= \partial_i A\exp(i\xi_1x_1 + i\xi_2x_2 - \gamma x_3 - i\omega t) \\ &\quad + e_{ijk}e_{kl3}\partial_j\partial_l B\exp(i\xi_1x_1 + i\xi_2x_2 - \nu x_3 - i\omega t) \end{aligned} \quad (3.1.62)$$

if $\xi_1^2 + \xi_2^2 = \kappa_R^2$, where the coefficients A and B satisfy Eq. (3.1.59).

We call the wave motion due to the solution of the elastic wave equation at $\xi_r = \kappa_R$ the *Rayleigh wave*. The phase velocity of the Rayleigh wave c_R at the free surface is defined by

$$c_R = \omega/\kappa_R < \omega/k_T \quad (3.1.63)$$

and the phase velocity of the Rayleigh wave is found to be slower than the S wave.

According to Eq. (3.1.62), the solution of Eq. (3.1.61) becomes as

$$\begin{aligned} u_r(\boldsymbol{x},t) &= i\kappa_R A\exp(i\kappa_R x_r - \gamma x_3 - i\omega t) \\ &\quad - i\kappa_R\nu B\exp(i\kappa_R x_r - \nu x_3 - i\omega t) \\ u_{r\perp}(\boldsymbol{x},t) &= 0 \\ u_3(\boldsymbol{x},t) &= -\gamma A\exp(i\kappa_R x_r - \gamma x_3 - i\omega t) \\ &\quad + \kappa_R^2\nu B\exp(i\kappa_R x_r - \nu x_3 - i\omega t) \end{aligned} \quad (3.1.64)$$

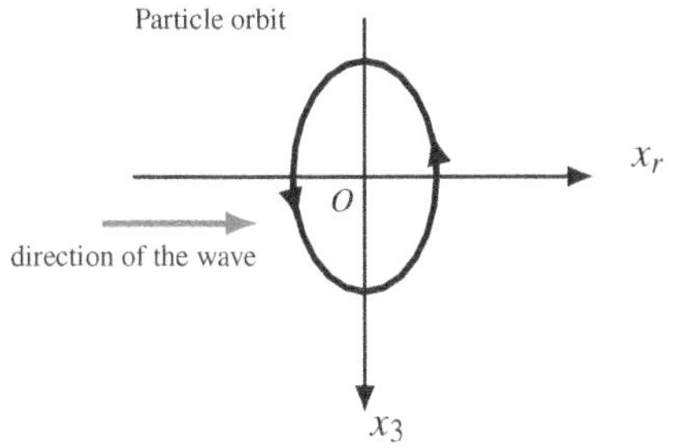

Figure 3.1.9 Direction of the Rayleigh wave propagation and its particle orbit at the free surface.

Substitution of $x_1 = x_2 = 0$ into Eq. (3.1.64) yields

$$\begin{aligned} u_r(0,0,x_3,t) &= iAU_r(\kappa_R,x_3)\exp(-i\omega t) \\ u_3(0,0,x_3,t) &= AU_3(\kappa_R,x_3)\exp(-i\omega t) \end{aligned} \quad (3.1.65)$$

where

$$U_r(\kappa_R,x_3) = \kappa_R\left(e^{-\gamma x_3} - \frac{B}{A}\nu e^{-\nu x_3}\right) \quad (3.1.66)$$

$$U_3(\kappa_R,x_3) = -\gamma e^{-\gamma x_3} + \frac{B}{A}\kappa_R^2 e^{-\nu x_3} \quad (3.1.67)$$

Figure 3.1.9 shows the trajectory of the real part of Eq. (3.1.65) at $x_3 = 0$, namely, we plot

$$\left(\mathrm{Re}\left(iU_r(\kappa_R,0)e^{-i\omega t}\right),\ \mathrm{Re}(U_3(\kappa_R,0)e^{-i\omega t})\right)$$

in $x_r - x_3$ plane. We call the trajectory the *particle orbit*. The particle orbit of the Rayleigh wave is found to form an ellipse and the direction of the particle velocity is the counterclockwise direction with respect to the right direction of the wave propagation. In addition, the major axis of the ellipse is orthogonal to the x_r axis.

Figure 3.1.10 shows $U_r(\kappa_R,x_3)$ and $U_3(\kappa_R,x_3)$ along the x_3 axis. We find large amplitude of the Rayleigh wave can be seen close to the free surface, which decreases rapidly toward the depth.

We constructed the Rayleigh wave as a non-trivial solution for the elastic wave propagation in an elastic half-space. This shows that the solution of the Rayleigh wave can be regarded as an eigenfunction. A homogeneous solution of elastic wave equation which satisfies the free boundary conditions exists for the root of the Rayleigh function, which can be regarded as an eigenvalue. We know that the Rayleigh wave is not caused by incident plane waves according to the discussions in this section. The cause of the Rayleigh wave will be clarified in the later discussions together with Green's function.

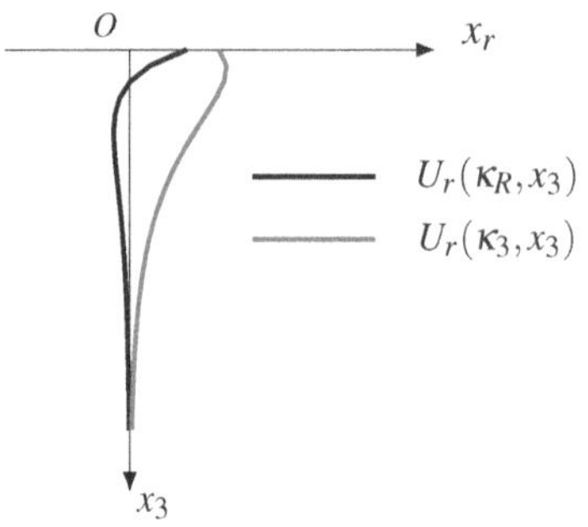

Figure 3.1.10 Amplitude of the Rayleigh wave with respect to depth from the free surface.

3.2 GREEN'S FUNCTION FOR ELASTIC HALF-SPACE

In this section, we present a derivation process for Green's function for an elastic half-space. The derivation takes place in the frequency domain. Consequently, the symbol $\hat{}$ is not employed to distinguish functions in the frequency domain from those in the time domain, as shown in Eqs. (2.4.7) and (2.4.3). Instead, the symbol is used to distinguish functions in the wavenumber domain from those in the spatial domain.

We must emphasize that both the derivation process and the representation of Green's function are not very simple. Moreover, expressing Green's function for an elastic half-space involves an infinite wavenumber integral. Nowadays, however, numerical computations for Green's function have become relatively straightforward. These computations allow us to analyze engineering problems, which will be discussed in the next chapter. Considering the significance of Green's function, as evidenced by the aforementioned facts, we proceed with our discussions in this section.

3.2.1 WEYL AND SOMMERFELD INTEGRALS

We start with a discussion of the Weyl and Sommerfeld integrals as a preparation for deriving Green's function for an elastic half-space. These integrals provide a representation of Green's function for the 3D Helmholtz equation in terms of the superposition of plane or cylindrical waves. As we proceed, we will find that Green's function for an elastic half-space can also be expressed using plane or cylindrical waves, similar to the representation found in the Weyl and/or the Sommerfeld integrals.

As discussed for Eq. (2.4.11) in Chapter 2, the definition of Green's function for the Helmholtz equation is given by

$$(\nabla^2 + k^2)G(\boldsymbol{x}, \boldsymbol{y}) = -\delta(\boldsymbol{x} - \boldsymbol{y}), \quad (\boldsymbol{x}, \boldsymbol{y} \in \mathbb{R}^3) \tag{3.2.1}$$

The solution is expressed as:

$$G(\boldsymbol{x},\boldsymbol{y}) = \frac{\exp(\pm ikR)}{4\pi R} \tag{3.2.2}$$

where

$$\begin{aligned} R &= |\boldsymbol{x}-\boldsymbol{y}| \\ &= \sqrt{(x_1-y_1)^2+(x_2-y_2)^2+(x_3-y_3)^2} \end{aligned} \tag{3.2.3}$$

To obtain the Weyl integral, we apply a Fourier transform with respect to the horizontal coordinate system to Eq. (3.2.1). As shown in Appendix B, the Fourier transform and its inverse transform with respect to the horizontal coordinate system are respectively given by

$$\widehat{G}(\hat{\boldsymbol{x}},\boldsymbol{y}) = \frac{1}{2\pi}\iint_{-\infty}^{\infty} G(\boldsymbol{x},\boldsymbol{y})\exp\big(-i(x_1\xi_1+x_2\xi_2)\big)dx_1dx_2 \tag{3.2.4}$$

$$G(\boldsymbol{x},\boldsymbol{y}) = \frac{1}{2\pi}\iint_{-\infty}^{\infty} \widehat{G}(\hat{\boldsymbol{x}},\boldsymbol{y})\exp\big(i(x_1\xi_1+x_2\xi_2)\big)d\xi_1d\xi_2 \tag{3.2.5}$$

where

$$\hat{\boldsymbol{x}} = (\xi_1,\xi_2,x_3) \tag{3.2.6}$$

and $\widehat{G}$ is the Fourier transform of G. Note that Eq. (3.2.5) is the expression of Green's function in terms of the superposition of plane waves propagating in the horizontal direction.

Our task at this point is to obtain the explicit form of $\widehat{G}$. Applying the Fourier transform in Eq. (3.2.4) to Eq. (3.2.1) yields the following result:

$$\left(\frac{d^2}{dx_3{}^2}-\nu^2\right)\widehat{G}(\hat{\boldsymbol{x}},\boldsymbol{y}) = -\frac{1}{2\pi}\delta(x_3-y_3)\exp\big(-i(\xi_1y_1+\xi_2y_2)\big) \tag{3.2.7}$$

where

$$\nu^2 = \xi_r^2-k^2,\ \ (\xi_r^2=\xi_1^2+\xi_2^2) \tag{3.2.8}$$

A method for the treatment of the Dirac delta function on the right-hand side of Eq. (3.2.7) is shown in Note 3.1. Based on the results of Note 3.1, the solution for Eq. (3.2.7) is expressed as:

$$\widehat{G}(\hat{\boldsymbol{x}},\boldsymbol{y}) = \frac{1}{4\pi\nu}\exp\big(-i(\xi_1y_1+\xi_2y_2)\big)\exp(-\nu|x_3-y_3|) \tag{3.2.9}$$

Given that ν has two branches when $k>\xi_r$, namely:

$$\nu = \pm i\bar{\nu},\ \ (\bar{\nu}=\sqrt{k^2-\xi_r^2}) \tag{3.2.10}$$

we employ the following notation:

$$\nu^{(\pm)} = \begin{cases} \nu & (\xi_r \geq k) \\ \pm i\bar{\nu} & (k>\xi_r\geq 0) \end{cases} \quad \text{(double-sign corresponds)} \tag{3.2.11}$$

Note that the branch $\nu = -i\bar{\nu}$ yields the out-going wave and the branch $\nu = +i\bar{\nu}$ yields the in-coming wave with a time factor of $\exp(-i\omega t)$ based on Eq. (3.2.9). Using the notation in Eq. (3.2.11), we can represent Green's function in the following form:

$$\begin{aligned}
&\frac{\exp(\pm ikR)}{4\pi R} \\
&= \frac{1}{8\pi^2}\iint_{-\infty}^{+\infty} \frac{e^{-\nu^{(\mp)}|x_3-y_3|}}{\nu^{(\mp)}} \exp\left(i\xi_1(x_1-y_1)+i\xi_2(x_2-y_2)\right)d\xi_1 d\xi_2 \\
&\quad \text{(double-sign corresponds)} \qquad (3.2.12)
\end{aligned}$$

This is called the *Weyl integral.*

It is also possible to represent Green's function in terms of the superposition of cylindrical waves. For this purpose, we apply the Fourier-Hankel transform to Eq. (3.2.1). As shown in Appendix B, the Fourier-Hankel transform and its inverse transform for Green's function are respectively expressed as

$$\widehat{G}_\xi^m(x_3-y_3) = \int_0^{2\pi} d\varphi \int_0^{\infty} rY_\xi^{m*}(r,\varphi)G(\boldsymbol{x},\boldsymbol{y})dr \qquad (3.2.13)$$

$$G(\boldsymbol{x},\boldsymbol{y}) = \frac{1}{2\pi}\sum_{m=-\infty}^{\infty}\int_0^{\infty} \xi Y_\xi^m(r,\varphi)\widehat{G}_\xi^m(x_3-y_3)d\xi \qquad (3.2.14)$$

where

$$\begin{aligned}
r &= \sqrt{(x_1-y_1)^2+(x_2-y_2)^2} \\
\varphi &= \tan^{-1}\frac{y_2-x_2}{y_1-x_1} \qquad (3.2.15)
\end{aligned}$$

and $\widehat{G}_\xi^m$ is the Fourier-Hankel transform of Green's function. Additionally, $Y_\xi^m(r,\varphi)$ and $Y_\xi^{m*}(r,\varphi)$ are defined by

$$Y_\xi^m(r,\varphi) = J_m(\xi r)\exp(im\varphi), \quad Y_\xi^{m*}(r,\varphi) = J_m(\xi r)\exp(-im\varphi) \qquad (3.2.16)$$

where m is the circumferential order number and J_m is the Bessel function of the first kind of the order m. The application of the Fourier-Hankel transform to Eq. (3.2.1) yields:

$$\left(\frac{d^2}{dx_3{}^2}-\nu^2\right)\widehat{G}_\xi^m(x_3-y_3) = -\hat{f}^m\delta(x_3-y_3) \qquad (3.2.17)$$

where

$$\begin{aligned}\hat{f}^m &= \int_0^{2\pi} e^{-im\varphi} d\varphi \int_0^{\infty} r J_m(\xi r) \frac{\delta(r)}{r} \delta(\varphi) \\ &= J_m(0) = \delta_{m0}\end{aligned} \tag{3.2.18}$$

For the derivation of Eq. (3.2.17), we used

$$\nabla^2 Y_\xi^m(r,\varphi) = -\xi^2 Y_\xi^m(r,\varphi) \tag{3.2.19}$$

and

$$\delta(\boldsymbol{x}-\boldsymbol{y}) = \frac{\delta(r)}{r} \delta(\varphi)\delta(x_3 - y_3) \tag{3.2.20}$$

Based on the results of Note 3.1, Eq. (3.2.17) yields the solution:

$$\widehat{G}_\xi^m(x_3 - y_3) = \frac{1}{2\nu} \hat{f}^m \exp(-\nu|x_3 - y_3|) \tag{3.2.21}$$

As a result, we get the following integral representation of Green's function

$$\frac{e^{\pm ikR}}{4\pi R} = \frac{1}{4\pi} \int_0^{\infty} \xi J_0(\xi r) \frac{e^{-\nu^{(\mp)}|x_3 - y_3|}}{\nu^{(\mp)}} d\xi, \text{ (double-sign corresponds)} \tag{3.2.22}$$

This is called the *Sommerfeld integral.*

We can specify the branches of ν by setting the branch cut and Riemann sheet for the complex wavenumber plane. Figure 3.2.1 shows the branch cut for $\nu = \sqrt{\xi^2 - k^2}$. The branch points for ν are $\xi = \pm k$. Therefore, we define the branch cut for ν as the line segment between $\xi = +k$ and $\xi = -k$, which connects the branch points. The argument of ξ around k in the Riemann sheet is set to

$$-\pi < \mathrm{Arg}(\xi - k) < \pi \tag{3.2.23}$$

Let ξ around the branch point k be

$$\xi = k + \delta e^{i\theta} \tag{3.2.24}$$

where δ is a very small real positive number and $-\pi < \theta < \pi$. Substitution of Eq. (3.2.24) into ν and neglecting the effects of δ^2 yields

$$\nu = \sqrt{2k\delta}\, e^{i\theta/2} \tag{3.2.25}$$

from which we see that

$$\nu \to \pm i\bar{\nu}, \ (\theta \to \pm\pi), \qquad \text{(double-sign corresponds)} \tag{3.2.26}$$

That is, below and above the branch cut, ν takes the branch $-i\bar{\nu}$ and $i\bar{\nu}$, respectively. In addition, we also see that $\mathrm{Re}(\nu) > 0$ in the Riemann sheet, which ensures the radiation condition of waves for $|x_3 - y_3| \to \infty$. As a result, we can set the paths of the integral, $W_{(+)}$ and $W_{(-)}$, in the complex wavenumber plane as shown in Fig. 3.2.2. The Sommerfeld integral can be defined by these paths as follows:

$$\frac{e^{\pm ikR}}{4\pi R} = \frac{1}{4\pi}\int_{W(\pm)} \xi J_0(\xi r)\frac{e^{-\nu|x_3-y_3|}}{\nu}d\xi \quad \text{(double-sign corresponds)} \tag{3.2.27}$$

The Sommerfeld integral is found to be the superposition of cylindrical waves for Green's function for the Helmholtz equation.

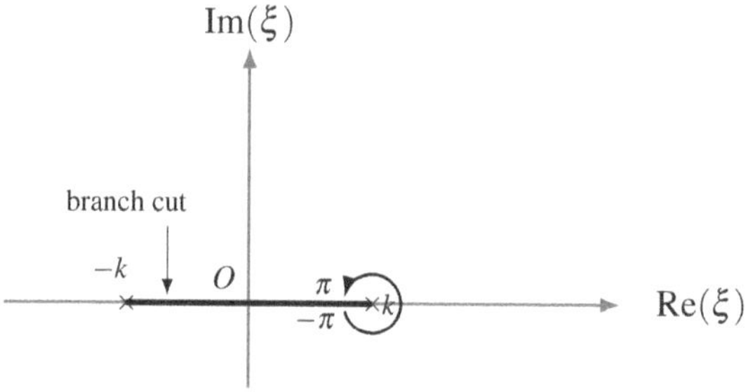

Figure 3.2.1 Argument of ξ in complex wavenumber plane for the upper sheet around the branch point of k to evaluate $\nu = \sqrt{\xi^2 - k^2}$.

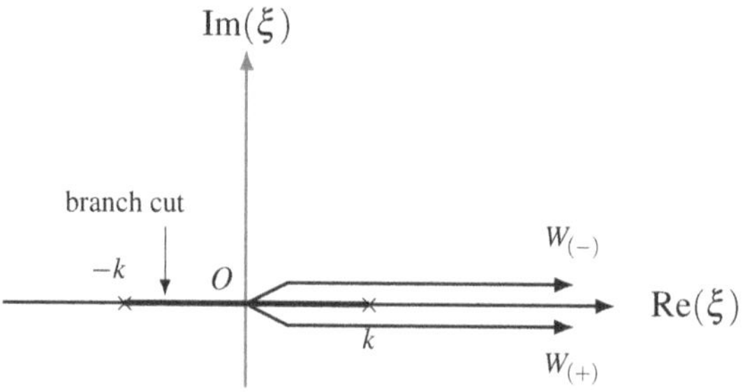

Figure 3.2.2 Paths of integrals for the Sommerfeld integral in the complex wavenumber plane.

Note 3.1 Treatment of the Dirac delta function in Eq. (3.2.7)

As mentioned earlier, our task is to solve Eq. (3.2.7). To address this, we first present a method for the following type of equation:

$$\left(\frac{d^2}{dx^2}-a^2\right)u(x,x')=-\delta(x-x'),\ (a>0) \tag{N3.1.1}$$

To solve Eq. (N3.1.1), $u(x,x')$ must satisfy:

$$\left(\frac{d^2}{dx^2}-a^2\right)u(x,x')=0,\ \ (x\neq x') \tag{N3.1.2}$$

To ensure a solution in the form:

$$\lim_{x\to\pm\infty}u(x,x')=0 \tag{N3.1.3}$$

$u(x,x')$ must be expressed as:

$$u(x,x')=\begin{cases}A\exp(-ax) & (x>x')\\ B\exp(ax) & (x<x')\end{cases} \tag{N3.1.4}$$

where A and B are coefficients to be determined from the singularities of the Dirac delta function, which are given by:

$$\lim_{\varepsilon\to 0}\left[\frac{du(x,x')}{dx}\right]_{x'-\varepsilon}^{x'+\varepsilon}=-1 \tag{N3.1.5}$$

$$\lim_{\varepsilon\to+0}\left(u(x'+\varepsilon,x')-u(x'-\varepsilon,x')\right)=0 \tag{N3.1.6}$$

Equations (N 3.1.5) and (N 3.1.6) are based on the fact that the derivative of the Heaviside unit step function $H(x)$ is the Dirac delta function $\delta(x)$ shown in Appendix-B, which can be expressed as:

$$\delta(x)=\frac{dH(x)}{dx} \tag{N3.1.7}$$

$$H(x)=\begin{cases}1 & (x>0)\\ 0 & (x<0)\end{cases} \tag{N3.1.8}$$

By applying Eqs. (N 3.1.5) and (N 3.1.6), we can determine the coefficients A and B in Eq. (N 3.1.4), leading to the solution:

$$u(x,x')=\frac{1}{2a}\exp(-a|x-x'|) \tag{N3.1.9}$$

Note 3.2 Concept of Riemann sheets and branch cut

To handle the $f(z) = \sqrt{z}, z \in \mathbb{C}$ as a complex-valued function, we must treat is as a two-valued function, with its values determined by $|z|$ and $\arg z$, where $\arg z$ represents the argument of z. For this particular case, the values of a function $f(z) = \sqrt{z}$ are determined as follows:

$$\sqrt{z} = |z|^{1/2} e^{i(\mathrm{Arg}\, z + 2n\pi)/2} = \begin{cases} |z|^{1/2} e^{i\mathrm{Arg}\, z} & n : \text{even} \\ -|z|^{1/2} e^{i\mathrm{Arg}\, z} & n : \text{odd} \end{cases} \tag{N3.2.1}$$

Here n is an integer and $\mathrm{Arg}\, z$ is the principal value of the argument of z. Equation (N3.2.1) shows that we can create two complex planes, with each plane defining the values of $\sqrt{z}$ uniquely. For instance, we can prepare two complex planes based on the argument of z as follows:

$$-\pi + 4n\pi < \arg z < \pi + 4n\pi \qquad (z \in \text{upper sheet}) \tag{N3.2.2}$$

$$\pi + 4n\pi < \arg z < 3\pi + 4n\pi \qquad (z \in \text{lower sheet}) \tag{N3.2.3}$$

These two complex planes are referred to as the upper and lower sheets concerning the argument of z. With the two complex planes, we can uniquely determine the two values of $\sqrt{z}$ by each plane, satisfying:

$$\mathrm{Re}\left(\sqrt{z}\right) > 0, \quad (z \in \text{upper sheet}) \tag{N3.2.4}$$

$$\mathrm{Re}\left(\sqrt{z}\right) < 0, \quad (z \in \text{lower sheet}) \tag{N3.2.5}$$

These upper and lower sheets are connected by the line $\mathrm{Re}\, z \leq 0$, allowing the argument z to move smoothly from the upper sheet to the lower sheet or vice versa, for example along the path of the integral. The line that connects the complex planes is called the *branch cut* and these complex planes are called *Riemann sheets*. Furthermore , the starting point of the branch cut $z = 0$, is referred to as the *branch point*. It's important to note that around the branch point, for instance, $z = \delta e^{i\theta}$, where δ is a very small positive real number, the value of $\sqrt{z}$ varies based on the movement of θ, which is a characteristic of the branch point. In general, a multi-valued complex function can be dealt with as a single-valued function by specifying the branch cut and Riemann sheet. Also, it's worth mentioning that the branch cut doesn't have to be a straight line. Later, we will explore cases where the branch cut is a curve.

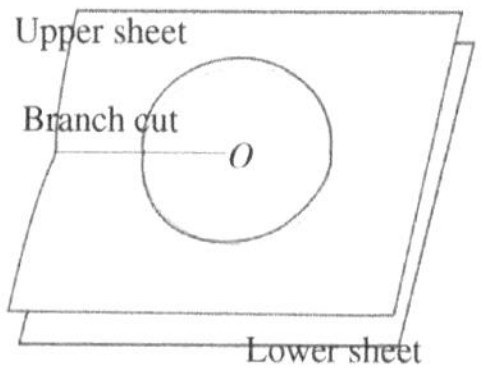

Figure: N.3.2.1 Riemann sheets and branch cut. On the branch cut, $\mathrm{Re}(z) \leq 0$.

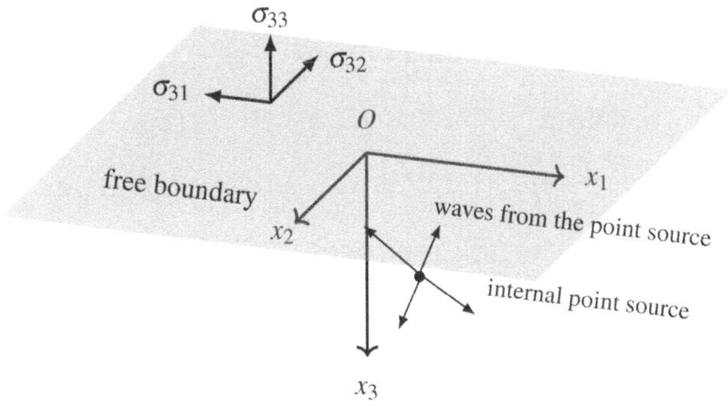

Figure 3.2.3 Point source in an elastic half-space used to define Green's function. Point source induces waves propagating in various directions, which are then reflected by the free boundary. Consequently, to satisfy the free boundary conditions for Green's function, it needs to be represented as a superposition of plane or cylindrical waves.

3.2.2 REPRESENTATION OF GREEN'S FUNCTION IN TERMS OF SUPERPOSITION OF PLANE WAVES

For wave problems in an elastic half-space, a point source induces waves propagating in various directions, which are then reflected by a plane boundary, as illustrated in Fig. 3.2.3. It is essential to recall that computation of wave reflection is not a simple task, and even for a single plane wave propagation, it involves considerable complexity, as discussed in §3.1.

The complexity becomes significantly greater when dealing with the problem of Green's function for an elastic half-space. To address this challenge, one possible solution is to express Green's function for an elastic half-space in terms of the superposition of plane or cylindrical waves, utilizing techniques such as the Weyl or Sommerfeld integral.

Now, we will proceed with our discussions to derive a representation of Green's function in terms of plane waves. We start by considering three scalar potentials expressed as plane waves propagating in the horizontal direction:

$$
\begin{aligned}
\phi(\boldsymbol{x}) &= \frac{1}{2\pi}\iint_{-\infty}^{\infty} \hat{\phi}(\hat{\boldsymbol{x}}) \exp(i\xi_1 x_1 + i\xi_2 x_2) d\xi_1 d\xi_2 \\
\psi(\boldsymbol{x}) &= \frac{1}{2\pi}\iint_{-\infty}^{\infty} \hat{\psi}(\hat{\boldsymbol{x}}) \exp(i\xi_1 x_1 + i\xi_2 x_2) d\xi_1 d\xi_2 \\
\chi(\boldsymbol{x}) &= \frac{1}{2\pi}\iint_{-\infty}^{\infty} \hat{\chi}(\hat{\boldsymbol{x}}) \exp(i\xi_1 x_1 + i\xi_2 x_2) d\xi_1 d\xi_2
\end{aligned}
\tag{3.2.28}
$$

Note that $\hat{\boldsymbol{x}}$ is defined in Eq. (3.2.6). Based on Eq. (3.1.4), the displacement field generated by the above potentials can be expressed as:

$$\begin{aligned} \boldsymbol{u}(\boldsymbol{x}) &= \nabla\phi(\boldsymbol{x}) + \nabla\times\nabla\times\psi(\boldsymbol{x})\boldsymbol{e}_3 + \nabla\times\chi(\boldsymbol{x})\boldsymbol{e}_3 \\ &= \frac{1}{2\pi}\iint_{-\infty}^{\infty}\Big[\big(i\xi_1\hat{\phi}(\hat{\boldsymbol{x}}) + i\xi_1\partial_3\hat{\psi}(\hat{\boldsymbol{x}}) + i\xi_2\hat{\chi}(\hat{\boldsymbol{x}})\big)\boldsymbol{e}_1 \\ &\quad + \big(i\xi_2\hat{\phi}(\hat{\boldsymbol{x}}) + i\xi_2\partial_3\hat{\psi}(\hat{\boldsymbol{x}}) - i\xi_1\hat{\chi}(\hat{\boldsymbol{x}})\big)\boldsymbol{e}_2 \\ &\quad + \big(\partial_3\hat{\phi}(\hat{\boldsymbol{x}}) + \xi_r^2\hat{\psi}(\hat{\boldsymbol{x}})\big)\boldsymbol{e}_3\Big]\exp(i\xi_1x_1 + i\xi_2x_2)d\xi_1d\xi_2 \\ &= \frac{1}{2\pi}\iint_{-\infty}^{\infty}\hat{u}_{i'}(\hat{\boldsymbol{x}})\boldsymbol{e}_{i'}\exp\big(i\xi_1x_1 + i\xi_2x_2\big)d\xi_1d\xi_2 \end{aligned} \tag{3.2.29}$$

where

$$\xi_r = \sqrt{\xi_1^2 + \xi_2^2} \tag{3.2.30}$$

and $\hat{u}_{i'}$ and $\boldsymbol{e}_{i'}$ are defined as

$$\begin{aligned} \hat{u}_{1'}(\hat{\boldsymbol{x}}) &= \partial_3\hat{\phi}(\hat{\boldsymbol{x}}) + \xi_r^2\hat{\psi}(\hat{\boldsymbol{x}}) \\ \hat{u}_{2'}(\hat{\boldsymbol{x}}) &= \xi_r\hat{\phi}(\hat{\boldsymbol{x}}) + \xi_r\partial_3\hat{\psi}(\hat{\boldsymbol{x}}) \\ \hat{u}_{3'}(\hat{\boldsymbol{x}}) &= \xi_r\hat{\chi}(\hat{\boldsymbol{x}}) \end{aligned} \tag{3.2.31}$$

$$\begin{aligned} \boldsymbol{e}_{1'} &= \boldsymbol{e}_3 \\ \boldsymbol{e}_{2'} &= ic\boldsymbol{e}_1 + is\boldsymbol{e}_2 \\ \boldsymbol{e}_{3'} &= is\boldsymbol{e}_1 - ic\boldsymbol{e}_2 \end{aligned} \tag{3.2.32}$$

Here, c and s in Eq. (3.2.32) are already defined by Eq. (3.1.16). They are respectively expressed as:

$$\mathrm{c} = \begin{cases} \xi_1/\xi_r & (\text{when } \xi_r \neq 0) \\ 1 & (\text{when } \xi_r = 0) \end{cases} \tag{3.2.33}$$

$$\mathrm{s} = \begin{cases} \xi_2/\xi_r & (\text{when } \xi_r \neq 0) \\ 0 & (\text{when } \xi_r = 0) \end{cases} \tag{3.2.34}$$

The set of base vectors $\{\boldsymbol{e}_{j'}\}_{j'=1'}^{3'}$ is an orthonormal basis since

$$\boldsymbol{e}_{i'}\cdot\boldsymbol{e}_{j'} = \boldsymbol{e}_{i'}^H\boldsymbol{e}_{j'} = \delta_{i'j'} \tag{3.2.35}$$

where $(\cdot)^H$ denotes the Hermitian conjugate. From Eq. (3.2.31), we observe that $\hat{u}_{1'}$ and $\hat{u}_{2'}$ represent the displacement field for the P-SV wave components, while $\hat{u}_{3'}$ describes the displacement field for the SH wave component. Thus, the orthonormal basis $\{\boldsymbol{e}_{j'}\}_{j'=1'}^{3'}$ holds significant physical meaning: $\boldsymbol{e}_{1'}$ and $\boldsymbol{e}_{2'}$ serve as the base vectors for the P-SV waves, and $\boldsymbol{e}_{3'}$ represents the base vector for the SH waves. To distinguish these base vectors from those of the Cartesian coordinate system, we use the subscript j', where $j = 1,2,3$, in Eq. (3.2.32).

It is important to establish the transformation rule for the vector components between two different coordinate systems, represented by $\boldsymbol{e}_j$ and $\boldsymbol{e}_{j'}$, respectively. As shown in Eq. (3.2.29), the displacement field can also be expressed as follows:

$$\boldsymbol{u}(\boldsymbol{x}) = \frac{1}{2\pi}\iint_{-\infty}^{\infty} \hat{u}_i(\hat{\boldsymbol{x}})\boldsymbol{e}_i \exp(i\xi_1 x_1 + i\xi_2 x_2)\,d\xi_1 d\xi_2 \tag{3.2.36}$$

Comparing Eqs. (3.2.29) and (3.2.36), we derive the following relationship:

$$\begin{aligned}
\hat{u}_j\boldsymbol{e}_j &= \hat{u}_{j'}\boldsymbol{e}_{j'} \\
&= \hat{u}_{1'}\boldsymbol{e}_3 + \hat{u}_{2'}(ic\boldsymbol{e}_1 + is\boldsymbol{e}_2) + \hat{u}_{3'}(is\boldsymbol{e}_1 - ic\boldsymbol{e}_2) \\
&= (ic\hat{u}_{2'} + is\hat{u}_{3'})\boldsymbol{e}_1 + (is\hat{u}_{2'} - is\hat{u}_{3'})\boldsymbol{e}_2 + \hat{u}_{1'}\boldsymbol{e}_3
\end{aligned} \tag{3.2.37}$$

which yields the transformation rule:

$$\begin{aligned}
\hat{u}_1 &= ic\hat{u}_{2'} + is\hat{u}_{3'} \\
\hat{u}_2 &= is\hat{u}_{2'} - ic\hat{u}_{3'} \\
\hat{u}_3 &= \hat{u}_{1'}
\end{aligned} \tag{3.2.38}$$

The transformation rule in Eq. (3.2.38) is represented as:

$$\hat{u}_i = T_{ij'}\hat{u}_{j'} \iff \hat{u}_{j'} = T^*_{ij'}\hat{u}_i \tag{3.2.39}$$

where the components of the matrix $T_{ij'}$ are defined as:

$$T_{ij'} = \begin{bmatrix} 0 & ic & is \\ 0 & is & -ic \\ 1 & 0 & 0 \end{bmatrix} \tag{3.2.40}$$

and $T^*_{ij'}$ represents the complex conjugate of $T_{ij'}$. It should be noted that $T_{ij'}$ is a unitary matrix.

Based on the preceding discussion, we can express the Fourier transform for an elastic half-space, which allows the decomposition of P-SV and SH waves, as follows:

$$\begin{aligned}
\hat{u}_{j'}(\hat{\boldsymbol{x}}) &= \frac{1}{2\pi}\iint_{-\infty}^{\infty} T^*_{ij'}u_i(\boldsymbol{x})\exp(-i(\xi_1 x_1 + \xi_2 x_2))\,dx_1\,dx_2 \\
u_i(\boldsymbol{x}) &= \frac{1}{2\pi}\iint_{-\infty}^{\infty} T_{ij'}\hat{u}_{j'}(\hat{\boldsymbol{x}})\exp(i(\xi_1 x_1 + \xi_2 x_2))\,d\xi_1\,d\xi_2
\end{aligned} \tag{3.2.41}$$

Our preparations for deriving Green's function are now complete. The definition of Green's function for an elastic half-space is given by the following equations:

$$\left[L_{ij}(\partial_1,\partial_2,\partial_3) + \rho\omega^2\delta_{ij}\right]G_{jk}(\boldsymbol{x},\boldsymbol{y}) = -\delta_{ik}\delta(\boldsymbol{x}-\boldsymbol{y}) \tag{3.2.42}$$

$$\lim_{x_3\to+0} P_{ij}(\partial_1,\partial_2,\partial_3)G_{jk}(\boldsymbol{x},\boldsymbol{y}) = 0, \quad (\boldsymbol{x},\boldsymbol{y}\in\mathbb{R}^3_+) \tag{3.2.43}$$

where L_{ij} and P_{ij} are respectively the Navier operator defined in Eq. (1.4.15) and the operator for the traction derived from Eq. (3.1.6). They are expressed as

$$L_{ij}(\partial_1,\partial_2,\partial_3) = (\lambda+\mu)\partial_i\partial_j + \delta_{ij}\mu\nabla^2 \tag{3.2.44}$$

$$P_{ij}(\partial_1,\partial_2,\partial_3) = \begin{bmatrix} \mu\partial_3 & 0 & \mu\partial_1 \\ 0 & \mu\partial_3 & \mu\partial_2 \\ \lambda\partial_1 & \lambda\partial_2 & (\lambda+2\mu)\partial_3 \end{bmatrix} \tag{3.2.45}$$

To proceed further, similar to what was done for the Weyl integral, we apply the Fourier transform with respect to the horizontal coordinate for Eqs. (3.2.42) and (3.2.43). In this case, we utilize the Fourier transform with respect to the horizontal coordinate system defined by Eq. (3.2.41).

Then, $\widehat{G}_{l'k}$ satisfies the following equations:

$$\left[-\mathscr{A}_{m'l'}(\xi_r,\partial_3)+\rho\omega^2\delta_{m'l'}\right]\widehat{G}_{l'k}(\hat{\mathbf{x}},\mathbf{y}) = -\frac{1}{2\pi}T^*_{km'}\,\delta(x_3-y_3)\exp(-i(\xi_1y_1+\xi_2y_2))$$
$$\lim_{x_3\to+0}\mathscr{P}_{m'l'}(\xi_r,\partial_3)\widehat{G}_{l'k}(\mathbf{x},\mathbf{y}) = 0 \tag{3.2.46}$$

where $\mathscr{A}$ and $\mathscr{P}$ are defined by

$$\begin{aligned} -\mathscr{A}_{m'l'}(\xi_r,\partial_3) &= T^*_{im'}L_{ij}(i\xi_1,i\xi_2,\partial_3)T_{jl'} \\ &= \begin{bmatrix} (\lambda+2\mu)\partial_3^2-\mu\xi_r^2 & -(\lambda+\mu)\xi_r\partial_3 & 0 \\ (\lambda+\mu)\xi_r\partial_3 & \mu\partial_3^2-(\lambda+2\mu)\xi_r^2 & 0 \\ 0 & 0 & \mu\partial_3^2-\mu\xi_r^2 \end{bmatrix} \end{aligned} \tag{3.2.47}$$

$$\begin{aligned} \mathscr{P}_{m'l'}(\xi_r,\partial_3) &= T^*_{im'}P_{ij}(i\xi_1,i\xi_2,\partial_3)T_{jl'} \\ &= \begin{bmatrix} (\lambda+2\mu)\partial_3 & -\lambda\xi_r & 0 \\ \mu\xi_r & \mu\partial_3 & 0 \\ 0 & 0 & \mu\partial_3 \end{bmatrix} \end{aligned} \tag{3.2.48}$$

The components of the operators $\mathscr{A}$ and $\mathscr{P}$ are separated into the effects of the P-SV wave and the SH wave, resulting in 2×2 and 1×1 matrices, respectively. Notably, ξ_1 and ξ_2 are eliminated from $\mathscr{A}$ and $\mathscr{P}$, and only ξ_r is needed, simplifying the treatment of these operators.

Once we have the solution to the following equations:

$$\left[-\mathscr{A}_{m'l'}(\xi_r,\partial_3)+\rho\omega^2\delta_{m'l'}\right]\hat{g}_{l'k'}(x_3,y_3,\xi_r) = -\delta_{k'm'}\,\delta(x_3-y_3) \tag{3.2.49}$$

$$\lim_{x_3\to+0}\mathscr{P}_{m'l'}(\xi_r,\partial_3)\hat{g}_{l'k'}(x_3,y_3,\xi_r) = 0 \tag{3.2.50}$$

then Eqs. (3.2.46) and (3.2.49) yield the following representation of the solution for Eq. (3.2.46):

$$\widehat{G}_{ik}(\hat{\mathbf{x}}:\mathbf{y}) = \frac{1}{2\pi}\exp(-i(\xi_1y_1+\xi_2y_2))T_{ip'}\hat{g}_{p'l'}(x_3,y_3,\xi_r)T^*_{kl'} \tag{3.2.51}$$

Therefore, we obtain a representation of Green's function for an elastic half-space in the following form:

$$G_{ij}(\boldsymbol{x},\boldsymbol{y}) = \frac{1}{4\pi^2}\iint_{-\infty}^{+\infty} T_{ip'}\,\hat{g}_{p'l'}(x_3,y_3,\xi_r)\,T^*_{jl'} \times \exp\big(i\xi_1(x_1-y_1)+i\xi_2(x_2-y_2)\big)d\xi_1 d\xi_2 \tag{3.2.52}$$

It is found from Eq. (3.2.52) that the construction of the solution for Eqs. (3.2.49) and (3.2.50) is necessary for an explicit form of Green's function for an elastic half-space. The detailed process of solving Eq. (3.2.49) with Eq. (3.2.50) is shown in Appendix C. The results are presented in Eq. (C.2.18). In the following, based on the results in Appendix C, we present $\hat{g}_{p'l'}(x_3,y_3,\xi)$ for the case $x_3=0$, which is found to have the following rather simple expressions:

$$\begin{aligned}
\hat{g}_{1'1'}(0,y_3,\xi_r) &= \frac{1}{\mu F(\xi_r)}\left[\gamma(\xi_r^2+\nu^2)e^{-\gamma y_3}-2\xi_r^2\gamma e^{-\nu y_3}\right]\\
\hat{g}_{1'2'}(0,y_3,\xi_r) &= \frac{\xi_r}{\mu F(\xi_r)}\left[-(\xi_r^2+\nu^2)e^{-\gamma y_3}+2\gamma\nu e^{-\nu y_3}\right]\\
\hat{g}_{2'1'}(0,y_3,\xi_r) &= \frac{\xi_r}{\mu F(\xi_r)}\left[2\gamma\nu e^{-\gamma y_3}-(\xi_r^2+\nu^2)e^{-\nu y_3}\right]\\
\hat{g}_{2'2'}(0,y_3,\xi_r) &= \frac{\nu}{\mu F(\xi_r)}\left[-2\xi_r^2 e^{-\gamma y_3}+(\xi_r^2+\nu^2)e^{-\nu y_3}\right]\\
\hat{g}_{3'3'}(0,y_3,\xi_r) &= \frac{1}{\mu\nu}e^{-\nu y_3}
\end{aligned} \tag{3.2.53}$$

where y_3 is the depth of the source point. In addition, recall that $F(\xi_r)$ is the Rayleigh function defined by Eq. (3.1.54).

3.2.3 REPRESENTATION OF GREEN'S FUNCTION IN TERMS OF SUPERPOSITION OF CYLINDRICAL WAVES

Subsequently, our focus shifts to seeking a representation of Green's function for an elastic half-space in terms of a cylindrical coordinate system. This representation holds significant importance as it allows us to apply approximation methods to Green's function in the subsequent sections. In the course of deriving this representation, we will discover that we can also use Green's function in the wavenumber domain, as presented in §3.2.2.

3.2.3.1 Relationship between Cartesian and cylindrical coordinate systems

We have already presented the notations for the cylindrical coordinate system in §2.6.2. Referring to Fig. 2.6.1 in §2.6.2, we represent the transformation rule of the base vectors between Cartesian and cylindrical coordinate systems in the following form:

$$\begin{pmatrix} \boldsymbol{e}_z \\ \boldsymbol{e}_r \\ \boldsymbol{e}_\varphi \end{pmatrix} = \begin{bmatrix} 0 & 0 & 1 \\ \cos\varphi & \sin\varphi & 0 \\ -\sin\varphi & \cos\varphi & 0 \end{bmatrix} \begin{pmatrix} \boldsymbol{e}_1 \\ \boldsymbol{e}_2 \\ \boldsymbol{e}_3 \end{pmatrix} \tag{3.2.54}$$

For a specific case, we express Eq. (3.2.54) in terms of the summation convention, which has the following form:

$$\boldsymbol{e}_\alpha = C_{\alpha j}(\varphi)\boldsymbol{e}_j \tag{3.2.55}$$

where the subscript α denotes the cylindrical coordinate system, with $\alpha = 1,2,3$ is used for expressing the components z, r, and φ, respectively. It is readily seen that Eq. (3.2.55) is equivalent to

$$\boldsymbol{e}_j = C_{\alpha j}(\varphi)\boldsymbol{e}_\alpha \tag{3.2.56}$$

Based on the transformation rule for base vectors shown in Eqs. (3.2.55) and (3.2.56), the transformation rule for the components of a vector field is expressed as

$$u_\alpha(\boldsymbol{x}) = C_{\alpha i}(\varphi)u_i(\boldsymbol{x}) \iff u_i(\boldsymbol{x}) = C_{\alpha i}(\varphi)u_\alpha(\boldsymbol{x}) \tag{3.2.57}$$

3.2.3.2 Fourier-Hankel transform for elastic wavefield

We employ the following three scalar potentials expressed in terms of the superposition of cylindrical waves:

$$\begin{aligned}
\phi(\boldsymbol{x}) &= \frac{1}{2\pi}\sum_{m=-\infty}^{\infty}\int_0^\infty \xi Y_\xi^m(r,\varphi)\hat{\phi}_\xi^m(z)d\xi \\
\psi(\boldsymbol{x}) &= \frac{1}{2\pi}\sum_{m=-\infty}^{\infty}\int_0^\infty \xi Y_\xi^m(r,\varphi)\hat{\psi}_\xi^m(z)d\xi \\
\chi(\boldsymbol{x}) &= \frac{1}{2\pi}\sum_{m=-\infty}^{\infty}\int_0^\infty \xi Y_\xi^m(r,\varphi)\hat{\chi}_\xi^m(z)d\xi
\end{aligned} \tag{3.2.58}$$

where $\hat{\phi}_\xi^m$, $\hat{\psi}_\xi^m$, and $\hat{\chi}_\xi^m$ are the Fourier-Hankel transform for ϕ, ψ, and χ, respectively. The displacement field generated by these potentials can be expressed as:

$$\begin{aligned}
\boldsymbol{u}(\boldsymbol{x}) &= \nabla\phi(\boldsymbol{x}) + \nabla\times\nabla\psi(\boldsymbol{x})\boldsymbol{e}_z + \nabla\times\chi(\boldsymbol{x})\boldsymbol{e}_z \\
&= \frac{1}{2\pi}\sum_{m=-\infty}^{\infty}\int_0^\infty \xi\Big[\boldsymbol{R}_\xi^m(r,\varphi)\big(\partial_3\hat{\phi}_\xi^m(z) + \xi^2\psi_\xi^m(z)\big) \\
&\quad + \boldsymbol{S}_\xi^m(r,\varphi)\big(\xi\hat{\phi}_\xi^m(z) + \xi\partial_3\hat{\psi}_\xi^m(z)\big) + \boldsymbol{T}_\xi^m(r,\varphi)\xi\hat{\chi}_\xi^m(z)\Big]d\xi
\end{aligned} \tag{3.2.59}$$

In the above expressions, $\boldsymbol{R}_\xi^m$, $\boldsymbol{S}_\xi^m$, and $\boldsymbol{T}_\xi^m$ are referred to as the *surface vector harmonics*. They are defined as:

$$\begin{aligned}
\boldsymbol{R}_\xi^m(r,\varphi) &= Y_\xi^m(r,\varphi)\boldsymbol{e}_z \\
\boldsymbol{S}_\xi^m(r,\varphi) &= \frac{1}{\xi}\partial_r Y_\xi^m(r,\varphi)\boldsymbol{e}_r + \frac{1}{\xi r}\partial_\varphi Y_\xi^m(r,\varphi)\boldsymbol{e}_\varphi \\
\boldsymbol{T}_\xi^m(r,\varphi) &= \frac{1}{\xi r}\partial_\varphi Y_\xi^m(r,\varphi)\boldsymbol{e}_r - \frac{1}{\xi}\partial_r Y_\xi^m(r,\varphi)\boldsymbol{e}_\varphi
\end{aligned} \tag{3.2.60}$$

We derived Eq. (3.2.58) using Eqs. (2.6.8) to (2.6.10).

To simplify the expression in Eq. (3.2.59), we introduce the following definitions:

$$\begin{aligned}
\hat{u}_{1'}(\xi,m,z) &= \partial_3\hat{\phi}_\xi^m(z) + \xi^2\hat{\psi}_\xi^m(z) \\
\hat{u}_{2'}(\xi,m,z) &= \xi\hat{\phi}_\xi^m(z) + \xi\partial_z\hat{\psi}_\xi^m(z) \\
\hat{u}_{3'}(\xi,m,z) &= \xi\hat{\chi}_\xi^m(z)
\end{aligned} \tag{3.2.61}$$

and

$$\begin{aligned}
\left[H_{\alpha l'}^m(\xi,r,\varphi)\right] &= \begin{bmatrix} \boldsymbol{R}_\xi^m(r,\varphi) & \boldsymbol{S}_\xi^m(r,\varphi) & \boldsymbol{T}_\xi^m(r,\varphi) \end{bmatrix} \\
&= \begin{bmatrix} 1 & 0 & 0 \\ 0 & \xi^{-1}\partial_r & (\xi r)^{-1}\partial_\varphi \\ 0 & (\xi r)^{-1}\partial_\varphi & -\xi^{-1}\partial_r \end{bmatrix} Y_\xi^m(r,\varphi)
\end{aligned} \tag{3.2.62}$$

Then, Eq. (3.2.59) can be expressed as

$$u_\alpha(\boldsymbol{x}) = \frac{1}{2\pi}\sum_{m=-\infty}^{\infty}\int \xi H_{\alpha l'}^m(\xi,r,\varphi)\hat{u}_{l'}(\xi,m,z)d\xi \tag{3.2.63}$$

Equation (3.2.63) provides the inverse Fourier-Hankel transform for an elastic wavefield, and $\left[H_{\alpha l'}^m(\xi,r,\varphi)\right]$ can be regarded as the kernel of this transform. Additionally, we can derive the forward Fourier-Hankel transform from Eq. (3.2.63). The forward transform, as derived in Note 3.3, is expressed as follows:

$$\hat{u}_{l'}(\xi,m,z) = \int_0^{2\pi} d\varphi \int_0^\infty rH_{\alpha l'}^{m*}(\xi,r,\varphi)\,u_\alpha(r,\varphi,z)\,dr \tag{3.2.64}$$

where $H_{\alpha l'}^{m*}(\xi,r,\varphi)$ represents the complex conjugate of $H_{\alpha l'}^m(\xi,r,\varphi)$.

Note 3.3 Fourier-Hankel transform for elastic wavefield

Equation (3.2.63) can be regarded as the inverse Fourier-Hankel transform for an elastic wavefield. By using this equation, we can derive the Fourier-Hankel transform for an elastic wavefield. Based on Eq. (3.2.63), we obtain the following expressions:

$$u_z(\boldsymbol{x}) = \frac{1}{2\pi}\sum_{m=-\infty}^{\infty} e^{im\varphi}\int_0^\infty \xi J_m(\xi r)\hat{u}_{1'}(m,\xi,z)d\xi \tag{N3.3.1}$$

$$u_r(\boldsymbol{x}) - iu_\varphi(\boldsymbol{x}) = \frac{1}{2\pi}\sum_{m=-\infty}^{\infty} e^{im\varphi}\int_0^\infty \xi J_{m-1}(\xi r)\left(\hat{u}_{2'}(\xi,m,z)+i\hat{u}_{3'}(\xi,m,z)\right)d\xi \tag{N3.3.2}$$

$$u_\varphi(\boldsymbol{x}) - iu_r(\boldsymbol{x}) = \frac{1}{2\pi}\sum_{m=-\infty}^{\infty} e^{im\varphi}\int_0^\infty \xi J_{m+1}(\xi r)\left(i\hat{u}_{2'}(\xi,m,z)+\hat{u}_{3'}(\xi,m,z)\right)d\xi \tag{N3.3.3}$$

In the derivation of Eqs. (N3.3.2) and (N3.3.3), we used the following formulas for Bessel functions:

$$\begin{aligned}\frac{1}{\xi}\partial_r J_m(\xi r) &= \frac{J_{m-1}(\xi r)-J_{m+1}(\xi r)}{2}\\ \frac{m}{\xi r}J_m(\xi r) &= \frac{J_{m-1}(\xi r)+J_{m+1}(\xi r)}{2}\end{aligned} \tag{N3.3.4}$$

The results from Eqs. (N3.3.1) to (N3.3.3) are:

$$u_z^m(r,z) = \int_0^\infty \xi J_m(\xi r)\hat{u}_{1'}(m,\xi,z)d\xi \tag{N3.3.5}$$

$$u_r^m(r,z) - iu_\varphi^m(r,z) = \int_0^\infty \xi J_{m-1}(\xi r)\left(\hat{u}_{2'}(m,\xi,z)+i\hat{u}_{3'}(m,\xi,z)\right)d\xi \tag{N3.3.6}$$

$$u_\varphi^m(r,z) - iu_r^m(r,z) = \int_0^\infty \xi J_{m+1}(\xi r)\left(i\hat{u}_{2'}(m,\xi,z)-\hat{u}_{3'}(m,\xi,z)\right)d\xi \tag{N3.3.7}$$

Here, we also introduced the following notations for azimuthal Fourier components:

$$u_z^m(r,z) = \int_0^{2\pi} e^{-im\varphi}u_z(r,\varphi,z)d\varphi \tag{N3.3.8}$$

$$u_r^m(r,z) = \int_0^{2\pi} e^{-im\varphi}u_r(r,\varphi,z)d\varphi \tag{N3.3.9}$$

$$u_\varphi^m(r,z) = \int_0^{2\pi} e^{-im\varphi}u_\varphi(r,\varphi,z)d\varphi \tag{N3.3.10}$$

Note 3.3 Fourier-Hankel transform for elastic wavefield (continued)

The application of the Hankel transform to Eqs. (N 3.3.5) to (N 3.3.7) leads to the following expressions:

$$\begin{aligned}\hat{u}_{1'}(m,\xi,z) &= \int_0^\infty rJ_m(\xi r)u_z^m(r,z)dr \\ &= \int_0^{2\pi} e^{-im\varphi}d\varphi \int_0^\infty rJ_m(\xi r)u_z(r,\varphi,z)dr \qquad \text{(N3.3.11)}\end{aligned}$$

$$\begin{aligned}\hat{u}_{2'}(m,\xi,z)+i\hat{u}_{3'}(m,\xi,z) &= \int_0^\infty rJ_{m-1}(\xi r)\left(u_r^m(r,z)-iu_\varphi^m(r,z)\right)dr \\ &= \int_0^{2\pi} e^{-im\varphi}d\varphi \int_0^\infty rJ_{m-1}(\xi r)\left(u_r(r,\varphi,z)-iu_\varphi(r,\varphi,z)\right)dr \qquad \text{(N3.3.12)}\end{aligned}$$

$$\begin{aligned}i\hat{u}_{2'}(m,\xi,z)+\hat{u}_{3'}(m,\xi,z) &= \int_0^\infty rJ_{m+1}(\xi r)\left(u_\varphi^m(r,z)-iu_r^m(r,z)\right)dr \\ &= \int_0^{2\pi} e^{-im\varphi}d\varphi \int_0^\infty rJ_{m+1}(\xi r)\left(u_\varphi(r,\varphi,z)-iu_r(r,\varphi,z)\right)dr \qquad \text{(N3.3.13)}\end{aligned}$$

By solving Eqs. (N3.3.11), (N3.3.12), and (N3.3.13), we find the following expressions:

$$\begin{aligned}\hat{u}_{2'}(m,\xi,z) &= \int_0^{2\pi} e^{-im\varphi}d\varphi \int_0^\infty r\left[\frac{1}{\xi}\partial_r J_m(\xi r)u_r(r,\varphi,z)+\frac{-im}{\xi r}J_m(\xi r)u_\varphi(r,\varphi,z)\right]dr \qquad \text{(N3.3.14)} \\ \hat{u}_{3'}(m,\xi,z) &= \int_0^{2\pi} e^{-im\varphi}d\varphi \int_0^\infty r\left[\frac{-im}{\xi r}J_m(\xi r)u_r(r,\varphi,z)-\frac{1}{\xi}\partial_r J_m(\xi r)u_\varphi(r,\varphi,z)\right]dr \qquad \text{(N3.3.15)}\end{aligned}$$

The coupling of Eqs. (N3.3.11), (N3.3.14), and (N3.3.15) yields the following Fourier-Hankel transform for an elastic wavefield:

$$\hat{u}_{l'}(\xi,m,z) = \int_0^{2\pi}\int_0^\infty rH_{\alpha l'}^{m*}(\xi,r,\varphi)\,u_\alpha(r,\varphi,z)\,dr \qquad \text{(N3.3.16)}$$

where $H_{\alpha l'}^{m*}(\xi,r,\varphi)$ is defined as

$$H_{\alpha l'}^{m*}(\xi,r,\varphi) = \begin{bmatrix} 1 & 0 & 0 \\ 0 & \xi^{-1}\partial_r & (\xi r)^{-1} \\ 0 & (\xi r)^{-1}\partial_\varphi & -\xi^{-1}\partial_r \end{bmatrix} e^{-im\varphi}J_m(\xi r) \qquad \text{(N3.3.17)}$$

3.2.3.3 Application of Fourier-Hankel transform to the definition equations of Green's function

Now, let us revisit the definition of Green's function as shown in Eqs. (3.2.42) and (3.2.43). Recall that these equations are expressed as:

$$\left(L_{ij}(\partial_1,\partial_2,\partial_3)+\rho\omega^2\delta_{ij}\right)G_{jk}(\boldsymbol{x},\boldsymbol{y}) = -\delta_{ik}\delta(\boldsymbol{x}-\boldsymbol{y}) \quad (3.2.65)$$

$$\lim_{x_3\to+0} P_{ij}(\partial_1,\partial_2,\partial_3)G_{jk}(\boldsymbol{x},\boldsymbol{y}) = 0, \ (\boldsymbol{x},\boldsymbol{y}\in\mathbb{R}^3_+) \quad (3.2.66)$$

where the subscripts for Green's function refer to a Cartesian coordinate system. We express Green's function in terms of the inverse Fourier-Hankel transform as follows:

$$G_{jk}(\boldsymbol{x},\boldsymbol{y}) = C_{\beta j}(\varphi)G_{\beta k}(\boldsymbol{x},\boldsymbol{y}) \quad (3.2.67)$$

$$G_{\beta k}(\boldsymbol{x},\boldsymbol{y}) = \frac{1}{2\pi}\sum_{m=-\infty}^{\infty}\int_0^{\infty}\xi H^m_{\beta l'}(\xi,r,\varphi)\hat{G}^m_{l'k}(\xi,z,\boldsymbol{y})d\xi \quad (3.2.68)$$

where $G_{\beta k}$ represents Green's function in terms of the cylindrical coordinate system, and $\hat{G}^m_{l'k}$ denotes the Fourier-Hankel transform of $G_{\beta k}$, given by

$$\hat{G}^m_{l'k}(\xi,z,\boldsymbol{y}) = \int_0^{2\pi}d\varphi\int_0^{\infty}rH^{m*}_{\beta l'}(\xi,r,\varphi)G_{\beta k}(\boldsymbol{x},\boldsymbol{y})dr \quad (3.2.69)$$

Note that we have to determine the explicit form of $\hat{G}^m_{l'k}$ by means of Eqs. (3.2.65) and (3.2.66). We also have to express the right-hand side of Eq. (3.2.65) in terms of the inverse Fourier-Hankel transform as

$$\delta_{ik}\delta(\boldsymbol{x}-\boldsymbol{y}) = C_{\beta i}(\varphi)C_{\beta k}(\varphi)\delta(\boldsymbol{x}-\boldsymbol{y}) \quad (3.2.70)$$

$$C_{\beta k}(\varphi)\delta(\boldsymbol{x}-\boldsymbol{y}) = \frac{1}{2\pi}\sum_{m=-\infty}^{\infty}\int_0^{\infty}\xi H^m_{\beta q'}(\xi,r,\varphi)\hat{\delta}^m_{q'k}(\xi,z,\boldsymbol{y})d\xi \quad (3.2.71)$$

In Eq. (3.2.70), we used

$$\delta_{ik} = C_{\beta i}(\varphi)C_{\beta k}(\varphi) \quad (3.2.72)$$

and $\hat{\delta}^m_{q'k}$ is the Fourier-Hankel transform of $C_{\beta k}(\varphi)\delta(\boldsymbol{x}-\boldsymbol{y})$, defined as

$$\hat{\delta}^m_{q'k}(\xi,z,\boldsymbol{y}) = \int_0^{2\pi}C_{\beta k}(\varphi)d\varphi\int_0^{\infty}rH^{m*}_{\beta q'}(\xi,r,\varphi)\delta(\boldsymbol{x}-\boldsymbol{y})dr \quad (3.2.73)$$

Based on Eqs. (3.2.65), (3.2.66), (3.2.67), and (3.2.70), we can establish the following equations to define Green's function:

$$C_{\alpha i}(\varphi)\left[L_{ij}(\partial_1,\partial_2,\partial_3)+\rho\omega^2\delta_{ij}\right]C_{\beta j}(\varphi)G_{\beta k}(\boldsymbol{x},\boldsymbol{y}) = -C_{\alpha i}(\varphi)C_{\beta i}(\varphi)C_{\beta k}(\varphi)\delta(\boldsymbol{x}-\boldsymbol{y})$$

$$\lim_{x_3\to 0}C_{\alpha i}(\varphi)P_{ij}(\partial_1,\partial_2,\partial_3)C_{\beta j}(\varphi)G_{\beta k}(\boldsymbol{x},\boldsymbol{y}) = 0 \ (\boldsymbol{x},\boldsymbol{y}\in\mathbb{R}^3_+) \quad (3.2.74)$$

which are equivalent to

$$
\begin{aligned}
(L_{\alpha\beta}(\partial_r,\partial_\varphi,\partial_z)+\delta_{\alpha\beta}\rho\omega^2)G_{\beta k}(\boldsymbol{x},\boldsymbol{y}) &= -\delta_{\alpha\beta}C_{\beta k}(\varphi)\delta(\boldsymbol{x}-\boldsymbol{y}) \quad (3.2.75)\\
\lim_{z\to 0} P_{\alpha\beta}(\partial_r,\partial_\varphi,\partial_z)G_{\beta k}(\boldsymbol{x},\boldsymbol{y}) &= 0, \ \ (\boldsymbol{x},\boldsymbol{y}\in\mathbb{R}^3_+) \quad (3.2.76)
\end{aligned}
$$

where $L_{\alpha\beta}$ and $P_{\alpha\beta}$ represent the Navier operator and the traction operator in the cylindrical coordinate system, respectively, and they are given by:

$$
L_{\alpha\beta}(\partial_r,\partial_\varphi,\partial_z) = \Big[(\lambda+2\mu)\nabla\nabla\cdot-\mu\nabla\times\nabla\times\Big]_{\alpha\beta} \quad (3.2.77)
$$

$$
P_{\alpha\beta}(\partial_r,\partial_\varphi,\partial_z) = \begin{bmatrix} (\lambda+2\mu)\partial_z & \lambda(\partial_r+1/r) & (\lambda/r)\partial_\varphi \\ \mu\partial_r & \mu\partial_z & 0 \\ (\mu/r)\partial_\varphi & 0 & \mu\partial_z \end{bmatrix} \quad (3.2.78)
$$

Note that the subscript on the right-hand side of Eq. (3.2.77) indicates the component of the operator. Additionally, we used Eqs. (2.6.8) to (2.6.10) and (2.6.15) to derive Eqs. (3.2.77) and (3.2.78). The substitution of the right-hand side of Eq. (3.2.68) into the left-hand side of Eq. (3.2.75) yields:

$$
\begin{aligned}
&\Big(L_{\alpha\beta}(\partial_r,\partial_\varphi,\partial_z)+\rho\omega^2\delta_{\alpha\beta}\Big)G_{\beta k}(\boldsymbol{x},\boldsymbol{y})\\
&= \Big(L_{\alpha\beta}(\partial_r,\partial_\varphi,\partial_z)+\rho\omega^2\delta_{\alpha\beta}\Big)\left[\frac{1}{2\pi}\sum_{m=-\infty}^{\infty}\int_0^\infty \xi H^m_{\beta l'}(\xi,r,\varphi)\hat{G}^m_{l'k}(\xi,z,\boldsymbol{y})d\xi\right]\\
&= \frac{1}{2\pi}\sum_{m=-\infty}^{\infty}\int_0^\infty \xi\Big[L_{\alpha\beta}(\partial_r,\partial_\varphi,\partial_z)+\rho\omega^2\delta_{\alpha\beta}\Big]H^m_{\beta l'}(\xi,r,\varphi)\hat{G}^m_{l'k}(\xi,z,\boldsymbol{y})d\xi\\
&= \frac{1}{2\pi}\sum_{m=-\infty}^{\infty}\int_0^\infty \xi H_{\alpha q'}(\xi,r,\varphi)\Big(-\mathscr{A}_{q'l'}(\xi,\partial_z)+\rho\omega^2\delta_{q'l'}\Big)\hat{G}^m_{l'k}(\xi,z,\boldsymbol{y})d\xi
\end{aligned}
\quad (3.2.79)
$$

where the operator $\mathscr{A}$, defined in Eq. (3.2.47), takes the following form:

$$
\mathscr{A}_{q'l'}(\xi,\partial_z) = \begin{bmatrix} -(\lambda+2\mu)\partial_z+\mu\xi^2 & (\lambda+\mu)\xi\partial_z & 0 \\ -(\lambda+\mu)\xi\partial_z & -\mu\partial_z^2+(\lambda+2\mu)\xi^2 & 0 \\ 0 & 0 & -\mu\partial_z^2+\mu\xi^2 \end{bmatrix}
$$

For the derivation of Eq. (3.2.79), we used the results from Note 3.4. Now, considering Eqs. (3.2.71), (3.2.75), and (3.2.79), we arrive at the following equation:

$$
\Big(-\mathscr{A}_{q'l'}(\xi,\partial_z)+\rho\omega^2\delta_{q'l'}\Big)\hat{G}^m_{l'k}(\xi,z,\boldsymbol{y}) = -\hat{\delta}^m_{q'k}(\xi,z,\boldsymbol{y}) \quad (3.2.80)
$$

We need to apply the Hankel transform to Eq. (3.2.76). Substituting the right-hand side of Eq. (3.2.68) into the left-hand side of Eq. (3.2.76) yields:

$$\begin{aligned}
& P_{\alpha\beta}(\partial_r,\partial_\varphi,\partial_z)\, G_{\beta k}(\boldsymbol{x},\boldsymbol{y}) \\
= & \; P_{\alpha\beta}(\partial_r,\partial_\varphi,\partial_z)\left[\frac{1}{2\pi}\sum_{m=-\infty}^{\infty}\int_0^\infty \xi H^m_{\beta l'}(\xi,r,\varphi)\hat{G}^m_{l'k}(\xi,z,\boldsymbol{y})d\xi\right] \\
= & \; \frac{1}{2\pi}\sum_{m=-\infty}\int_0^\infty \xi P_{\alpha\beta}(\partial_r,\partial_\varphi,\partial_z)\, H^m_{\beta l'}(\xi,r,\varphi)\hat{G}^m_{l'k}(\xi,z,\boldsymbol{y}d\xi \\
= & \; \frac{1}{2\pi}\sum_{m=-\infty}\int_0^\infty \xi H^m_{\alpha q'}(\xi,r,\varphi)\mathscr{P}_{q'l'}(\xi,\partial_z)\hat{G}^m_{l'k}(\xi,z,\boldsymbol{y})d\xi
\end{aligned} \tag{3.2.81}$$

Hence, Eq. (3.2.76) is modified as

$$\lim_{z\to 0}\mathscr{P}_{q'l'}(\xi,\partial_z)\hat{G}^m_{l'k}(\xi,z,\boldsymbol{y}) = 0 \tag{3.2.82}$$

In Eq. (3.2.82), the operator $\mathscr{P}$, defined in Eq. (3.2.48), takes the following form:

$$\mathscr{P}_{q'l'}(\xi,\partial_z) = \begin{bmatrix} (\lambda+2\mu)\partial_z & -\lambda\xi & 0 \\ \mu\xi & \mu\partial_z & 0 \\ 0 & 0 & \mu\partial_z \end{bmatrix}$$

For the derivation of Eq. (3.2.81), we used the results from Note 3.4.

Now, we have Eqs. (3.2.80) and (3.2.82) to determine $\hat{G}^m_{l'k}(\xi,x_3,\boldsymbol{y})$. It is worth noting that these equations have similarities to Eqs. (3.2.49) and (3.2.50), which define Green's function in the wavenumber domain for a Cartesian coordinate system. To proceed with our discussion, we need to explicitly define $\hat{\delta}^m_{q'k}$ as defined in Eq. (3.2.73).

Let us consider the components of the field and source points for Green's function as follows:

$$\begin{aligned}
\boldsymbol{x} &= (r,\varphi,z) \\
\boldsymbol{y} &= (0,0,y_3)
\end{aligned} \tag{3.2.83}$$

Then, the expression for the Dirac delta function in terms of the cylindrical coordinate system becomes

$$\delta(\boldsymbol{x}-\boldsymbol{y}) = \frac{1}{r}\delta(r)\delta(\varphi)\delta(z-y_3) \tag{3.2.84}$$

As a result, the Fourier-Hankel transform for $\hat{\delta}(\boldsymbol{x}-\boldsymbol{y})$ can be expressed as

$$\begin{aligned}
& \hat{\delta}^m_{q'k}(\xi,z,\boldsymbol{y}) \\
= & \int_0^{2\pi} d\varphi \int_0^\infty rH^{m*}_{\beta q'}(\xi,r,\varphi)dr\, C_{\beta k}(\varphi)\frac{\delta(r)}{r}\delta(\varphi)\delta(z-y_3) \\
= & \; H^{m*}_{\beta q'}(\xi,0,0)C_{\beta k}(0)\delta(z-y_3) := \hat{F}^m_{q'k}\delta(z-y_3)
\end{aligned} \tag{3.2.85}$$

where $\hat{F}^m_{q'k}$ is expressed as

$$\hat{F}^m_{q'k} = \begin{cases} \begin{bmatrix} 0 & 0 & 1 \\ 0 & 0 & 0 \\ 0 & 0 & 0 \end{bmatrix} & (m=0) \\ (1/2)\begin{bmatrix} 0 & 0 & 0 \\ \pm 1 & -i & 0 \\ -i & \mp 1 & 0 \end{bmatrix} & (m=\pm 1) \\ \begin{bmatrix} 0 & 0 & 0 \\ 0 & 0 & 0 \\ 0 & 0 & 0 \end{bmatrix} & (\text{otherwise}) \end{cases} \tag{3.2.86}$$

The substitution of Eq. (3.2.85) into the right-hand side of Eq. (3.2.80) yields:

$$\left[-\mathscr{A}_{q'l'}(\xi,\partial_z)+\rho\omega^2\delta_{q'l'}\right]\hat{G}^m_{l'k}(\xi,z,:\boldsymbol{y}) = -\hat{F}^m_{q'k}\delta(z-y_3) \tag{3.2.87}$$

Therefore, if we define $\hat{g}_{l'p'}(z,y_3)$ as follows:

$$\hat{G}^m_{l'k}(\xi,z,\boldsymbol{y}) = \hat{g}_{l'p'}(z,y_3,\xi)\hat{F}^m_{p'k} \tag{3.2.88}$$

then Eqs. (3.2.87) and (3.2.82) are modified as

$$\left[-\mathscr{A}_{q'l'}(\xi,\partial_z)+\rho\omega^2\delta_{q'l'}\right]\hat{g}_{l'p'}(z,y_3,\xi) = -\delta_{q'p'}\delta(z-y_3) \tag{3.2.89}$$

$$\lim_{z\to+0}\mathscr{P}_{q'l'}(\xi,\partial_z)\hat{g}_{l'p'}(z,y_3,\xi) = 0 \tag{3.2.90}$$

These equations are the same as Eqs. (3.2.49) and (3.2.50), respectively. We have already constructed the solutions for Eqs. (3.2.49) and (3.2.50) in Eq. (3.2.53). Therefore, we arrive at the following representation of Green's function for an elastic half-space in terms of the superposition of cylindrical waves:

$$G_{ij}(\boldsymbol{x},\boldsymbol{y}) = \frac{1}{2\pi}C_{i\alpha}(\varphi)\sum_{m=-1}^{+1}\int_0^\infty \xi H^m_{\alpha l'}(\xi,r,\varphi)\hat{g}_{l'p'}(x_3,y_3,\xi)d\xi\hat{F}^m_{p'j} \tag{3.2.91}$$

Here, for the sake of convenience, we have used x_3 instead of z in the representation. It's important to note that the infinite series for m shown in Eq. (3.2.68) is reduced to a finite sum for Green's function, as can be seen in Eq. (3.2.91). This reduction is a consequence of the horizontal range of the source point being located at the origin of the coordinate system, as shown in Eq. (3.2.83). In conclusion, the representation of Green's function given in Eq. (3.2.91) is convenient for investigating the properties of Green's function in this specific cylindrical coordinate system setting.

Note 3.4 Properties of surface vector harmonics

Recall that the kernel of the Fourier-Hankel transform for an elastic wavefield was defined in Eq. (3.2.62), which can be expressed as

$$\left[H^{m}_{\alpha k'}(\xi,r,\varphi)\right]=\left[\begin{array}{ccc} \boldsymbol{R}^{m}_{\xi}(r,\varphi) & \boldsymbol{S}^{m}_{\xi}(r,\varphi) & \boldsymbol{T}^{m}_{\xi}(r,\varphi) \end{array}\right] \tag{N3.4.1}$$

Note that $\boldsymbol{R}^{m}_{\xi}$, $\boldsymbol{S}^{m}_{\xi}$, and $\boldsymbol{T}$ are defined in Eq. (3.2.60) and and are generally referred to as *surface vector harmonics*. Using Eqs. (2.6.8) to (2.6.10), we can deduce the following properties for surface vector harmonics:

$$\begin{aligned} \nabla\times\nabla\times\boldsymbol{R}^{m}_{\xi}(r,\varphi)u_{1'}(z) &= \boldsymbol{R}^{m}_{\xi}(r,\varphi)\xi^{2}u_{1'}(z)+\boldsymbol{S}^{m}_{\xi}(r,\varphi)\xi\partial_{z}u_{1'}(z) \\ \nabla\nabla\cdot\boldsymbol{R}^{m}_{\xi}(r,\varphi)u_{1'}(z) &= \boldsymbol{R}^{m}_{\xi}(r,\varphi)\partial_{z}^{2}u_{1'}(z)+\boldsymbol{S}^{m}_{\xi}(r,\varphi)\xi\partial_{z}u_{1'}(z) \end{aligned} \tag{N3.4.2}$$

$$\begin{aligned} \nabla\times\nabla\times\boldsymbol{S}^{m}_{\xi}(r,\varphi)u_{2'}(z) &= -\boldsymbol{R}^{m}_{\xi}(r,\varphi)\xi\partial_{z}u_{2'}(z)-\boldsymbol{S}^{m}_{\xi}(r,\varphi)\partial_{z}^{2}u_{2'}(z) \\ \nabla\nabla\cdot\boldsymbol{S}^{m}_{\xi}(r,\varphi)u_{2'}(z) &= -\boldsymbol{R}^{m}_{\xi}(r,\varphi)\xi\partial_{z}u_{2'}(z)-\boldsymbol{S}^{m}_{\xi}(r,\varphi)\xi^{2}u_{2'}(z) \end{aligned} \tag{N3.4.3}$$

$$\begin{aligned} \nabla\times\nabla\times\boldsymbol{T}^{m}_{\xi}(r,\varphi)u_{3'}(z) &= -\boldsymbol{T}^{m}_{\xi}(r,\varphi)(\partial_{z}^{2}-\xi^{2})u_{3'}(z) \\ \nabla\nabla\cdot\boldsymbol{T}^{m}_{\xi}(r,\varphi)u_{3'}(z) &= 0 \end{aligned} \tag{N3.4.4}$$

Consequently, we can express the action of the Navier operator $L_{\alpha\beta}(\partial_r,\partial_\varphi,\partial_z)$ on $H^{m}_{\beta k'}(\xi,r,\varphi)u_{k'}(z)$ as follows:

$$\begin{aligned} &L_{\alpha\beta}(\partial_r,\partial_\varphi,\partial_z)H^{m}_{\beta k'}(\xi,r,\varphi)u_{k'}(z) \\ &= H^{m}_{\alpha k'}(\xi,r,\varphi)\left(-\mathscr{A}_{k'l'}(\xi,\partial_z)\right)u_{k'}(z) \end{aligned} \tag{N3.4.5}$$

where $L_{\alpha\beta}$ is the Navier operator presented in the cylindrical coordinate system given by Eq. (3.2.77). The components of the operator $\mathscr{A}_{k'l'}$ are given by

$$\mathscr{A}_{k'l'}(\xi,\partial_z)=\left[\begin{array}{ccc} -(\lambda+2\mu)\partial_z^2+\mu\xi^2 & (\lambda+\mu)\xi\partial_z & 0 \\ -(\lambda+\mu)\xi\partial_z & -\mu\partial_z^2+(\lambda+2\mu)\xi^2 & 0 \\ 0 & 0 & -\mu\partial_z^2+\mu\xi^2 \end{array}\right] \tag{N3.4.6}$$

Similarly, the action of the traction operator $P_{\alpha\beta}(\partial_r,\partial_\varphi,\partial_z)$ on $H^{m}_{\beta k'}(\xi,r,\varphi)u_{k'}(z)$ can be expressed as:

$$\begin{aligned} &P_{\alpha\beta}(\partial_r,\partial_\varphi,\partial_z)H^{m}_{\beta k'}(\xi,r,\varphi)u_{k'}(z) \\ &= H^{m}_{\alpha k'}(\xi,r,\varphi)\left(\mathscr{P}_{k'l'}(\xi,\partial_z)\right)u_{k'}(z) \end{aligned} \tag{N3.4.7}$$

where $P_{\alpha\beta}$ is the traction operator in the cylindrical coordinate system given by Eq. (3.2.78), and the components of the operator $\mathscr{P}_{k'l'}$ are given by:

$$\mathscr{P}_{k'l'}(\xi,\partial_z)=\left[\begin{array}{ccc} (\lambda+2\mu)\partial_z & -\lambda\xi & 0 \\ \mu\xi & \mu\partial_z & 0 \\ 0 & 0 & \mu\partial_z \end{array}\right] \tag{N3.4.8}$$

These expressions and properties are crucial for further analyses and computations in the context of elastic wavefields in cylindrical coordinate systems.

3.3 EVALUATION OF GREEN'S FUNCTION IN THE COMPLEX WAVENUMBER PLANE BY USE OF BRANCH LINE INTEGRALS

3.3.1 GENERAL REMARKS

We found that Green's function for an elastic half-space can be expressed in the form of an integral representation, as shown in Eq. (3.2.52) or (3.2.91). As is mentioned before, it is not difficult to compute these integral representations, nowadays. In the past, however, only approximation methods were available for the representations. Approximation methods, which are thought to be historical, have limitations in the sense that they do not provide precise quantitative results for the integral. Nevertheless, their results can be used to determine the properties of wave propagation. In addition, the historical approximation methods still have possibilities for enhancement of computational methods, as shown in the next chapter. Here, we investigate two historical approximation methods, one based on the branch line integral and the other based on the steepest descent path in the complex wavenumber plane. In this section, we consider the method based on the branch line integral.

3.3.2 METHOD BASED ON BRANCH LINE INTEGRAL FOR SOMMERFELD INTEGRAL

3.3.2.1 Review of Sommerfeld integral and its modification

As a starting point for the discussion, we investigate the Sommerfeld integral, which was given in Eq. (3.2.22). We discuss the method based on the branch line integral for the Sommerfeld integral using the following expression:

$$S(r,x_3,y_3) = \frac{1}{4\pi}\int_0^{\infty} \xi J_0(\xi r)\frac{e^{-\nu|x_3-y_3|}}{\nu}d\xi \left(= \frac{e^{ikR}}{4\pi R}\right) \tag{3.3.1}$$

where

$$R = \sqrt{r^2+(x_3-y_3)^2} \tag{3.3.2}$$

and

$$\nu = \begin{cases} \sqrt{\xi^2-k^2} & (\xi \geq k) \\ -i\sqrt{k^2-\xi^2} & (\xi < k) \end{cases} \tag{3.3.3}$$

We extend the interval of the integral in Eq. (3.3.1) from $[0,\infty)$ to $(-\infty,\infty)$ by the use of

$$\begin{aligned} J_0(\xi r) &= (1/2)\left(H_0^{(1)}(\xi r)+H_0^{(2)}(\xi r)\right) \\ H_0^{(1)}(-\xi r) &= -H_0^{(2)}(\xi r) \end{aligned} \tag{3.3.4}$$

where $H_0^{(1)}$ and $H_0^{(2)}$ are Hankel functions of the first and second kind, respectively, of order zero. Then, Eq. (3.3.1) is modified as

$$S(r,x_3,y_3) = \frac{1}{8\pi}\int_{-\infty}^{\infty} \xi H_0^{(1)}(\xi r)\frac{e^{-\nu|x_3-y_3|}}{\nu}d\xi \tag{3.3.5}$$

In addition, we apply the asymptotic form of the Hankel function

$$H_0^{(1)}(\xi r) \sim \sqrt{\frac{2}{\pi \xi r}} \exp\left(i\xi r - i\frac{\pi}{4}\right), \ (r \to \infty) \tag{3.3.6}$$

to Eq. (3.3.5). This yields

$$S(r,x_3,y_3) \sim \frac{1}{8\pi}\sqrt{\frac{2}{\pi r}} e^{-i\pi/4} \int_{-\infty}^{\infty} \sqrt{\xi} \frac{e^{-\nu|x_3-y_3|}}{\nu} e^{i\xi r} d\xi, \ (r \to \infty) \tag{3.3.7}$$

3.3.2.2 Riemann sheet and branch cut for Eq. (3.3.7)

We discuss Eq. (3.3.7) to apply the distortion of the path of the integral in the complex wavenumber plane to obtain the branch line integral. For the distortion of the path of the integral, we have to be aware that ν and $\sqrt{\xi}$ in the integrand in Eq. (3.3.7) are multi-valued functions in the complex wavenumber plane and as a result, Riemann cuts are required to determine the branches for ν and $\sqrt{\xi}$. The basic concepts of Riemann sheets and cuts are given in Note 3.2. At this point, we have to determine the branch cuts according to the physical requirement. The physical requirement is the radiation condition, by which we can avoid the divergence of the integrand in Eq. (3.3.7). It is expressed by $\mathrm{Re}(\nu) > 0$. The Riemann sheet that satisfies the physical requirement is called the *permissible sheet*. The Riemann cuts that realize the Riemann sheet for $\mathrm{Re}(\nu) > 0$ are the lines $\mathrm{Re}(\nu) = 0$, which are shown in Fig. 3.3.1. It is found from Fig. 3.3.1 that the Riemann cuts are on the whole imaginary axis and on the real axis in the range of $-k \leq \mathrm{Re}(\xi) \leq k$. Note that the distorted path of the integral in the complex wavenumber plane has to be closed in the permissible sheet without crossing the branch cuts. In this sense, the distortion of the path of the integral in the complex wavenumber plane seems to be impossible if one employs the branch cut shown in Fig. 3.3.1.

To overcome this problem, let us assume that the wavenumber of the wavefield k is a complex value; that is,

$$k = k_R + i\varepsilon, \ (\varepsilon > 0) \tag{3.3.8}$$

Then, $\mathrm{Re}(\nu) = 0$ requires

$$\begin{aligned} \mathrm{Re}(\xi)\,\mathrm{Im}(\xi) &= k_R\varepsilon \\ \left(\mathrm{Re}(\xi)\right)^2 - \left(\mathrm{Im}(\xi)\right)^2 &\leq k_R^2 - \varepsilon^2 \end{aligned} \tag{3.3.9}$$

and the locations of the branch cuts are shifted, as shown in Fig. 3.3.2, from which we see that the two branch cuts are separated. Due to the separation of the branch cuts, we can establish a closed path of the integral for Eq. (3.3.7).

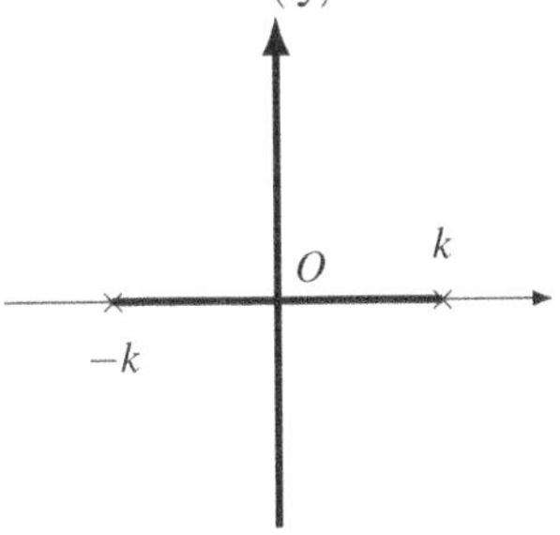

Figure 3.3.1 Riemann cuts for ν for case where k is real-valued parameter.

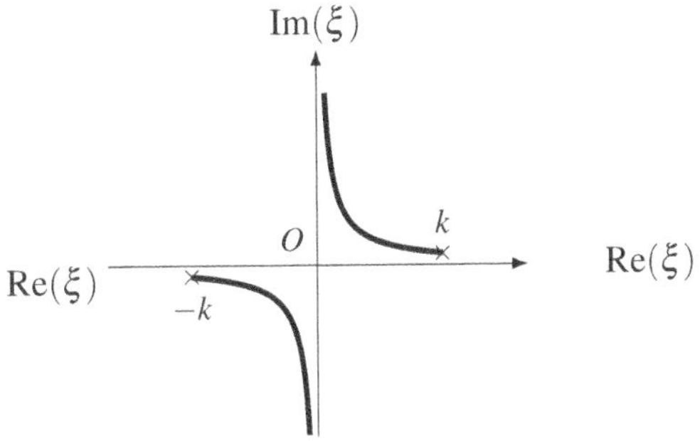

Figure 3.3.2 Riemann cuts for ν for case where k is complex-valued parameter.

3.3.2.3 Representation of Eq. (3.3.7) in terms of branch line integral

Now, we derive the expression in Eq. (3.3.7) in terms of the branch line integral. We set the closed path of the integral for Eq. (3.3.7) in the permissible sheet as shown in Fig. 3.3.3. For the path of the integral, the branch cut for $\sqrt{\xi}$, which is on the negative real axis, is also added. In the permissible sheet, the branch of $\sqrt{\xi}$ is chosen such that $\mathrm{Re}\left(\sqrt{\xi}\right) > 0$. The paths of integral M_1 and M_2 in Fig. 3.3.3 are arcs of radius R_d from the origin of the complex wavenumber plane. The integrand of Eq. (3.3.7) has no poles, so we have

$$\int_{C+M_1+B_1+B_2+M_2} \sqrt{\xi}\frac{\exp(-\nu|x_3-y_3|}{\nu}\exp(i\xi r)d\xi = 0 \tag{3.3.10}$$

At this point, we would like to show that

$$I_{M_1+M_2} = \int_{M_1+M_2} \sqrt{\xi}\frac{e^{-\nu|x_3-y_3|}}{\nu}e^{i\xi r}d\xi \to 0,\ (R_d \to \infty) \tag{3.3.11}$$

To show this, we set $\xi = R_d\exp(i\theta)$ and use

$$\left|e^{-\nu|x_3-y_3|}\right| < 1,\ \ (\because \mathrm{Re}(\nu) > 0) \tag{3.3.12}$$

$$|\nu| = \sqrt{\left|R_d^2e^{2i\theta} - (k_R+i\varepsilon)^2\right|} \geq \sqrt{R_d^2 - k_R^2}\left[1-\alpha\right]^{1/4} \tag{3.3.13}$$

$$\sin\theta \geq \frac{2}{\pi}\theta,\ (0 \leq \theta \leq \pi/2) \tag{3.3.14}$$

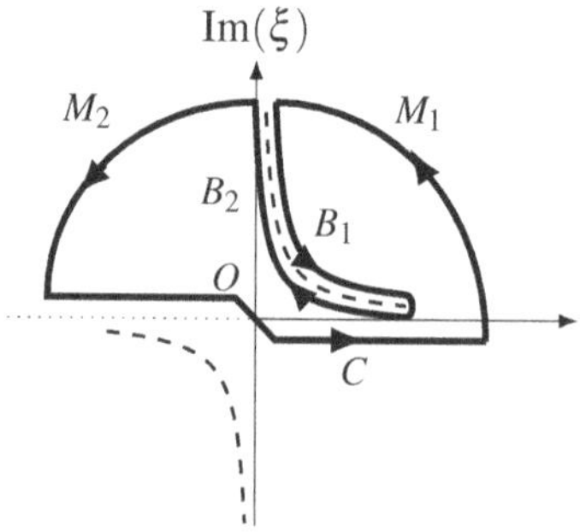

Figure 3.3.3 Path of integral for Eq. (3.3.7) in complex wavenumber plane for case where k is complex-valued parameter.

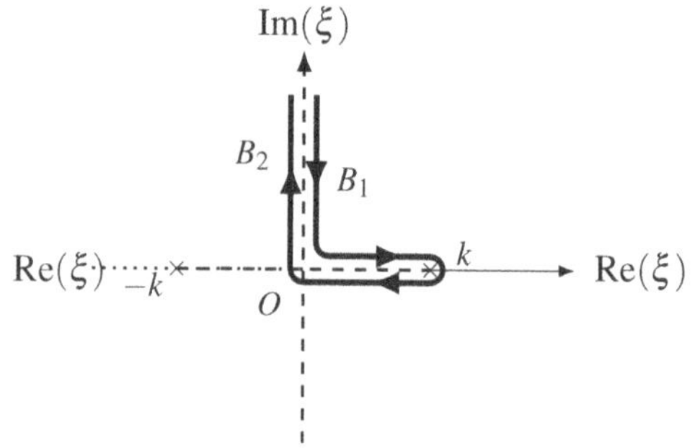

Figure 3.3.4 Path of branch line integral for Eq. (3.3.7) in complex wavenumber plane for case where k is real-valued parameter.

where

$$\alpha = \frac{4\varepsilon k_R R_d^2}{(R_d^2 - k_R^2)^2} \to 0, \ \ (R_d \to \infty) \tag{3.3.15}$$

Based on Eqs. (3.3.12) to (3.3.14), we have

$$\begin{aligned} |I_{M_1+M_2}| &\leq 2\frac{R_d\sqrt{R_d}}{\sqrt{R_d^2 - k_R^2}(1-\alpha)^{1/4}} \int_0^{\pi/2} e^{-R_d(2/\pi)\theta} d\theta \\ &\to 0, \ (R_d \to \infty) \end{aligned} \tag{3.3.16}$$

Therefore, based on Eqs. (3.3.10) and (3.3.16), we have the branch line integral representation for Eq. (3.3.7) in the following form:

$$S(r, x_3, y_3) \sim -\frac{1}{8\pi}\sqrt{\frac{2}{\pi r}} e^{-i\pi/4} \int_{B_1+B_2} \sqrt{\xi} \frac{e^{-\nu|x_3-y_3|}}{\nu} e^{i\xi r} d\xi \tag{3.3.17}$$

3.3.2.4 Approximation technique for branch line integral of Sommerfeld integral

Now, our task is to evaluate Eq. (3.3.17) under the assumption that k is a real-valued parameter. For this case, the paths of the branch line integral become as shown in Fig. 3.3.4. For the integral, we use the following notation for ν:

$$\begin{aligned} \nu_1 &= \left[\nu\right]_{\xi \in B_1} \\ \nu_2 &= \left[\nu\right]_{\xi \in B_2} \end{aligned} \tag{3.3.18}$$

ν_1 and ν_2 are found to have the following properties:

$$\begin{aligned} \mathrm{Im}\,\nu_1 &> 0 \\ \mathrm{Im}\,\nu_2 &< 0, \text{ (unless } \xi \in \text{third quadrant)} \end{aligned} \tag{3.3.19}$$

In addition, we also use the notation $\xi = \zeta + i\eta$. Equation (3.3.17) is expressed as

$$\begin{aligned} S(r,x_3,y_3) \sim & -\frac{1}{8\pi}\sqrt{\frac{2}{\pi r}}e^{-i\pi/4}\int_0^k \sqrt{\zeta}\left[\frac{e^{-\nu_1|x_3-y_3|}}{\nu_1} - \frac{e^{-\nu_2|x_3-y_3|}}{\nu_2}\right]\exp(i\zeta r)d\zeta \\ & +\frac{1}{8\pi}\sqrt{\frac{2}{\pi r}}e^{-i\pi/4}\int_0^\infty \sqrt{i\eta}\left[\frac{e^{-\nu_1|x_3-y_3|}}{\nu_1} - \frac{e^{-\nu_2|x_3-y_3|}}{\nu_2}\right]\exp(-\eta r)id\eta \end{aligned} \tag{3.3.20}$$

For the first term on the right-hand side of Eq. (3.3.20), we notice that

$$\begin{aligned} \nu_1 &= i\sqrt{k^2-\zeta^2} \\ \nu_2 &= -i\sqrt{k^2-\zeta^2} \end{aligned} \tag{3.3.21}$$

and we approximate this first term by the properties of the integrand close to the region $\zeta \sim k$. That is, the integrand is approximated by

$$\begin{aligned} & \sqrt{\zeta}\left[\frac{e^{-\nu_1|x_3-y_3|}}{\nu_1} - \frac{e^{-\nu_2|x_3-y_3|}}{\nu_2}\right]\exp(i\zeta r) \\ & = \frac{\sqrt{k}}{i}\left[\frac{2}{\sqrt{2k}}u^{-1/2} - \left(\frac{1}{\sqrt{2k}^3} + \sqrt{2k}\,|x_3-y_3|^2\right)u^{1/2}\right]e^{ikr}e^{-iku} \\ & \quad +O(u^{3/2}), \quad (u = k-\zeta \to 0) \end{aligned} \tag{3.3.22}$$

Then, the first term on the right-hand side of Eq. (3.3.20) can be approximated by

$$\begin{aligned} & -\frac{1}{8\pi}\sqrt{\frac{2}{\pi r}}e^{-i\pi/4}\int_0^k \sqrt{\zeta}\left[\frac{e^{-\nu_1|x_3-y_3|}}{\nu_1} - \frac{e^{-\nu_2|x_3-y_3|}}{\nu_2}\right]\exp(i\zeta r)d\zeta \\ & \sim -\frac{1}{8\pi}\sqrt{\frac{2}{\pi r}}e^{-i\pi/4}e^{ikr}\int_0^k \frac{\sqrt{k}}{i}\frac{2}{\sqrt{2k}}\left[u^{-1/2} + Bu^{1/2} + O(u^{3/2})\right]e^{-iur}du \\ & \sim \frac{1}{4\pi}\sqrt{\frac{1}{\pi r}}e^{i\pi/4}e^{ikr}\int_0^\infty \left[u^{-1/2} + Bu^{1/2} + O(u^{3/2})\right]e^{-iur}du \end{aligned} \tag{3.3.23}$$

where

$$B = -\frac{1}{4k} - k|x_3-y_3|^2 \tag{3.3.24}$$

For the computation of the last term of the integral, we again distort the path of the integral, which yields

$$\begin{aligned}\int_0^{\infty} u^{-1/2} e^{-iur} du &= \int_0^{-i\infty} u^{-1/2} e^{-iur} du \\ &= \int_0^{\infty} (-iu')^{-1/2} e^{-u'r} (-i) du' \\ &= e^{-i\pi/4} \int_0^{\infty} (u')^{-1/2} e^{-u'r} du' = \sqrt{\frac{\pi}{r}} e^{-i\pi/4} \end{aligned} \tag{3.3.25}$$

$$\begin{aligned}\int_0^{\infty} u^{1/2} e^{-iur} du &= \int_0^{-i\infty} u^{1/2} e^{-iur} du \\ &= \int_0^{\infty} (-iu')^{1/2} e^{-u'r} (-i) du' \\ &= e^{-3i\pi/4} \int_0^{\infty} (u')^{1/2} e^{-u'r} du' = \frac{1}{2}\sqrt{\frac{\pi}{r^3}} e^{-3i\pi/4} \end{aligned} \tag{3.3.26}$$

Therefore, we have the result for Eq. (3.3.23) in the following form:

$$\begin{aligned} &-\frac{1}{8\pi}\sqrt{\frac{2}{\pi r}} e^{-i\pi/4} \int_0^{k} \sqrt{\zeta} \left[\frac{e^{-\nu_1|x_3-y_3|}}{\nu_1} - \frac{e^{-\nu_2|x_3-y_3|}}{\nu_2} \right] \exp(i\zeta r) d\zeta \\ &\sim \frac{1}{4\pi r} \exp(ikr) \left(1 - \frac{i}{2}\frac{B}{r} + O(r^{-2}) \right), \ (r \to \infty) \end{aligned} \tag{3.3.27}$$

For the second term on the right-hand side of Eq. (3.3.20), we notice that

$$\begin{aligned} \nu_1 &= i\sqrt{\eta^2 + k^2} \\ \nu_2 &= -i\sqrt{\eta^2 + k^2} \end{aligned} \tag{3.3.28}$$

and we approximate the integral by the properties of the integrand close to the range $\eta \sim 0$. Therefore, we have

$$\begin{aligned} &\frac{1}{8\pi}\sqrt{\frac{2}{\pi r}} e^{-i\pi/4} \int_0^{\infty} \sqrt{i\eta} \left[\frac{e^{-\nu_1|x_3-y_3|}}{\nu_1} - \frac{e^{-\nu_2|x_3-y_3|}}{\nu_2} \right] \exp(-\eta r) i d\eta \\ &\sim \frac{1}{8\pi}\sqrt{\frac{2}{\pi r}} \frac{2\cos(k|x_3-y_3|)}{k} \int_0^{\infty} \sqrt{\eta} e^{-\eta r} d\eta \\ &\sim \frac{\cos(k|x_3-y_3|)}{4\sqrt{2}k\pi r^2} + O(r^{-3}), \ (r \to \infty) \end{aligned} \tag{3.3.29}$$

Now, based on the results for Eqs. (3.3.27) and (3.3.29), we have the approximation of the Sommerfeld integral by means of the branch line integral in the following form:

$$
\begin{aligned}
\frac{1}{4\pi R}\exp(ikR) &= \frac{1}{4\pi}\int_0^{\infty}\frac{\xi}{\nu}\exp(-\nu|x_3-y_3|)J_0(\xi r)d\xi \\
&\sim \frac{1}{4\pi r}\exp(ikr)\left[1+\frac{i}{2r}\left(\frac{1}{4k}+k|x_3-y_3|^2\right)\right] \\
&\quad +\frac{\cos(k|x_3-y_3|)}{4\sqrt{2}\pi k r^2}+O(r^{-3}),\quad (r\to\infty)
\end{aligned}
\tag{3.3.30}
$$

Equation (3.3.30) is the approximation of the Sommerfeld integral obtained using the method based on the branch line integral. The principal part of the approximated result for Eq. (3.3.30), namely

$$
\frac{1}{4\pi R}\exp(ikR) \sim \frac{1}{4\pi r}\exp(ikr),\quad (r\to\infty)
$$

reveals the properties of the approximation method based on the branch line integral.

The principal part of the approximation result itself is quite natural in the sense that there cannot be other than the result. The second term on the right-hand side of Eq. (3.3.30) is due to the result from the branch line integral along the imaginary axis in the complex wavenumber plane. The decrease in amplitude is on the order of r^{-2}, which is faster than that for the principal part. In addition, we find that the second term does not have a factor $\exp(ikr)$, meaning that it does not exhibit wave propagation toward the horizontal direction and is characterized as a standing wave.

3.3.3 APPROXIMATION OF GREEN'S FUNCTION FOR ELASTIC HALF-SPACE BY METHOD BASED ON BRANCH LINE INTEGRAL

3.3.3.1 Review of Green's function in terms of Fourier-Hankel transform and its modification

Now, let's revisit the expression of Green's function for an elastic half-space, which can be represented in terms of the Fourier-Hankel transform. As shown in Eq. (3.2.91), Green's function can be expressed as

$$
G_{ij}(\boldsymbol{x},\boldsymbol{y}) = \frac{1}{2\pi}C_{i\alpha}(\varphi)\sum_{m=-1}^{+1}\int_0^{\infty}\xi H^m_{\alpha l'}(\xi,r,\varphi)\hat{g}_{l'p'}(x_3,y_3,\xi)d\xi\,\hat{F}^m_{p'j} \tag{3.3.31}
$$

where $C_{i\alpha}$, $H^m_{\alpha l'}$, $\hat{g}_{l'p'}$, and $\hat{F}^m_{p'j}$ are given in Eqs. (3.2.55), (3.2.62), (C.2.18), and (3.2.86), respectively.

For the following discussion, we restrict ourselves to Green's function for the surface response due to an interior point source for simplicity. That is, we use Eq. (3.2.53) for Green's function in the wavenumber domain denoted, as $\hat{g}_{l'p'}$.

To derive the branch line integral for Eq. (3.3.31), our tasks are twofold: first, we need to extend the interval of the integral in Eq. (3.3.31) from $[0,+\infty)$ to $(-\infty,+\infty)$; secondly, we must introduce the far field properties of $H^m_{\alpha l'}$.

To extend the interval of the integral, we use the following properties of the Hankel functions:

$$\begin{aligned} J_m(\xi r) &= \frac{1}{2}\left[H_m^{(1)}(\xi r) + H_m^{(2)}(\xi r)\right] \\ H_m^{(1)}(-\xi r) &= -H_m^{(2)}(\xi r), \ (m = 0, \pm 2) \\ H_m^{(1)}(-\xi r) &= H_m^{(2)}(\xi r), \ (m = \pm 1) \end{aligned} \tag{3.3.32}$$

With these properties, Eq. (3.3.31) is modified as follows:

$$G_{ij}(\boldsymbol{x}, \boldsymbol{y}) = \frac{1}{4\pi} C_{i\alpha}(\varphi) \sum_{m=-1}^{+1} \int_{-\infty}^{\infty} \xi H_{\alpha l'}^{m(1)}(\xi, r, \varphi) \hat{g}_{l'p'}(x_3, y_3, \xi) d\xi \, \hat{F}_{p'j}^m \tag{3.3.33}$$

where $H_{\alpha l'}^{m(1)}(\cdot)$ is defined by the matrix:

$$H_{\alpha l'}^{m(1)}(\xi, r, \varphi) = \begin{bmatrix} 1 & 0 & 0 \\ 0 & \xi^{-1}\partial_r & (\xi r)^{-1}\partial_\varphi \\ 0 & (\xi r)^{-1}\partial_\varphi & -\xi^{-1}\partial_r \end{bmatrix} H_m^{(1)}(\xi r) \exp(im\varphi) \tag{3.3.34}$$

Using the following asymptotic form of the Hankel function:

$$H_m^{(1)}(\xi r) \sim \sqrt{\frac{2}{\pi \xi r}} \exp\left(i\xi r - \frac{(2m+1)\pi i}{4}\right), \ (r \to \infty) \tag{3.3.35}$$

we can express Eq. (3.3.34) as:

$$H_{\alpha l'}^{m(1)}(\xi, r, \varphi) \sim \sqrt{\frac{2}{\pi \xi r}} \exp\left(i\xi r - \frac{(2m+1)\pi i}{4}\right) h_{\alpha l'} \exp(im\varphi), \ (r \to \infty) \tag{3.3.36}$$

where $h_{\alpha l'}$ is given by the matrix:

$$h_{\alpha l'} = \begin{bmatrix} 1 & 0 & 0 \\ 0 & i & 0 \\ 0 & 0 & -i \end{bmatrix} \tag{3.3.37}$$

The substitution of Eq. (3.3.36) into Eq. (3.3.33) yields the following representation of Green's function:

$$\begin{aligned} G_{ij}(\boldsymbol{x}, \boldsymbol{y}) \sim \ & \frac{1}{4\pi} \sqrt{\frac{2}{\pi r}} C_{i\alpha}(\varphi) h_{\alpha l'} \sum_{m=-1}^{+1} e^{-(2m+1)\pi i/4} e^{im\varphi} \\ & \times \int_{-\infty}^{\infty} \sqrt{\xi} \exp(i\xi r) \, \hat{g}_{l'p'}(x_3, y_3, \xi) d\xi \, \hat{F}_{p'j}^m, \ (r \to \infty) \end{aligned} \tag{3.3.38}$$

This expression serves as the starting point for distorting the path of the integral in the complex wavenumber plane.

3.3.3.2 Distortion of path of integral of Eq. (3.3.38) in complex wavenumber plane

As shown in Eq. (3.2.53), Green's function in the wavenumber domain $\hat{g}_{l'p'}$ has multi-valued functions γ and ν defined by

$$\begin{aligned}\gamma &= \sqrt{\xi^2 - k_L^2}\\ \nu &= \sqrt{\xi^2 - k_T^2}\end{aligned}$$

Therefore, we need three kinds of branch cut for ν, γ, and $\sqrt{\xi}$ for the evaluation of Eq. (3.3.38) in the complex wavenumber plane. To set up the branch cuts, we assume that k_T and k_L are complex-valued parameters:

$$\begin{aligned}k_T &= \mathrm{Re}(k_T) + i\varepsilon\\ k_L &= \mathrm{Re}(k_L) + i\varepsilon, \quad (\varepsilon > 0)\end{aligned} \qquad (3.3.39)$$

Then, the branch cut for ν, which is $\mathrm{Re}(\nu) = 0$, requires

$$\mathrm{Re}(\xi)\,\mathrm{Im}(\xi) = \mathrm{Re}(k_T)\varepsilon \qquad (3.3.40)$$

$$\Big(\mathrm{Re}(\xi)\Big)^2 - \Big(\mathrm{Im}(\xi)\Big)^2 \leq \mathrm{Re}(k_T)^2 - \varepsilon^2 \qquad (3.3.41)$$

Likewise, the branch cut $\mathrm{Re}(\gamma) = 0$ requires

$$\mathrm{Re}(\xi)\,\mathrm{Im}(\xi) = \mathrm{Re}(k_L)\varepsilon \qquad (3.3.42)$$

$$\Big(\mathrm{Re}(\xi)\Big)^2 - \Big(\mathrm{Im}(\xi)\Big)^2 \leq \mathrm{Re}(k_L)^2 - \varepsilon^2 \qquad (3.3.43)$$

Figure 3.3.5 shows the branch cuts and the path of the integral for Eq. (3.3.38) in the permissible sheet of the complex wavenumber plane. As can be seen, the branch cuts for ν and γ are separated, and as a result, a closed path of the integral for Eq. (3.3.38) in the permissible sheet can be established. An important fact that we have to take into account at this point is the presence of the root of the Rayleigh function.

Due to the introduction of small parameter ε for Eq. (3.3.39), the roots of the Rayleigh function are shifted to the first and third quadrants of the complex wavenumber plane. Note that the closed path of the integral includes the root of the Rayleigh function in the first quadrant, which affects Green's function.

The paths of the integral M_1, M_2, and M_3 in Fig. 3.3.5 are arcs of radius R_d from the origin of the complex wavenumber plane. Since the effects of the path of the integral along M_1, M_2, and M_3 become negligible as $R_d \to \infty$, Green's function shown in Eq. (3.3.38) is expressed as the branch line integrals and the residue term in the

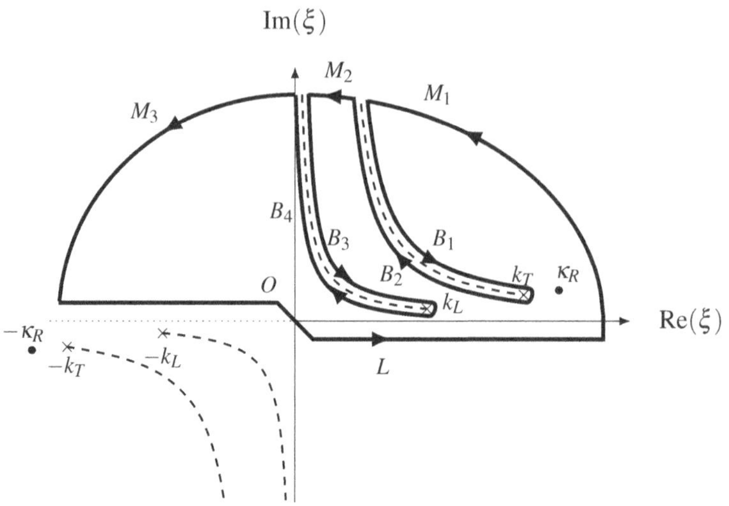

Figure 3.3.5 Path of integral for Eq. (3.3.38) in complex wavenumber plane for case where k_T and k_L are complex-valued parameters. The path of the branch line integral $B_1 + B_2$ is due to the presence of ν, and $B_3 + B_4$ is due to γ, where ν and γ are defined for Green's function in the wavenumber domain. The branch cut for $\sqrt{\xi}$, which is on the negative real axis, is also required. Note that the presence of the root of the Rayleigh function κ_R in the first quadrant affects Green's function.

following form:

$$
\begin{aligned}
G_{ij}(\boldsymbol{x},\boldsymbol{y}) \sim & -\frac{1}{4\pi}\sqrt{\frac{2}{\pi r}}C_{i\alpha}(\varphi)h_{\alpha l'}\sum_{m=-1}^{+1} e^{-(2m+1)\pi i/4}e^{im\varphi} \\
& \times\left[\int_{B_1+B_2}\sqrt{\xi}\exp(i\xi r)\,\hat{g}_{l'p'}(x_3,y_3,\xi)d\xi \right.\\
& \left. + \int_{B_3+B_4}\sqrt{\xi}\exp(i\xi r)\,\hat{g}_{l'p'}(x_3,y_3,\xi)d\xi\right]\hat{F}^m_{p'j} \\
& + \frac{1}{4\pi}\sqrt{\frac{2}{\pi r}}C_{i\alpha}(\varphi)h_{\alpha l'}\sum_{m=-1}^{+1} e^{-(2m+1)\pi i/4}e^{im\varphi} \\
& \times\left[2\pi i\sqrt{\kappa_R}\exp(i\kappa_R r)\operatorname*{Res}_{\xi=\kappa_R}\ \hat{g}_{l'p'}(x_3,y_3,\xi)\right]\hat{F}^m_{p'j}, \\
& (r\to\infty)
\end{aligned}
\tag{3.3.44}
$$

Note that the contribution of the residue due to the root of the Rayleigh function is found from the last term on the right-hand side of Eq. (3.3.44), which corresponds to the Rayleigh wave. We discover that the Rayleigh wave can be induced by a point source as its residue contribution.

3.3.3.3 Approximation of integral of Eq. (3.3.44)

Now, let us obtain the approximation of Eq. (3.3.44) by assuming that k_T, k_L, and κ_R are real-valued parameters. Figure 3.3.6 shows the paths of the branch line integral and the path around the root of the Rayleigh function for the case where k_T and k_L are real-valued parameters. For this case, the paths B_1 and B_3 as well as B_2 and B_4 contract. Nevertheless, we have to distinguish the paths B_1 and B_3 as well as B_2 and B_4 since the branch cuts for ν and γ define different Riemann sheets. Based on a similar discussion for the Sommerfeld integral, we have the following branches of ν and γ along the paths B_1 to B_4:

$$\mathrm{Im}(\nu_1) = \mathrm{Im}([\nu]_{\xi \in B_1}) > 0, \quad \mathrm{Im}(\nu_2) = \mathrm{Im}([\nu]_{\xi \in B_2}) < 0 \tag{3.3.45}$$

$$\mathrm{Im}(\gamma_3) = \mathrm{Im}([\gamma]_{\xi \in B_3}) > 0, \quad \mathrm{Im}(\gamma_4) = \mathrm{Im}([\gamma]_{\xi \in B_4}) < 0 \tag{3.3.46}$$

We also have to be aware of the branch of γ on the path B_1 and B_2 as well as ν on the path B_3 and B_4. We summarize the branches of ν and γ on the paths of B_1 and B_2 as well as B_3 and B_4 in Tables 3.3 and 3.4

Table 3.3
Branch of ν and γ on the paths of B_1 and B_2

ξ	$i\infty$ ··· 0 ··· k_L	··· k_T
ν_1	$\mathrm{Im}(\nu_1) > 0$	
ν_2	$\mathrm{Im}(\nu_2) < 0$	
γ	$\mathrm{Im}(\gamma) > 0$	$\mathrm{Re}(\gamma) > 0$

Table 3.4
Branch of ν and γ on the paths of B_3 and B_3

ξ	$i\infty$ ··· 0 ··· k_L
γ_3	$\mathrm{Im}(\gamma_3) > 0$
γ_4	$\mathrm{Im}(\gamma_4) < 0$
ν	$\mathrm{Im}(\nu) < 0$

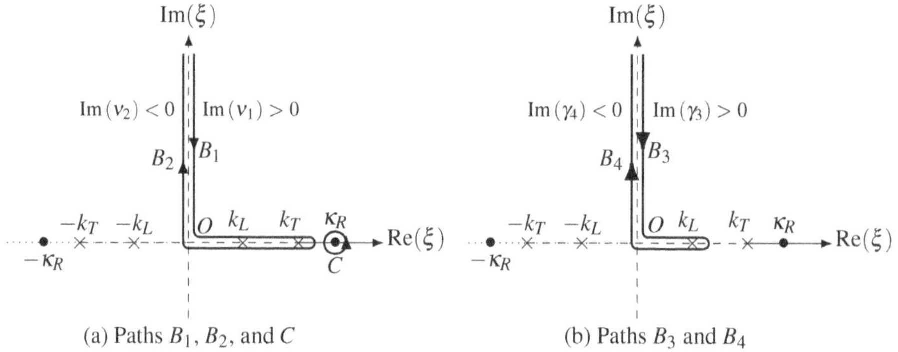

Figure 3.3.6 Paths of branch line integral and around pole in complex wavenumber plane for Green's function shown in Eq. (3.3.38) for case where k_T, k_L, and κ_R are real-valued parameters.

We express Eq. (3.3.44) in the following form to obtain its approximation:

$$G_{ij}(\boldsymbol{x},\boldsymbol{y}) \sim \frac{1}{4\pi}C_{i\alpha}(\varphi)h_{\alpha l'}\sum_{m=-1}^{+1} e^{-(2m+1)\pi i/4}e^{im\varphi} \times \left[J_{l'p'}^{(B_1B_2)}(r,x_3,y_3)+J_{l'p'}^{(B_3B_4)}(r,x_3,y_3)+2\pi i J_{l'p'}^{(C)}(r,x_3,y_3)\right]\hat{F}_{p'j}^{(m)} \tag{3.3.47}$$

where $J_{l'p'}^{(B_1B_2)}$, $J_{l'p'}^{(B_3B_4)}$, and $J_{l'p'}^{(C)}$ are defined by

$$\begin{aligned} J_{l'p'}^{(B_1B_2)}(r,x_3,y_3) &= -\sqrt{\frac{2}{\pi r}}\int_{B_1+B_2}\sqrt{\xi}\exp(i\xi r)\,\hat{g}_{l'p'}(x_3,y_3,\xi)d\xi \\ &= -\sqrt{\frac{2}{\pi r}}\int_0^{k_T}\sqrt{\xi}\exp(i\xi r)\left[\hat{g}_{l'p'}(x_3,y_3,\xi)\right]_{B_2}^{B_1}d\xi \\ &\quad +\sqrt{\frac{2}{\pi r}}\int_0^{i\infty}\sqrt{\xi}\exp(i\xi r)\left[\hat{g}_{l'p'}(x_3,y_3,\xi)\right]_{B_2}^{B_1}d\xi \end{aligned} \tag{3.3.48}$$

$$\begin{aligned} J_{l'p'}^{(B_3B_4)}(r,x_3,y_3) &= -\sqrt{\frac{2}{\pi r}}\int_{B_3+B_4}\sqrt{\xi}\exp(i\xi r)\,\hat{g}_{l'p'}(x_3,y_3,\xi)d\xi \\ &= -\sqrt{\frac{2}{\pi r}}\int_0^{k_L}\sqrt{\xi}\exp(i\xi r)\left[\hat{g}_{l'p'}(x_3,y_3,\xi)\right]_{B_4}^{B_3}d\xi \\ &\quad +\sqrt{\frac{2}{\pi r}}\int_0^{i\infty}\sqrt{\xi}\exp(i\xi r)\left[\hat{g}_{l'p'}(x_3,y_3,\xi)\right]_{B_4}^{B_3}d\xi \end{aligned} \tag{3.3.49}$$

$$J_{l'p'}^{(C)}(r,x_3,y_3) = \sqrt{\frac{2}{\pi r}}\sqrt{\kappa_R}\exp(i\kappa_R r)\operatorname*{Res}_{\xi=\kappa_R}\hat{g}_{l'p'}(x_3,y_3,\xi) \tag{3.3.50}$$

where the notation $[\,\cdot\,]_{B_j}^{B_i}$ in Eqs. (3.3.48) and (3.3.49) is expressed as

$$\left[\hat{g}_{l'p'}(x_3,y_3,\xi)\right]_{B_j}^{B_i} = \left[\hat{g}_{l'p'}(x_3,y_3,\xi)\right]_{\xi\in B_i} - \left[\hat{g}_{l'p'}(x_3,y_3,\xi)\right]_{\xi\in B_j} \quad (3.3.51)$$

Based on the same procedure as that used for the Sommerfeld integral, the approximation of the integral of Eq. (3.3.48) becomes

$$\begin{aligned} J_{l'p'}^{(B_1B_2)}(r,x_3,y_3) \sim\; & \frac{e^{ik_Tr}}{r^2}\hat{g}_{l'p'}^{(T_V)}(x_3,y_3) \\ & + \frac{e^{ik_Tr}}{r}\hat{g}_{l'p'}^{(T_H(1))}(x_3,y_3) + \frac{e^{ik_Tr}}{r^2}\hat{g}_{l'p'}^{(T_H(2))}(x_3,y_3) + \\ & + \frac{1}{r^2}\hat{g}_{l'p'}^{(T_O)}(x_3,y_3) + O(r^{-3}), \quad (r\to\infty) \end{aligned} \quad (3.3.52)$$

where $\hat{g}_{l'p'}^{(T_V)}$, $\hat{g}_{l'p'}^{(T_H(1))}$, $\hat{g}_{l'p'}^{(T_H(2))}$ and $\hat{g}_{l'p'}^{(T_O)}$ represent the effects of the SV, SH, and standing waves, respectively, which are obtained from

$$\begin{aligned} \hat{g}_{l'p'}^{(T_H(1))} &= \lim_{\xi\to k_T-0}(k_T-\xi)^{1/2}\left(\sqrt{2}e^{3\pi i/4}\sqrt{\xi}\left[\hat{g}_{l'p'}(x_3,y_3,\xi)\right]_{B_2}^{B_1}\right) \\ \hat{g}_{l'p'}^{(T_H(2))} &= \lim_{\xi\to k_T-0}(k_T-\xi)^{-1/2}e^{\pi i/4}\frac{1}{\sqrt{2}}\left(\sqrt{\xi}\left[\hat{g}_{l'p'}(x_3,y_3,\xi)\right]_{B_2}^{B_1}\right. \\ & \quad \left. -\frac{1}{\sqrt{2}}e^{-3\pi i/4}(k_T-\xi)^{-1/2}\hat{g}_{l'p'}^{(T_H(1))}\right), \quad (l'=3,p'=3) \\ \hat{g}_{l'p'}^{(T_V)} &= \lim_{\xi\to k_T-0}(k_T-\xi)^{-1/2}e^{\pi i/4}\frac{1}{\sqrt{2}}\sqrt{\xi}\left[\hat{g}_{l'p'}(x_3,y_3,\xi)\right]_{B_2}^{B_1}, \quad (1\le l',p'\le 2) \\ \hat{g}_{l'p'}^{(T_O)} &= \lim_{\xi\to+i0}\frac{e^{3\pi i/4}}{\sqrt{2}}\left[\hat{g}_{l'p'}(x_3,y_3,\xi)\right]_{B_2}^{B_1} \end{aligned} \quad (3.3.53)$$

Likewise, the approximation of Eq. (3.3.49) becomes

$$J_{l'p'}^{(B_3B_4)}(r,x_3,y_3) \sim \frac{e^{ik_Lr}}{r^2}\hat{g}_{l'p'}^{(L)}(x_3,y_3) + \frac{1}{r^2}\hat{g}_{l'p'}^{(L_O)}(x_3,y_3), \quad (r\to\infty) \quad (3.3.54)$$

where $\hat{g}_{l'p'}^{(L)}$ and $\hat{g}_{l'p'}^{L_O}$ represent the effects of the P wave and the contribution from the integral along the imaginary axis obtained from

$$\begin{aligned} \hat{g}_{l'p'}^{(L)} &= \lim_{\xi\to k_L-0}(k_L-\xi)^{-1/2}e^{\pi i/4}\frac{1}{\sqrt{2}}\sqrt{\xi}\left[\hat{g}_{l'p'}(x_3,y_3,\xi)\right]_{B_4}^{B_4}, \quad (1\le l'p'\le 2) \\ \hat{g}_{l'p'}^{(L_O)} &= \lim_{\xi\to+i0}\frac{e^{3\pi i/4}}{\sqrt{2}}\left[\hat{g}_{l'p'}(x_3,y_3,\xi)\right]_{B_4}^{B_3} \end{aligned} \quad (3.3.55)$$

Based on Eqs. (3.3.47), (3.3.50), (3.3.52), and (3.3.54), the approximated result for Green's function for an elastic half-space is in the following form:

$$\begin{aligned} G_{ij}(\boldsymbol{x},\boldsymbol{y}) \sim & \frac{1}{4\pi} C_{i\alpha}(\varphi) h_{\alpha l'} \sum_{m=-1}^{1} e^{-(2m+1)\pi i/4} e^{im\varphi} \\ & \times \left[\frac{e^{i\kappa_R r}}{\sqrt{r}} \hat{g}^{(R)}_{l'p'} + \frac{e^{ik_T r}}{r} \left(\hat{g}^{(T_H(1))}_{l'p'} + \frac{1}{r} \hat{g}^{(T_H(2))}_{l'p'} \right) + \frac{e^{ik_T r}}{r^2} \hat{g}^{(T_V)}_{l'p'} \right. \\ & \left. + \frac{e^{ik_L r}}{r^2} \hat{g}^{(L)}_{l'p'} + \frac{1}{r^2} \hat{g}^{(O)}_{l'p'} \right] \hat{F}^m_{p'j} + O(r^{-3}), \quad (r \to \infty) \end{aligned} \tag{3.3.56}$$

where

$$\begin{aligned} \hat{g}^{(O)}_{l'p'} &= \hat{g}^{(T_O)}_{l'p'} + \hat{g}^{(L_O)}_{l'p'} \\ \hat{g}^{(R)}_{l'p'} &= 2\sqrt{2\pi\kappa_R}\, i \operatorname*{Res}_{\xi=\kappa_R} \hat{g}_{l'p'}(x_3, y_3, \xi) \end{aligned} \tag{3.3.57}$$

Equation (3.3.56) shows the geometrical decay factor for each component of a wave. As can be seen, the Rayleigh, SH, and P-SV waves have decay factors of $r^{-1/2}$, r^{-1}, and r^{-2}. It is interesting that the decay of the P-SV wave (r^{-2}) is faster than that of the SH wave (r^{-1}), which is on the same order as that of a scalar wave in 3D full space. The Rayleigh wave has the slowest decay, which shows that it tends to propagate to the far field. The contribution from the integral along the branch cuts on the imaginary axis has a decay factor of r^{-2}. Note that this contribution shows non-propagating waves toward the horizontal direction.

At the end of this section, we have to be aware that the method of the branch line integral clarified wave propagation toward the horizontal direction together with decaying factor of the P, S and the Rayleigh waves. Important points are that the P and S waves are due to the integral along the branch line, while the Rayleigh wave is due to the residue contribution from the root of the Rayleigh function. The explicit forms of $\hat{g}^{(T_V)}_{l'p'}$, $\hat{g}^{(T_H)}_{l'p'}$, $\hat{g}^{(L)}_{l'p'}$, $\hat{g}^{(R)}_{l'p'}$, $\hat{g}^{(T_O)}_{l'p'}$ and $\hat{g}^{(L_O)}_{l'p'}$ are summarized in Note 3.5, for the case of $x_3 = 0$. The derivation of the explicit forms of $\hat{g}^{(R)}_{l'p'}$ is not very complicated for the case of $x_3 = 0$. For example, according to Eq. (3.2.53), we have

$$\operatorname*{Res}_{\xi=\kappa_R} \hat{g}_{1'1'}(0, y_3, \xi) = \frac{1}{\mu F'(\kappa_R)} \left[\gamma(\xi^2 + \nu^2) e^{-\gamma y_3} - 2\xi^2 \gamma e^{-\nu y_3} \right]_{\xi=\kappa_R} \tag{3.3.58}$$

Therefore, we can omit the details of the explicit forms of $\hat{g}^{(R)}_{l'p'}$ herein.

Note 3.5 Explicit forms of $\hat{g}^{(T_V)}_{l'p'}, \hat{g}^{(T_H)}_{l'p'}, \hat{g}^{(L)}_{l'p'}, \hat{g}^{(T_O)}_{l'p'}$ and $\hat{g}^{L_O}_{l'p'}$

We summarize the explicit forms of $\hat{g}^{(T_V)}_{l'p'}, \hat{g}^{(T_H)}_{l'p'}, \hat{g}^{(L)}_{l'p'}, \hat{g}^{(T_O)}_{l'p'}$ and $\hat{g}^{(L_O)}_{l'p'}$ given in Eqs. (3.3.53) and (3.3.55) for the case of $x_3 = 0$ as

$$
\begin{aligned}
\hat{g}^{(T_H(1))}_{3'3'} &= \frac{2e^{\pi i/4}}{\mu} \\
\hat{g}^{(T_H(2))}_{3'3'} &= \frac{e^{3\pi i/4}}{\mu}\left(\frac{1}{4k_T} + k_T y_3^2\right) \\
\hat{g}^{(T_V)}_{1'1'} &= \frac{4e^{3\pi i/4}}{\mu k_T^3}\left(\gamma k_T^2 y_3 + 2\gamma^2 e^{-\gamma y_3} - 4\gamma^2\right) \\
\hat{g}^{(T_V)}_{1'2'} &= \frac{4e^{3\pi i/4}}{\mu k_T^2}\left(\gamma - 2\gamma e^{-\gamma y_3}\right) \\
\hat{g}^{(T_V)}_{2'1'} &= \frac{2e^{3\pi i/4}}{\mu k_T^2}\left(2\gamma e^{-\gamma y_3} + k_T^2 y_3 - 4\gamma\right) \\
\hat{g}^{(T_V)}_{2'2'} &= \frac{2e^{3\pi i/4}}{\mu k_T}\left(-2e^{-\gamma y_3} + 1\right) \\
\hat{g}^{(T_O)}_{2'2'} &= e^{\pi i/4}\frac{\sqrt{2}\cos_T y_3}{\mu k_T} \\
\hat{g}^{(T_O)}_{3'3'} &= e^{\pi i/4}\frac{\sqrt{2}\cos k_T y_3}{\mu k_T} \\
\hat{g}^{(L)}_{1'1'} &= \frac{2e^{3\pi i/4}}{\mu(2k_L^2 - k_T^2)^2}\left(k_L(2k_L^2 - k_T^2) - 2k_L^3 e^{-\nu y_3}\right) \\
\hat{g}^{(L)}_{1'2'} &= \frac{2e^{3\pi i/4}}{\mu(2k_L^2 - k_T)^3}\left((2k_L^2 - k_T^2)^2 k_L^2 y_3 + 2k_L^2 \nu(2k_L^2 - k_T^2)e^{-\gamma y_3} - 4k_L^4 \nu\right) \\
\hat{g}^{(L)}_{2'1'} &= \frac{4e^{3\pi i/4}}{\mu(2k_L^2 - k_T)^3}\left(\nu k_L^2(2k_L^2 - k_T^2) - 2\nu k_L^4 e^{-\nu y_3}\right) \\
\hat{g}^{(L)}_{2'2'} &= \frac{4e^{3\pi i/4}}{\mu(2k_L^2 - k_T^2)^4}\left(\nu k_L^3(2k_L^2 - k_T^2)^2 y_3\right. \\
&\qquad \left. - 4\nu^2 k_L^5 + 2k_L^3 \nu^2(2k_L^2 - k_T^2)e^{-\gamma y_3}\right) \\
\hat{g}^{(L_O)}_{2'2'} &= e^{\pi i/4}\frac{\sqrt{2}k_L \cos k_L y_3}{\mu k_T^2}
\end{aligned}
$$

3.4 EVALUATION OF GREEN'S FUNCTION IN COMPLEX WAVENUMBER PLANE USING STEEPEST DESCENT PATH METHOD

3.4.1 CONCEPT OF STEEPEST DESCENT PATH

As an alternative to the approximation method based on the branch line integral, we apply the steepest descent path method. To provide a brief overview of this method, we first investigate an integral of the following type:

$$J = \int_{-\infty}^{+\infty} Q(\xi)\exp(f(\xi))d\xi \tag{3.4.1}$$

where f and Q are regular functions in the complex wavenumber plane. We watch a point in the wavenumber space that satisfies

$$f'(\xi_s) = 0 \tag{3.4.2}$$

where ξ_s is called the *saddle point* for $f(\xi)$ according to the theory of analytic functions. To facilitate the discussion, we first describe the basic concept of the saddle point. Figure 3.4.1 shows the contour lines for the real and imaginary parts of a regular function defined by

$$f_a(\xi) = A(\xi - \xi_s)^2 e^{-i\delta} \tag{3.4.3}$$

where A and δ are real-valued constants and it is assumed that $A > 0$. It is clear that $\xi = \xi_s$ is the saddle point for $f_a(\xi)$. For the saddle point, we can define the following two sets:

$$\begin{aligned} S_1 &= \{\xi \in \mathbb{C}\,|\mathrm{Re}\left(f_a(\xi)\right) \le \mathrm{Re}\left(f_a(\xi_s)\right) = 0,\ \ \mathrm{Im}\left(f_a(\xi)\right) = 0\} \\ S_2 &= \{\xi \in \mathbb{C}\,|\mathrm{Re}\left(f_a(\xi)\right) \ge \mathrm{Re}\left(f_a(\xi_s)\right) = 0,\ \ \mathrm{Im}\left(f_a(\xi)\right) = 0\} \end{aligned} \tag{3.4.4}$$

which are found to be straight lines that cross the saddle point and are orthogonal to each other. Along the line of the set S_1 (S_2), $\mathrm{Re}(f_a)$ takes the maximum (minimum) value at the saddle point. That is, $\xi = \xi_a$ does not provide the local maximum or minimum for $\mathrm{Re}(f_a)$ or $\mathrm{Im}(f_a)$ even though $f_a'(\xi_s) = 0$. This is a significant property of the saddle point for a complex-valued analytic function. The contour lines for the real and imaginary parts of f_a show that the line S_1 can be called the *steepest descent path* since $\mathrm{Re}(f_a)$ decreases most rapidly from the saddle point along the line S_1.

The approximation method for Eq. (3.4.1) is developed using the steepest descent path that crosses the saddle point for $f(\xi)$. We use a Taylor series expansion around the saddle point ξ_s and approximate $f(\xi)$ by truncating the series as:

$$f(\xi) \sim f(\xi_s) + (1/2)(\xi - \xi_s)^2 f''(\xi_s) \tag{3.4.5}$$

which means that the approximation method uses the information of the integrand close to the saddle point. We approximate Eq. (3.4.1) as

$$\begin{aligned} J &= \int_{-\infty}^{\infty} Q(\xi)\exp(f(\xi))d\xi \\ &\sim Q(\xi_s)e^{f(\xi_s)}\int_{S_1} \exp((1/2)(\xi - \xi_s)^2 f''(\xi_s))d\xi \end{aligned} \tag{3.4.6}$$

where S_1 is the steepest descent path defined by

$$\begin{aligned} \mathrm{Re}((\xi - \xi_s)^2 f''(\xi_s)) &\le 0 \\ \mathrm{Im}((\xi - \xi_s)^2 f''(\xi_s)) &= 0 \end{aligned} \tag{3.4.7}$$

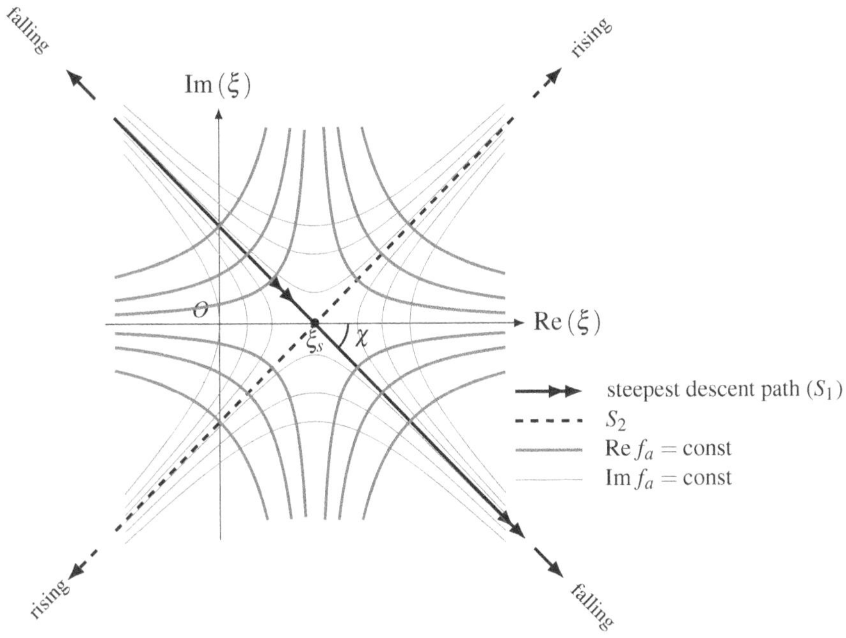

Figure 3.4.1 Contour lines for Re $f_a(\xi)$ and Im $f_a(\xi)$, where $f_a(\xi) = A(\xi - \xi_s)^2 e^{-i\delta}$. We have Re $\left(f_a(\xi)\right) \leq$ Re $\left(f_a(\xi_s)\right)$ for $\xi \in S_1$ and Re $\left(f_a(\xi)\right) \geq$ Re$\left(f_a(\xi_s)\right)$ for $\xi \in S_2$. This characterizes the properties of the saddle point for $\xi = \xi_s$. The line S_1 is called the steepest descent path. The angle χ between Re(ξ) axis and S_1 axis, which characterizes the forward direction of the steepest descent path, has to be chosen such that $-\pi/2 < \chi < \pi/2$.

Note that the steepest descent path becomes a straight line, as discussed for Fig. 3.4.1. To carry out the integral of Eq. (3.4.6), we need to determine the direction of the steepest descent path. For this purpose, let u be a real parameter and assume that

$$\xi - \xi_s = u\exp(i\chi), \quad f''(\xi_s) = |f''(\xi_s)|\exp(-i\delta) \tag{3.4.8}$$

Based on the condition of the steepest descent path defined by Eq. (3.4.7), χ and δ have to satisfy

$$2\chi - \delta = \pm\pi \tag{3.4.9}$$

We choose the solution for χ in the range of $-\pi/2 < \chi < \pi/2$, as explained in Fig. 3.4.1. We then obtain the following approximation of Eq. (3.4.6):

$$\begin{aligned} J &\sim Q(\xi_s)e^{f(\xi_s)}e^{i\chi}\int_{-\infty}^{+\infty}\exp\Big(-(1/2)u^2|f''(\xi_s)|\Big)du \\ &= \sqrt{\frac{2\pi}{|f''(\xi_s)|}}Q(\xi_s)e^{f(\xi_s)}e^{i\chi} \end{aligned} \tag{3.4.10}$$

The above procedure used to obtain the approximation of the integral is called the steepest decent path method. Equation (3.4.10) is the result of its application.

3.4.2 APPLICATION OF STEEPEST DESCENT PATH METHOD TO SOMMERFELD INTEGRAL

Our task now is to apply the steepest descent path method to the Sommerfeld integral. Our starting point is Eq. (3.3.7), which is derived from the Sommerfeld integral. We rewrite Eq. (3.3.7) in the following form:

$$F_s(\xi) \sim \frac{1}{8\pi}\sqrt{\frac{2}{\pi r}}e^{-i\pi/4}\int_{-\infty}^{+\infty} Q(\xi)\exp(f(\xi))d\xi \tag{3.4.11}$$

where $Q(\xi)$ and $f(\xi)$ are expressed as

$$\begin{aligned} Q(\xi) &= \frac{\sqrt{\xi}}{\nu} \\ f(\xi) &= -\nu|x_3 - y_3| + i\xi r \end{aligned} \tag{3.4.12}$$

Recall that ν is defined by

$$\nu = \begin{cases} \sqrt{\xi^2 - k^2} & (|\xi| \geq k) \\ -i\sqrt{k^2 - \xi^2} & (|\xi| < k) \end{cases} \tag{3.4.13}$$

Since the derivative of $f(\xi)$ becomes

$$f'(\xi) = -\frac{\xi}{\sqrt{\xi^2 - k^2}}|x_3 - y_3| + ir \tag{3.4.14}$$

the saddle point for $f(\xi)$ is expressed as

$$\xi_s = \pm\frac{r}{R}k = \pm k\sin\theta \tag{3.4.15}$$

where θ and R are defined by

$$\sin\theta = \frac{r}{R} \tag{3.4.16}$$

$$R = \sqrt{r^2 + |x_3 - y_3|^2} \tag{3.4.17}$$

The angle θ defined by R and r is shown in Fig. 3.4.2.

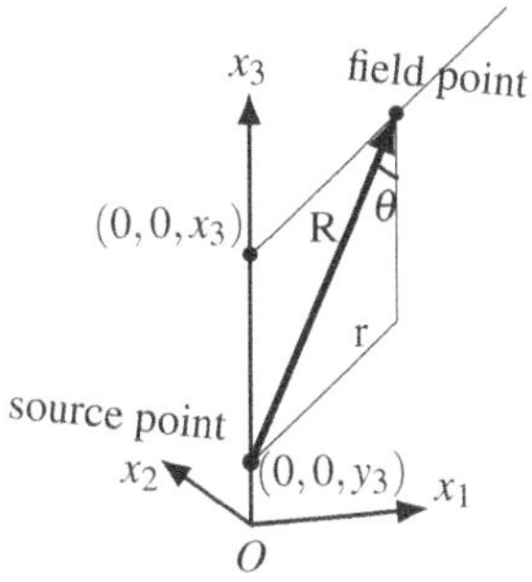

Figure 3.4.2 Geometry of source point and field point for the Sommerfeld integral. Utilizing Eq. (3.4.15), the saddle point is determined as $\xi_s = k\sin\theta$, with $\sin\theta = r/R$. Additionally, it is evident that $R = r\sin\theta + |x_3 - y_3|\cos\theta$, where R and r are the distance and horizontal range between the source and field point, respectively.

Based on Eq. (3.4.10), we need $f(\xi_s)$, $f''(\xi_s)$, and $Q(\xi_s)$, which are expressed as

$$\begin{aligned} f(\xi_s) &= i\sqrt{k^2 - k^2\sin\theta}\,|x_3 - y_3| + ikr\sin\theta \\ &= ikR \\ f''(\xi_s) &= \frac{k^2}{\sqrt{\xi_s^2 - k^2}^3}|x_3 - y_3| \\ &= (-i)\frac{|x_3 - y_3|}{k\sqrt{1 - r^2/R^2}^3} \\ Q(\xi_s) &= \frac{1}{-ik\cos\theta}\sqrt{\frac{r}{R}k} \end{aligned} \tag{3.4.18}$$

As a result, Eq. (3.4.11) becomes

$$\begin{aligned} F_s(\xi) &\sim \frac{1}{8\pi}\sqrt{\frac{2}{\pi r}}e^{-i\pi/4}Q(\xi_s)\exp(f(\xi_s)) \\ &\quad \times \int_{S_1} \exp\left((1/2)(\xi - \xi_s)^2 f''(\xi_s)\right)d\xi \\ &\sim \frac{1}{8\pi}\sqrt{\frac{2}{\pi r}}e^{-i\pi/4}\frac{e^{ikR}}{-ik\cos\theta}\sqrt{\frac{r}{R}k} \\ &\quad \times \int_{S_1} \exp\left((1/2)(\xi - \xi_s)^2 f''(\xi_s)\right)d\xi \end{aligned} \tag{3.4.19}$$

Let us determine the direction of the steepest descent path, which is characterized by the angle χ in Fig. 3.4.2. Let $\xi - \xi_s = u\exp(i\chi)$, where u is a real-valued parameter. Then, we have

$$\mathrm{Arg}\left((\xi - \xi_s)^2 f''(\xi_s)\right) = 2\chi - \frac{\pi}{2} = \pm\pi \tag{3.4.20}$$

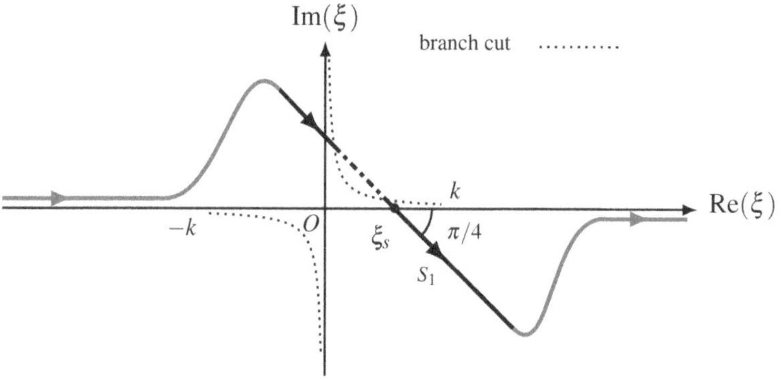

Figure 3.4.3 Steepest descent path for Sommerfeld integral for Eq. (3.4.19). Distortion is applied to the path of the integral so that it includes the path S_1.

since $\mathrm{Arg}f''(\xi_s) = -\pi/2$ based on Eq. (3.4.18). Equation (3.4.20) yields $\chi = -\pi/4$ or $\chi = (3/4)\pi$. Here, we employ $\chi = -\pi/4$. We can approximate the result of the integral shown in Eq. (3.4.19). The steepest descent path for Eq. (3.4.19) is shown in Fig. 3.4.3. The path of the integral is set so that it crosses the saddle point at an angle of $\chi = -\pi/4$ and includes S_1. The distorted path is found to cross the Riemann cut, and as a result the whole path cannot be in the permissible sheet. Nevertheless, it is clear the integrand does not diverge. In addition, the saddle point is located below the branch cut and we find that

$$\mathrm{Im}(\sqrt{\xi_s^2 - k^2}) < 0 \tag{3.4.21}$$

which agrees with Eq. (3.4.13). Now, we have the approximated result for the Sommerfeld integral as

$$\begin{aligned} F_s(\xi) &\sim \frac{1}{8\pi}\sqrt{\frac{2}{\pi r}}e^{-i\pi/4}\frac{e^{ikR}}{-ik\cos\theta}\sqrt{\frac{r}{R}}k \\ &\quad \times \int_{-\infty}^{\infty} \exp\left(-(1/2)u^2\frac{R^3}{k|x_3 - y_3|^2}\right)e^{-\pi i/4}du \\ &= \frac{1}{4\pi R}\exp(ikR) \end{aligned} \tag{3.4.22}$$

The application of the steepest descent path method to the Sommerfeld integral yields $\exp(ikR)/(4\pi R)$. Although this result might be trivial, it tells us the properties of the steepest descent path method. Later, we discuss the result for Green's function for an elastic half-space, which has a much more complicated form than that for the Sommerfeld integral.

The above procedure is based on the truncation of the Taylor series of $f(\xi)$ up to the second term. The steepest descent path can also be set without truncating the

Taylor series. At this point, let us investigate the steepest descent path for the case where the Taylor series is not truncated since the discussion will be used for Green's function for an elastic half-space with respect to the treatment of the Rayleigh pole.

In general, the steepest descent path has to satisfy the following equation:

$$f(\xi) - f(\xi_s) = -X^2 \tag{3.4.23}$$

where X is a real-valued parameter if the Taylor series is not truncated. Equation (3.4.23) shows that by using the real-valued parameter X, a point ξ on the steepest descent path can be determined. For the Sommerfeld integral, the equation for the steepest descent path becomes

$$-\sqrt{\xi^2 - k^2}\,|x_3 - y_3| + ir\xi = -X^2 + ikR \tag{3.4.24}$$

We can solve the equation with respect to ξ using the given parameter X in the following form:

$$\xi = (ip + k)\sin\theta \pm \sqrt{p^2 - 2pik}\,\cos\theta \tag{3.4.25}$$

where the parameter p is defined by

$$p = \frac{X^2}{R}\left(= \frac{X^2}{\sqrt{r^2 + |x_3 - y_3|^2}}\right) \tag{3.4.26}$$

Recall that θ is defined by $\sin\theta = r/R$. It is found from Eq. (3.4.25) that the steepest descent path has an asymptote of

$$\xi = \pm p\cos\theta + ip\sin\theta, \quad (p > 0) \tag{3.4.27}$$

which is obtained from Eq. (3.4.25) for $p \to \infty$. Figure 3.4.4 shows the steepest descent path and its asymptote for the Sommerfeld integral. As can be seen, the angle between the asymptote and the Re(ξ) axis is θ, which is also found from Eq. (3.4.25). In addition, we find that the steepest descent is no longer a straight line. We can trace the steepest descent path using various values of p. The relationship among the values of p, ξ and points on the steepest descent path is summarized in Table 3.5. As can be seen in Fig. 3.4.4, the steepest descent path crosses the imaginary axis at point A, where $\xi = ik\tan\theta$. After crossing the imaginary axis, the path crosses the Riemann cut. The steepest descent path crosses the real axis of the complex wavenumber plane at point B, where $\xi = k\sin\theta$. This point is the saddle point. After crossing the real axis at point B, the steepest descent path returns to the real axis at point C, where $\xi = k/\sin\theta$. After crossing point C, the path approaches the asymptote. Point C is important for a later discussion to examine the effects of the Rayleigh pole on Green's function for an elastic half-space.

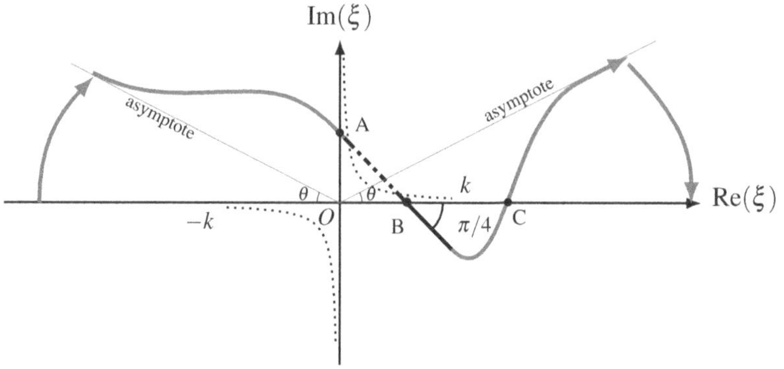

Figure 3.4.4 Steepest descent path for Sommerfeld integral for case where Taylor series is not truncated. The steepest descent path is no longer a straight line. It crosses the real axis of the complex wavenumber plane two times and then approaches the asymptote.

Table 3.5

Relationship among values p, ξ and points on the steepest descent path shown in Fig. 3.4.4. Note that the point B is the saddle point.

p	ξ	Fig. 3.4.4
$k\sin\theta\tan\theta$	$ik\tan\theta$	A
0	$k\sin\theta$	B
$k\cos\theta\cot\theta$	$k/\sin\theta$	C

3.4.3 APPLICATION OF STEEPEST DESCENT PATH METHOD TO GREEN'S FUNCTION FOR ELASTIC HALF-SPACE

3.4.3.1 Preparation

Now, let us apply the steepest descent path method to Green's function for an elastic half-space [2]. We restrict our discussion to Green's function for the response at the free surface due to an interior point source. Figure 3.4.5 shows the location of the point source and the field point for Green's function, which are used for the application of the steepest descent path method to Green's function. We modify the

[2]The current discussion incorporates findings from the published article in 'Pseudo-projection approach to reconstruct locations of point-like scatterers characterized by Lamé parameters and mass densities in an elastic half-space' by Touhei, T., published in the International Journal of Solids and Structures (2019, Vol.169, pages 187-204, Copyright Elsevier). Several modifications have been made to the content of this published article.

representation of Green's function presented in Eq. (3.3.38) as follows:

$$\begin{aligned} G_{ij}(\boldsymbol{x},\boldsymbol{y}) \sim & \frac{1}{4\pi}\sqrt{\frac{2}{\pi r}}C_{i\alpha}(\varphi)h_{\alpha l'}\sum_{m=-1}^{+1} e^{-(2m+1)\pi i/4}e^{im\varphi} \\ & \times \int_{-\infty}^{+\infty}\left[\frac{\sqrt{\xi}}{\nu}Q^{(T_V)}_{l'p'}(\xi)\exp(f_T(\xi)) + \frac{\sqrt{\xi}}{\nu}Q^{(T_H)}_{l'p'}(\xi)\exp(f_T(\xi)) \right. \\ & \left. +\frac{\sqrt{\xi}}{\gamma}Q^{(L)}_{l'p'}(\xi)\exp(f_L(\xi))\right] d\xi\, F^{(m)}_{p'j} \end{aligned} \tag{3.4.28}$$

where f_T and f_L are defined by

$$\begin{aligned} f_T(\xi) &= -\nu y_3 + i\xi r \\ f_L(\xi) &= -\gamma y_3 + i\xi r \end{aligned} \tag{3.4.29}$$

and $Q^{(T_V)}_{l'p'}(\xi)$, $Q^{(T_H)}_{l'p'}(\xi)$, and $Q^{(L)}_{l'p'}(\xi)$ are the arrays for the functions for the SV, SH, and P waves, respectively, derived from Green's function in the wavenumber domain. They are expressed as

$$\begin{aligned} Q^{(T_V)}_{l'p'}(\xi) &= \frac{1}{\mu F(\xi)}\begin{bmatrix} -2\nu\xi^2\gamma & 2\gamma\nu^2\xi & 0 \\ -\xi\nu(\xi^2+\nu^2) & \nu^2(\xi^2+\nu^2) & 0 \\ 0 & 0 & 0 \end{bmatrix} \\ Q^{(T_H)}_{l'p'}(\xi) &= \frac{1}{\mu}\begin{bmatrix} 0 & 0 & 0 \\ 0 & 0 & 0 \\ 0 & 0 & 1 \end{bmatrix} \\ Q^{(L)}_{l'p'}(\xi) &= \frac{1}{\mu F(\xi)}\begin{bmatrix} \gamma^2(\xi^2+\nu^2) & -\xi\gamma(\xi^2+\nu^2) & 0 \\ 2\xi\gamma^2\nu & -2\nu\gamma\xi^2 & 0 \\ 0 & 0 & 0 \end{bmatrix} \end{aligned} \tag{3.4.30}$$

Recall that ν and γ in Eqs. (3.4.29) and (3.4.30) are given by

$$\begin{aligned} \nu &= \sqrt{\xi^2 - k_T^2} \\ \gamma &= \sqrt{\xi^2 - k_L^2} \end{aligned}$$

Equation (3.4.30) is the result obtained from Eq. (3.2.53).

Equation (3.4.28) shows that the steepest descent paths and the saddle points for Green's function have to be constructed for S and P waves separately. The saddle points for the S and P waves have to satisfy the following equations:

$$f'_T(\xi_T) = 0 \tag{3.4.31}$$

$$f'_L(\xi_L) = 0 \tag{3.4.32}$$

where ξ_T and ξ_L are the saddle points for the S and P waves, respectively. We have

$$\xi_T = k_T\sin\theta, \quad \xi_L = k_L\sin\theta \tag{3.4.33}$$

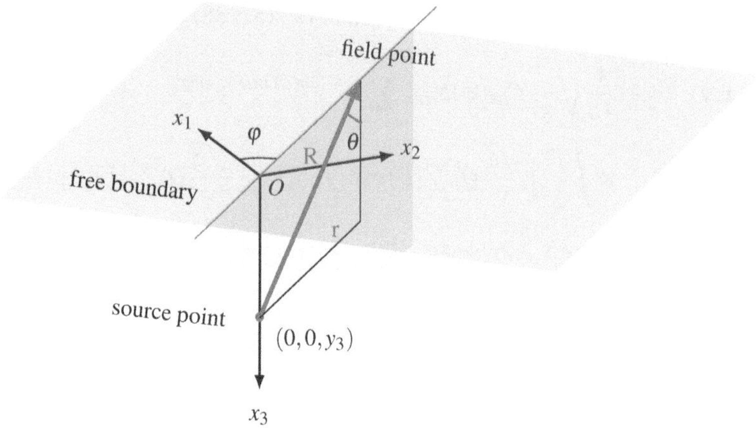

Figure 3.4.5 Interior source point and field point at free surface for Green's function. The depth of the source point is y_3 and the horizontal range from the source point to the field point is r, from which the incidence angle θ is defined. The saddle points for the P and S waves, ξ_L and ξ_L, are expressed as $\xi_L = k_L \sin\theta$ and $\xi_T = k_T \sin\theta$, respectively.

Note that the angle θ is defined by

$$\sin\theta = r/R = r/\sqrt{r^2 + y_3^2} \tag{3.4.34}$$

as shown in Fig. 3.4.5. For the construction of the steepest descent paths for the approximation of the integral, we employ Eqs. (3.4.5) and (3.4.6). Therefore, the steepest descent paths become straight lines. For the steepest descent paths defined above, it is not very important to define the permissible sheets of the complex wavenumber plane since it is not possible to set the whole steepest descent path in the permissible sheet, as discussed for the Sommerfeld integral. We also have to be aware that the approximated result is determined by information on the integrand close to the saddle point. Therefore, the branch cuts we employ are those shown in Fig. 3.4.6, to which the steepest descent paths are added.

The saddle point for the P wave ξ_L is always $\xi_L < k_L$. Therefore, the steepest descent path for the P wave we employ is always that shown in Fig. 3.4.6 (a). On the other hand, there is a case where the saddle point for the S wave is located at $k_L < \xi_T < k_T$. For this case, we have to employ the steepest descent path shown in Fig. 3.4.6 (b), where the branch line integral around k_L becomes necessary. Note that the branch line integral around k_L is only for the SV wave since the SH wavefunction does not require the branch cut for the P wave. The condition $\xi_T > k_L$ is equivalent to

$$\theta > \theta_c = \sin^{-1}\frac{k_L}{k_T} = \tan^{-1}\frac{k_L}{\sqrt{k_T^2 - k_L^2}} \tag{3.4.35}$$

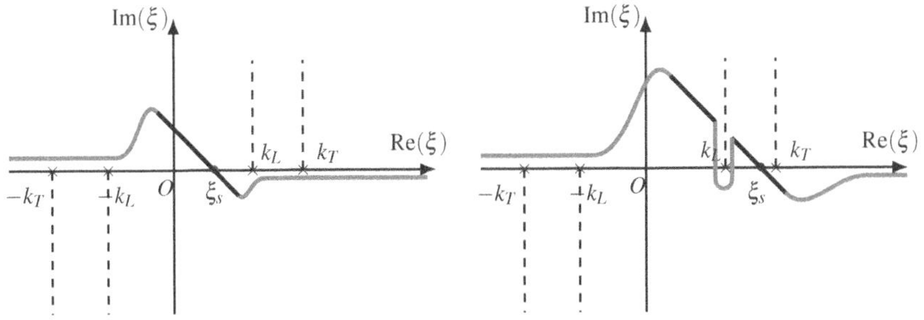

(a) Path of integral (case 1) (b) Path of integral (case 2)

Figure 3.4.6 Steepest descent paths. ξ_s, which represents ξ_L or ξ_T, is the saddle point. The path of the integral for cases 1 and 2 describes the case where $\xi_s < k_L$ and $k_L < \xi_s (= \xi_T)$, respectively. The path for case 2 is only for the SV wave, for which the branch line integral is required.

where θ_c is the critical angle, which is defined by Eq. (3.1.45). As discussed for Eq. (3.1.45), the S-P wave is generated for the case where the incidence angle of the SV wave exceeds the critical angle, as explained in Fig. 3.4.7.

At this point, note that the treatment of the Rayleigh pole is unclear as long as we employ the paths of the integral in the complex wavenumber plane shown in Fig. 3.4.6. The treatment of the Rayleigh pole using the method for the steepest descent path based on Eq. (3.4.23) is discussed later.

3.4.3.2 Results of application of steepest descent path method

Let the results of the integration along the steepest decent path for Green's function be denoted by $G_{ij}^{(S)}$. The integration process along the steepest descent path is almost the same as that for the Sommerfeld integral we have previously discussed. Therefore, instead of presenting the details of the process, we summarize the results. We

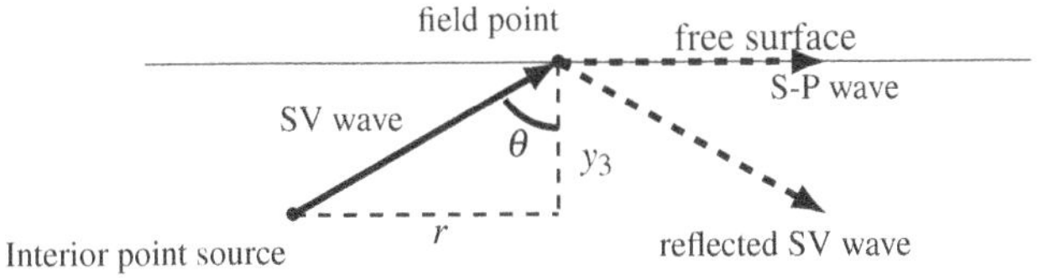

Figure 3.4.7 S-P wave propagating along free surface for case where incidence angle θ for SV wave is larger than critical angle θ_c. This wave corresponds to the branch line integral along k_L shown in Fig. 3.4.6.

can express the results of the steepest descent path method in the following form:

$$\begin{aligned} G_{ij}^{(S)}(\boldsymbol{x},\boldsymbol{y}) &= \frac{\exp(ik_T R)}{4\pi R}\left(D_{ij}^{(T_V)}(\theta,\varphi)+D_{ij}^{(T_H)}(\theta,\varphi)\right)+\frac{\exp(ik_L R)}{4\pi R}D_{ij}^{(L)}(\theta,\varphi) \\ &\quad +O(R^{-2}) \end{aligned} \tag{3.4.36}$$

where θ and φ are the angles defined in Fig. 3.4.5 and R is the distance between $\boldsymbol{x}$ and $\boldsymbol{y}$. $D_{ij}^{(T_V)}$, $D_{ij}^{(T_H)}$, and $D_{ij}^{(L)}$ are the directivity tensors for the SV, SH, and P waves, respectively. We can factorize the directivity tensors in the following form:

$$D_{ij}^{(\#)}(\theta,\varphi) = A^{(\#)}(\theta)\,W_i^{(\#)}(\theta,\varphi)\,V_j^{(\#)}(\theta,\varphi), \quad (\# = T_V,\ T_H,\ L) \tag{3.4.37}$$

where the superscript # takes the symbol T_V, T_H, or L. In addition, $V_j^{\#}$ and $W_i^{\#}$ are vectors that constitute the directivity tensor and $A^{\#}$ describes the amplitude. The components of the vector $V_j^{\#}$ are

$$\begin{aligned} V_j^{(T_V)}(\theta,\varphi) &= \left(\cos\varphi\cos\theta\sin\theta,\ \ \sin\varphi\cos\theta\sin\theta,\ \ \sin^2\theta \right) \\ V_j^{(T_H)}(\theta,\varphi) &= \left(\sin\varphi\sin\theta,\ \ -\cos\varphi\sin\theta,\ \ 0 \right) \\ V_j^{(L)}(\theta,\varphi) &= \left(\cos\varphi\sin\theta,\ \ \sin\varphi\sin\theta,\ \ -\cos\theta \right) \end{aligned} \tag{3.4.38}$$

and those of the vector $W_i^{\#}$ are

$$\begin{aligned} W_i^{(T_V)}(\theta,\varphi) &= \left(\cos\varphi\cos\theta\sin\theta,\ \ \sin\varphi\cos\theta\sin\theta,\ \ \kappa^{(T_V)}(\theta) \right) \\ W_i^{(T_H)}(\theta,\varphi) &= \left(\sin\varphi\sin\theta,\ \ -\cos\varphi\sin\theta,\ \ 0 \right) \\ W_i^{(L)}(\theta,\varphi) &= \left(\cos\varphi\sin\theta,\ \ \sin\varphi\sin\theta,\ \ \kappa^{(L)}(\theta) \right) \end{aligned} \tag{3.4.39}$$

where κ^{T_V} and $\kappa^{(L)}$ are given as

$$\begin{aligned} \kappa^{(T_V)}(\theta) &= -i\frac{2\sin^2\theta\cos\theta\sqrt{\sin^2\theta-(k_L/k_T)^2}}{2\sin^2\theta-1} \\ \kappa^{(L)}(\theta) &= \frac{2\sin^2\theta-(k_T/k_L)^2}{2\sqrt{(k_T/k_L)^2-\sin^2\theta}} \end{aligned} \tag{3.4.40}$$

The amplitudes $A^{(\#)}$ are

$$\begin{aligned} A^{(T_V)}(\theta) &= \frac{-2k_T^4(2\sin^2\theta-1)}{\mu\sin^2\theta F(k_T\sin\theta)} \\ A^{(TH)}(\theta) &= \frac{2}{\mu\sin^2\theta} \\ A^{(L)}(\theta) &= \frac{4k_L^3\cos\theta\sqrt{k_T^2-k_L^2\sin^2\theta}}{\mu F(k_L\sin\theta)} \end{aligned} \tag{3.4.41}$$

To obtain physical interpretations for the vectors $V_j^{(\#)}$ and $W_i^{(\#)}$, let us return to Fig. 3.4.5, where we find that the direction vector for the wave from the point source to the field point at the free surface is expressed as

$$d_j(\theta,\varphi)=\left(\begin{array}{ccc}\cos\varphi\sin\theta, & \sin\varphi\sin\theta, & -\cos\theta\end{array}\right) \tag{3.4.42}$$

Using the direction vector d_j, we can show that $V_j^{(T_V)}$, $V_j^{(T_H)}$, and $V_j^{(L)}$ are the polarizations of the plane SV, SH, and P waves, respectively, from the point source to the field point. In addition, it is readily seen that

$$\left(V_j^{(L)}(\theta,\varphi)\right)\perp\left(V_j^{(TV)}(\theta,\varphi)\right)\perp\left(V_j^{(TH)}(\theta,\varphi)\right) \tag{3.4.43}$$

Plane SV, SH, and P waves with a direction vector d_j are reflected at the free surface. Based on the discussion in Section 3.1, we can show that $W_i^{(T_V)}$, $W_i^{(T_H)}$, and $W_i^{(L)}$ are the polarizations of waves at the free surface due to the plane incident SV, SH, and P waves, respectively. We see that $W_i^{(T_V)}$ and $W_i^{(L)}$ are not orthogonal to each other due to coupling of the SV and P waves at the free surface. On the other hand, the reflection of the plane incident SH wave is only the SH wave, and thus

$$W_i^{(T_H)}(\theta,\varphi)=V_i^{(T_H)}(\theta,\varphi) \tag{3.4.44}$$

3.4.3.3 Effects of S-P wave due to branch line integral

Next, we investigate the effects of the S-P wave due to the branch line integral shown in Fig. 3.4.6(b). We express the contribution of the S-P wave to Green's function in the following form:

$$\begin{aligned}G_{ij}^{(S-P)}(\boldsymbol{x},\boldsymbol{y}) &\sim \frac{1}{4\pi}\sqrt{\frac{2}{\pi r}}C_{i\alpha}(\varphi)h_{\alpha l'}\sum_{m=-1}^{+1}e^{-i(2m+1)\pi/4}e^{im\varphi}\\ &\times\int_{B_-+B_+}\frac{\sqrt{\xi}}{\nu}Q_{l'p'}^{(T_V)}(\xi)e^{i\xi r-\nu y_3}d\xi\, F_{p''j}^{(m)}\end{aligned} \tag{3.4.45}$$

where the paths of the integral B_- and B_+ are shown in Fig. 3.4.8. Recall that $\gamma=\sqrt{\xi^2-k_L^2}$ and let $\xi=k_L+i\zeta$ along the branch line. Then, we have

$$\begin{aligned}\gamma_{\xi\in B_+} &\sim \sqrt{2k_L\zeta}e^{i\pi/4}\\ \gamma_{\xi\in B_-} &\sim -\sqrt{2k_L\zeta}e^{i\pi/4}\end{aligned} \tag{3.4.46}$$

where $|\zeta|\ll 1$. In addition, we also have

$$\begin{aligned}&\left[\frac{Q_{l'p'}^{(T_V)}(\xi)}{\nu}\right]_{\xi\in B_+}-\left[\frac{Q_{l'p'}^{(T_V)}(\xi)}{\nu}\right]_{\xi\in B_-}\\ &=\ \gamma_+Q_{l'p'}^{(T_{VB})}(k_L))+O(\gamma_+^3),\ \ (|\zeta|\ll 1)\end{aligned} \tag{3.4.47}$$

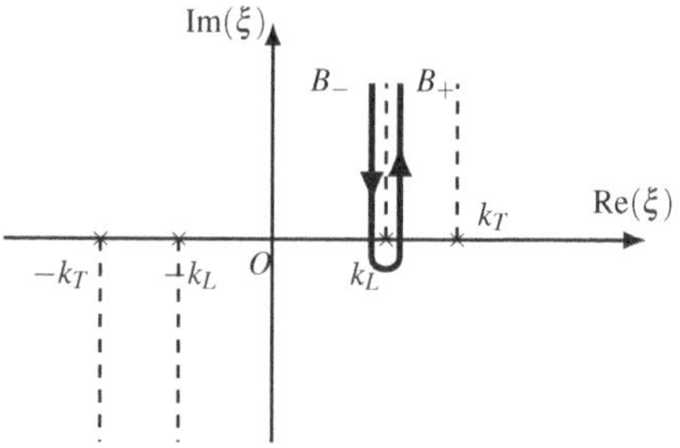

Figure 3.4.8 Paths of integral B_- and B_+ around branch cut.

where $\gamma_+ = \sqrt{2k_L\zeta}\exp(i\pi/4)$ and

$$
\begin{aligned}
Q_{1'1'}^{(T_VB)}(\xi) &= \frac{-4\xi^2}{\mu(2\xi^2-k_T^2)^2} \\
Q_{1'2'}^{(T_VB)}(\xi) &= \frac{4\nu\xi}{\mu(2\xi^2-k_T^2)^2} \\
Q_{2'1'}^{(T_VB)}(\xi) &= \frac{-8\xi^3\nu}{\mu(2\xi^2-k_T^2)^3} \\
Q_{2'2'}^{(T_VB)}(\xi) &= \frac{8\xi^2\nu^2}{\mu(2\xi^2-k_T^2)^3}
\end{aligned}
\tag{3.4.48}
$$

For the evaluation of Eq. (3.4.45), we have to apply the Taylor series expansion for $f_T(\xi)$ defined in Eq. (3.4.29) around $\xi = k_L$ as

$$
\begin{aligned}
& f_T(\xi)\left(= i\xi r - \nu y_3\right) \\
&= f_T(k_L) + f_T'(k_L)(\xi - k_L) + \cdots
\end{aligned}
\tag{3.4.49}
$$

where

$$
\begin{aligned}
f_T(k_L) &= ik_L r + i\sqrt{k_T^2 - k_L^2}y_3 \\
&= ik_T\sin\theta_c R\sin\theta + ik_T\cos\theta_c R\cos\theta \\
&= ik_T R\cos(\theta-\theta_c)
\end{aligned}
\tag{3.4.50}
$$

$$
\begin{aligned}
f_T'(k_L) &= ir - \frac{k_L}{\sqrt{k_L^2 - k_T^2}}y_3 \\
&= iR\sin\theta - \frac{ik_T\sin\theta_c}{k_T\cos\theta_c}R\cos\theta \\
&= \frac{iR}{\cos\theta_c}\sin(\theta-\theta_c) = i\beta
\end{aligned}
\tag{3.4.51}
$$

Note that θ_c is the critical angle that satisfies $k_L = k_T \sin\theta_c$ and β is defined by

$$\beta = \frac{R}{\cos\theta_c}\sin(\theta-\theta_c) \tag{3.4.52}$$

We know that the S-P wave is generated for the case $\theta > \theta_c$, so $\beta > 0$.

Now, we can evaluate Eq. (3.4.45). We have the following result for the integral:

$$\begin{aligned}
&\int_{B_-+B_+} \frac{\sqrt{\xi}}{\nu} Q_{l'p'}^{(T_V)}(\xi) e^{i\xi r - \nu y_3}\, d\xi \\
&\sim \int_0^{\infty} \sqrt{k_L}\gamma_+ Q_{l'p'}^{(T_{VB})}(k_L) e^{ik_T R\cos(\theta-\theta_c)} e^{i\beta i\zeta}\, i d\zeta \\
&= \sqrt{2} k_L^{3/2} e^{3\pi i/4} e^{ik_T R\cos(\theta-\theta_c)} Q_{l''p''}^{(T_{VB})}(k_L) \int_0^{\infty} \sqrt{\zeta} e^{-\beta\zeta} d\zeta \\
&= \sqrt{2} k_L^{3/2} e^{3\pi i/4} e^{ik_T R\cos(\theta-\theta_c)} Q_{l'p'}^{(T_{VB})}(k_L) \frac{\sqrt{\pi}}{2}\left(\frac{\cos\theta_c}{R\sin(\theta-\theta_c)}\right)^{3/2}
\end{aligned} \tag{3.4.53}$$

As a result, we have the result for the S-P wave in the following form:

$$\begin{aligned}
G_{ij}^{(S-P)}(\boldsymbol{x},\boldsymbol{y}) &\sim \frac{1}{4\pi R^2} e^{ik_T R\cos(\theta-\theta_c)} \sqrt{\frac{k_L^3\cos^3\theta_c}{\sin\theta\sin^3(\theta-\theta_c)}} \\
&\times \sum_{m=-1}^{1} e^{-i(m-1)\pi/2} e^{im\varphi} C_{i\alpha}(\varphi) h_{\alpha l'} Q_{l'p'}^{(T_{VB})}(k_L) F_{p'j}^{m}
\end{aligned} \tag{3.4.54}$$

The S-P wave has a decay factor of R^{-2}.

3.4.3.4 Effects of Rayleigh pole on Green's function

We have discussed the steepest descent path method for Green's function. At this point, we have to be aware that the effects of the Rayleigh pole on Green's function are still unclear. In order to clarify the effects of the Rayleigh pole, we employ the steepest descent path defined in Eq. (3.4.23), which is a curve that crosses the real axis two times. Figure 3.4.9 shows this descent path. Note that the steepest descent path has to be defined for the P and S waves separately. The points that cross the real axis of the complex wavenumber are $k_\# \sin\theta$ and $k_\#/\sin\theta$, where # takes the symbol T or L. From Table 3.4.1, recall that $k_\# \sin\theta$ is the saddle point.

Since θ is defined by $\sin\theta = r/R$, there are two cases for which $\kappa_R > k_\#/\sin\theta$ and $\kappa_R < k_\#/\sin\theta$, namely the far field case and the near field case. The Rayleigh pole κ_R has to be located above the path of the integral, and thus the contribution of the Rayleigh pole is necessary for the far field case, as shown in Fig. 3.4.9. For the case $k_L/\sin\theta < \kappa_R < k_T/\sin\theta$, the contribution of the residue term to Green's

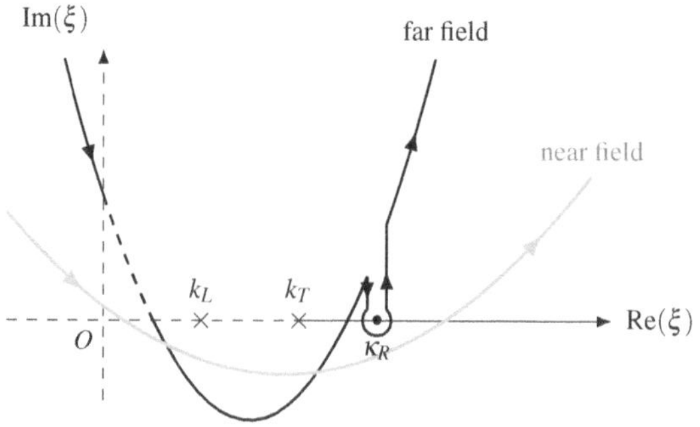

Figure 3.4.9 Steepest descent path defined in Eq. (3.4.23) for Green's function for elastic half-space. As shown in Table 3.4.1, the steepest decent path returns to the real axis of the complex wavenumber plane where $k_T/\sin\theta$ for the S wave and $k_L/\sin\theta$ for the P wave.

function is

$$G_{ij}^{(R)}(\boldsymbol{x},\boldsymbol{y}) \sim \frac{1}{4\pi}\sqrt{\frac{2}{\pi R}}\sqrt{\sin\theta}C_{i\alpha}(\varphi)h_{\alpha l'}\sum_{m=-1}^{+1} e^{-(2m+1)\pi i/4}e^{im\varphi} \times 2\pi i\sqrt{\kappa_R}\exp(i\kappa_R r)\operatorname*{Res}_{\xi=\kappa_R}\left[\frac{e^{-\gamma y_3}}{\gamma}Q_{l'p'}^{(L)}(\xi)\right]F_{p'j}^{(m)} \tag{3.4.55}$$

For the case $k_L/\sin\theta < k_T/\sin\theta < \kappa_R$, it is

$$G_{ij}^{(R)}(\boldsymbol{x},\boldsymbol{y}) \sim \frac{1}{4\pi}\sqrt{\frac{2}{\pi R}}\sqrt{\sin\theta}C_{i\alpha}(\varphi)h_{\alpha l'}\sum_{m=-1}^{+1} e^{-(2m+1)\pi i/4}e^{im\varphi} \times 2\pi i\sqrt{\kappa_R}\exp(i\kappa_R r)\operatorname*{Res}_{\xi=\kappa_R}\left[\frac{e^{-\nu y_3}}{\nu}Q_{l'p'}^{(T_V)}(\xi)+\frac{e^{-\gamma y_3}}{\gamma}Q_{l'p'}^{(L)}(\xi)\right]F_{p'j}^{(m)} \tag{3.4.56}$$

We see that the decay factor for the Rayleigh wave is $R^{-1/2}$, which agrees with the discussion in the previous section.

3.5 SPECTRAL REPRESENTATION OF GREEN'S FUNCTION FOR ELASTIC HALF-SPACE

So far, we have dealt with approximation methods for Green's function based on the branch line integral and the steepest descent path method, which are thought to be historical ones. In this section, we apply a modern viewpoint of mathematics to the representation of Green's function [3]. We show the effectiveness of the obtained

[3]This section is a part of a published article in 'Generalized Fourier transform and its application to the volume integral equation for elastic wave propagation in a half-space', by Touhei, T., published in the International Journal of Solids and Structures (2009, Vol. 46, pages 52-73, Copyright Elsevier). Several modifications have been made to the content of this published article.

representation for the computation of a scattering problem in the next chapter. The discussion in this section is based on operator theory. Therefore, the starting point of the discussion is the characterization of the operator for the elastic wave equation and the boundary conditions.

3.5.1 REVIEW OF ELASTIC WAVE EQUATION AND BOUNDARY CONDITIONS

We start with the following elastic wave equation and boundary conditions at the free surface:

$$\begin{aligned} \left(L_{ij}(\partial_1, \partial_2, \partial_3) + \delta_{ij}\rho\omega^2\right) u_j(\boldsymbol{x}) &= -f_i(\boldsymbol{x}) \\ P_{ij}(\partial_1, \partial_2, \partial_3) u_j(\boldsymbol{x}) &= 0, \quad (\text{at } x_3 = 0) \end{aligned} \tag{3.5.1}$$

where the operators L_{ij} and P_{ij} are defined in Eqs. (3.2.44) and (3.2.45), respectively. In this section, the force density f_i and the displacement field u_i are assumed to be in $L_2(\mathbb{R}^3_+)$. The scalar product of the function in $L_2(\mathbb{R}^3_+)$ is defined by

$$\left(u_i, v_i\right)_{L_2(\mathbb{R}^3_+)} = \int_{\mathbb{R}^3_+} u_i^*(\boldsymbol{x})\, v_i(\boldsymbol{x}) d\boldsymbol{x} \tag{3.5.2}$$

where the symbol $*$ denotes the complex conjugate. We introduce the following Fourier transform with respect to the horizontal coordinate system:

$$\begin{aligned} \left(\mathscr{F}^{(h)} u_i\right)(\hat{\boldsymbol{x}}) &= \frac{1}{2\pi} \iint_{-\infty}^{\infty} u_i(\boldsymbol{x}) \exp\left(-i(x_1\xi_1 + x_2\xi_2)\right) dx_1 dx_2 \\ \left(\mathscr{F}^{(h)^{-1}} \hat{u}_i\right)(\boldsymbol{x}) &= \frac{1}{2\pi} \iint_{-\infty}^{\infty} \hat{u}_i(\hat{\boldsymbol{x}}) \exp\left(i(x_1\xi_1 + x_2\xi_2)\right) d\xi_1 d\xi_2 \end{aligned} \tag{3.5.3}$$

which was previously given in Eqs. (3.2.4) and (3.2.5). Remember that $\hat{\boldsymbol{x}}$ is defined by

$$\hat{\boldsymbol{x}} = (\ \xi_1,\ \ \xi_2,\ \ x_3\ ) \tag{3.5.4}$$

according to Eq. (3.2.6). Using the above Fourier transform, Eq. (3.5.1) is transformed into

$$\begin{aligned} \left(L_{ij}(i\xi_1, i\xi_2, \partial_3) + \rho\omega^2\delta_{ij}\right) \hat{u}_j(\hat{\boldsymbol{x}}) &= -\hat{f}_i(\hat{\boldsymbol{x}}) \\ P_{ij}(i\xi_1, i\xi_2, \partial_3) \hat{u}_j(\hat{\boldsymbol{x}}) &= 0, \ \text{at } x_3 = 0 \end{aligned} \tag{3.5.5}$$

where $\hat{u}_j$ and $\hat{f}_j$ are respectively defined by

$$\hat{u}_j = \mathscr{F}^{(h)} u_j, \ \ \hat{f}_i = \mathscr{F}^{(h)} f_i \tag{3.5.6}$$

We also introduce the component of the displacement field denoted by $u_{i'}$, for which the transformation rule between u_i and $u_{i'}$ is given in Eq. (3.2.39). Recall that the transformation rule is expressed as

$$\hat{u}_i = T_{ij'} \hat{u}_{j'} \Longleftrightarrow \hat{u}_{j'} = T^*_{ij'} u_i \tag{3.5.7}$$

where $T_{ij'}$ is defined by

$$T_{ij'} = \begin{bmatrix} 0 & ic & is \\ 0 & is & -ic \\ 1 & 0 & 0 \end{bmatrix} \tag{3.5.8}$$

Note that c and s in Eq. (3.5.8) have been already provided in Eqs. (3.2.33) and (3.2.34), which are as follows:

$$\begin{aligned} c &= \begin{cases} \xi_1/\xi_r & (\text{when } \xi_r \neq 0) \\ 1 & (\text{when } \xi_r = 0) \end{cases} \\ s &= \begin{cases} \xi_2/\xi_r & (\text{when } \xi_r \neq 0) \\ 0 & (\text{when } \xi_r = 0) \end{cases} \end{aligned} \tag{3.5.9}$$

where $\xi_r = \sqrt{\xi_1^2 + \xi_2^2}$. In addition, recall that $T_{ij'}$ is a unitary matrix; that is,

$$T_{ij'} T^*_{kj'} = \delta_{ik} \tag{3.5.10}$$

The application of the Fourier transform in Eq. (3.5.3) and the unitary matrix $T_{ij'}$ to Eq. (3.5.1) yields

$$\begin{aligned} (-\mathscr{A}_{k'l'}(\xi_r, \partial_3) + \rho\omega^2 \delta_{k'l'})\hat{u}_{l'}(\hat{\boldsymbol{x}}) &= -T^*_{ik'} \hat{f}_i(\boldsymbol{x}) \\ \mathscr{P}_{k'l'}(\xi_r, \partial_3)\hat{u}_{l'}(\hat{\boldsymbol{x}}) &= 0, \ (\text{at } x_3 = 0) \end{aligned} \tag{3.5.11}$$

where the operators $\mathscr{A}_{k'l'}$ and $P_{k'l'}$ are defined by

$$\begin{aligned} \mathscr{A}_{k'l'}(\xi_r, \partial_3) &= -T^*_{ik'} L_{ij}(i\xi_1, i\xi_2, \partial_3) T_{jl'} \\ \mathscr{P}_{k'l'}(\xi_r, \partial_3) &= T^*_{ik'} P_{ij}(i\xi_1, i\xi_2, \partial_3) T_{jl'} \end{aligned} \tag{3.5.12}$$

The components of the operators $\mathscr{A}_{k'l'}$ and $\mathscr{P}_{k'l'}$ are

$$\begin{aligned} &\mathscr{A}_{k'l'}(\xi_r, \partial_3) \\ &= \begin{bmatrix} -(\lambda + 2\mu)\partial_3^2 + \mu\xi_r^2 & (\lambda + \mu)\xi_r\partial_3 & 0 \\ -(\lambda + \mu)\xi_r\partial_3 & -\mu\partial_3^2 + (\lambda + 2\mu)\xi_r^2 & 0 \\ 0 & 0 & -\mu\partial_3^2 + \mu\xi_r^2 \end{bmatrix} \end{aligned} \tag{3.5.13}$$

$$\begin{aligned} &\mathscr{P}_{k'l'}(\xi_r, \partial_3) \\ &= \begin{bmatrix} (\lambda + 2\mu)\partial_3 & -\lambda\xi_r & 0 \\ \mu\xi_r & \mu\partial_3 & 0 \\ 0 & 0 & \mu\partial_3 \end{bmatrix} \end{aligned} \tag{3.5.14}$$

according to Eqs. (3.2.47) and (3.2.48).

In this section, we begin by examining the self-adjointness of the operator $\mathscr{A}_{k'l'}$. Next, we explore the spectral representation of this operator, drawing upon its self-adjoint properties. Finally, we derive the spectral representation of Green's function for an elastic half-space based on the aforementioned discussions.

Note 3.6 Concept of symmetry operator and self-adjoint operator

Let A be a linear operator whose domain is denoted as $D(A) \subset \mathscr{H}$, where $\mathscr{H}$ is a Hilbert space. Adjoint operator A^* for A is defined as

$$\left(g, Af\right) = \left(A^* g, f\right), \quad (f \in D(A)) \tag{N3.6.1}$$

For the case where the domain of $D(A^*)$ satisfies

$$D(A) \subseteq D(A^*) \tag{N3.6.2}$$

that is, for the case where

$$\left(g, Af\right) = \left(Ag, f\right), \quad (f, g \in D(A)) \tag{N3.6.3}$$

we say that A is a symmetry operator. For the case where a symmetry operator satisfies the following:

$$\left(f, Af\right) \geq 0, \quad (\forall f \in D(A) \tag{N3.6.4}$$

A is non-negative. If the following equation is valid:

$$D(A) = D(A^*) \tag{N3.6.5}$$

we say that A is a self-adjoint operator. It is known that a spectral representation is possible only for a self-adjoint operator. Therefore, it is necessary to prove the self-adjointness of the operator in order to use the spectral representation of the operator. To prove that a symmetry operator A is self-adjoint, we can check whether the following statement is valid:

$$\forall f \in \mathscr{H}, \ \exists u \in D(A) \text{ such that } (A \pm is)u = f \tag{N3.6.6}$$

where s is a real number and $s \neq 0$.

3.5.2 SELF-ADJOINTNESS OF OPERATOR $\mathscr{A}_{K'L'}$

Now, we examine the self-adjoint properties of the operator $\mathscr{A}_{k'l'}$. The following discussion is in accordance with Note 3.6. Specifically, we will first clarify the symmetry of the operator and then proceed to discuss its self-adjointness.

For our discussions, it is necessary to define the domain of the operator, as well as the scalar product within the function space under consideration. We define the scalar product as follows:

$$(u_{i'}, v_{i'})_{L_2(\mathbb{R}_+)} = \int_{\mathbb{R}_+} u_{i'}^*(x_3) v_{i'}(x_3) dx_3 \tag{3.5.15}$$

The domain of the operator $\mathscr{A}_{k'l'}$ is specified as:

$$D(\mathscr{A}_{k'l'}) = \{u_{i'} \in L_2(\mathbb{R}_+) \,|\, \mathscr{A}_{k'l'} u_{l'} \in L_2(\mathbb{R}_+), \ \ \mathscr{P}_{k'l'} u_{l'} = 0 \text{ at } x_3 = 0\} \tag{3.5.16}$$

Now, we can assert the following:

Lemma 3.1 *The operator $\mathscr{A}_{k'l'}$ is symmetric and non-negative.*

Proof: Let $u_{i'}$ and $v_{i'} \in D(\mathscr{A}_{i'j'})$. Then,

$$\begin{aligned}
&\left(u_{i'},\, \mathscr{A}_{i'j'}v_{j'}\right)_{L_2(\mathbb{R}_+)} \\
&= -\left[u_{i'}^{*}\left(\mathscr{P}_{i'j'}v_{j'}\right)\right]_0^{\infty} + \left[\left(\mathscr{P}_{j'i'}u_{i'}^{*}\right)v_{j'}\right]_0^{\infty} + \left(\mathscr{A}_{j'i'}u_{i'},\, v_{j'}\right)_{L_2(\mathbb{R}_+)} \\
&= \left(\mathscr{A}_{j'i'}u_{i'},\, v_{j'}\right)_{L_2(\mathbb{R}_+)} \qquad (3.5.17)
\end{aligned}$$

$$\begin{aligned}
&\left(u_{i'},\, \mathscr{A}_{i'j'}u_{j'}\right)_{L_2(\mathbb{R}_+)} \\
&= \int_0^{\infty}\left[(\lambda+2\mu)\left|\partial_3 u_{1'}\right|^2 + \mu\left|\partial_3 u_{2'}\right|^2 + \mu\left|\partial_3 u_{3'}\right|^2\right]dx_3 \\
&\quad + \int_0^{\infty}\left[\mu\xi_r^2\left|u_{1'}\right|^2 + (\lambda+2\mu)\xi_r^2\left|u_{2'}\right|^2 + \mu\xi_r^2\left|u_{3'}\right|^2\right]dx_3 \\
&\quad + \int_0^{\infty}\left[-\lambda\xi_r\left(\partial_3 u_{1''}^{*}u_{2''} + u_{2'}\partial_3 u_{1'}\right) + \mu\xi_r\left(\partial_3 u_{2'}^{*}u_{1'} + u_{1'}^{*}\partial_3 u_{2'}\right)\right]dx_3 \\
&= \int_0^{\infty}\left[\lambda\left|\partial_3 u_{1'} - \xi_r u_{2'}\right|^2 + \mu\left|\partial_3 u_{2'} + \xi_r u_{1'}\right|^2\right]dx_3 \\
&\quad + 2\mu\int_0^{\infty}\left[\left|\partial_3 u_{1'}\right|^2 + \xi_r^2\left|u_{2'}\right|^2\right]dx_3 + \mu\int_0^{\infty}\left[\left|\partial_3 u_{3'}\right|^2 + \xi_r^2\left|u_{3'}\right|^2\right]dx_3 \\
&\geq 0 \qquad \square \qquad (3.5.18)
\end{aligned}$$

To investigate the self-adjointness of the operator, we employ the resolvent kernel introduced in Appendix C, which is also referred to in Note 3.7. This kernel, as defined in Appendix C, satisfies the following equation:

$$\left(\mathscr{A}_{i'j'} - \eta^2\mu\delta_{i'j'}\right)g_{j'k'}(x_3, y_3, \xi_r, \eta) = \delta_{i'k'}\delta(x_3 - y_3), \quad (\eta \in \mathbb{C}) \qquad (3.5.19)$$

together with the boundary condition

$$\mathscr{P}_{i'j'}g_{j'k'}(x_3, y_3, \xi_r, \eta) = 0, \quad (\text{at } x_3 = 0) \qquad (3.5.20)$$

where $\mathbb{C}$ is the set of the complex numbers. As detailed in Appendix C, the resolvent kernel can be constructed for $\eta \in \mathbb{C} \setminus (B_p \cup B_c)$, with the properties

$$\begin{aligned}
\sup_{x_3 \in \mathbb{R}_+} \int_{\mathbb{R}_+} |g_{i'j'}(x_3, y_3, \xi_r, \eta)|\,dy_3 &< \infty \\
\sup_{y_3 \in \mathbb{R}_+} \int_{\mathbb{R}_+} |g_{i'j'}(x_3, y_3, \xi_r, \eta)|\,dx_3 &< \infty \qquad (3.5.21)
\end{aligned}$$

where B_p and B_c are defined by

$$\begin{aligned}
B_p &= \{\eta \in \mathbb{R} \,|\, F_R(\xi_r, \eta) = 0\} \\
B_c &= \{\eta \in \mathbb{R} \,|\, \eta \geq \xi_r\} \qquad (3.5.22)
\end{aligned}$$

Note that F_R in Eq. (3.5.22) is defined by Eq. (C.3.2), which is expressed as:

$$F_R(\xi_r,\eta) = (2\xi_r^2-\eta^2)^2 - 4\xi_r^2\sqrt{\xi_r^2-\eta^2}\sqrt{\xi_r^2-(c_T/c_L)^2\eta^2} \tag{3.5.23}$$

With respect to the resolvent kernel, we present the following lemma:

Lemma 3.2 *For $f_{i'} \in L_2(\mathbb{R}_+)$ and $\eta \in \mathbb{C}\setminus(B_p \cup B_c)$, the following equation holds:*

$$u_{i'}(x_3) = \int_{\mathbb{R}_+} g_{i'j'}(x_3,y_3,\xi_r,\eta) f_{j'}(y_3)dy_3 \in L_2(\mathbb{R}_+) \tag{3.5.24}$$

Proof: We define

$$v_{i'}(x_3) = \int_{\mathbb{R}_+} g_{i'j'}(x_3,y_3,\xi_r,\eta) f_{j'}(y_3)dy_3 \tag{3.5.25}$$

for fixed i' and j'. According to the Schwarz inequality, we have

$$\begin{aligned} |v_{i'}(x_3)| &\le \left[\int_{\mathbb{R}_+} |g_{i'j'}(x_3,y_3,\xi_r,\eta)|\,|f_{j'}(y_3)|^2 dy_3\right]^{1/2} \left[\int_{\mathbb{R}_+} |g_{i'j'}(x_3,y_3,\xi_r,\eta)| dy_3\right]^{1/2} \\ &\le \left[\int_{\mathbb{R}_+} |g_{i'j'}(x_3,y_3,\xi_r,\eta)|\,|f_{j'}(y_3)|^2 dy_3\right]^{1/2} M_1 \end{aligned} \tag{3.5.26}$$

where

$$M_1 = \sup_{x_3\mathbb{R}_+}\left[\int_{\mathbb{R}_+} |g_{I'j'}(x_3,y_3,\xi_r,\eta)| dy_3\right]^{1/2} \tag{3.5.27}$$

Therefore, the following inequality:

$$\begin{aligned} \int_{\mathbb{R}_+} |v_{i'}(x_3)|^2 dx_3 &\le M_1^2 \int_{\mathbb{R}_+}\int_{\mathbb{R}_+} |g_{i'j'}(x_3,y_3,\xi_r,\eta)|\,|f_{j'}(y_3)|^2 dy_3 dx_3 \\ &\le M_1^2 M_2 \|f_{j'}\|^2_{L_2(\mathbb{R}_+} \end{aligned} \tag{3.5.28}$$

where

$$M_2 = \sup_{y_3\in\mathbb{R}_+}\int_{\mathbb{R}_+} |g_{i'j'}(x_3,y_3,\xi_r,\eta)| dx_3 \tag{3.5.29}$$

is possible. Equation (3.5.28) concludes the proof. □

In accordance with Note 3.6 as well as the lemma presented above, we can state the following theorem:

Theorem 3.1 *The operator $\mathscr{A}_{i'j'}$ with the domain $D(\mathscr{A}_{i'j'})$ is self-adjoint.*

Proof: To establish the self-adjointness, it suffices to demonstrate that $\forall f_{I'} \in L_2(\mathbb{R}+)$, there exist $ui'^{(+)}$ and $u_{i'}^{(-)}$ in $D(\mathscr{A}_{i'j'})$ satisfying: It is sufficient to prove that

$$(\mathscr{A}_{i'j'} + ip\mu\delta_{i'j'})u_{j'}^{(+)}(x_3) = f_{i'}(y_3) \tag{3.5.30}$$

$$(\mathscr{A}_{i'j'} - ip\mu\delta_{i'j'})u_{j'}^{(-)}(x_3) = f_{i'}(x_3) \tag{3.5.31}$$

where p is a positive real number. For the construction of $u_{j'}^{(+)}$, define

$$u_{i'}^{(+)}(x_3) = \int_{\mathbb{R}_+} g_{i'j'}(x_3, y_3, \xi_r, \eta) f_j(y_3) dy_3 \tag{3.5.32}$$

where η is chosen such that $\eta^2 = ip$. Note that $\eta \in \mathbb{C} \setminus (B_p \cup B_c)$.

The following equation:

$$\begin{aligned}
&\int_{\mathbb{R}_+} \varphi_{i'}(x_3)(\mathscr{A}_{i'j'} + ip\mu\delta_{i'j'})u_{j'}^{(+)}(x_3)dx_3 \\
&= \int_{\mathbb{R}_+} \left[(\mathscr{A}_{j'i'} + ip\mu\delta_{i'j'})\varphi_{i'}(x_3)\right] u_{j'}^{(+)}(x_3)dx_3 \\
&= \int_{\mathbb{R}_+} (\mathscr{A}_{j'i'} + ip\mu\delta_{i'j'})\varphi_{i'}(x_3) \int_{\mathbb{R}_+} g_{j'k'}(x_3, y_3, \xi_r, \eta) f_{k'}(y_3) dy_3 dx_3 \\
&= \int_{\mathbb{R}_+} \left[\int_{\mathbb{R}_+} ((\mathscr{A}_{j'i'} + ip\mu\delta_{j'i'})\varphi_{i'}(x_3)) g_{j'k'}(x_3, y_3, \xi_r, \eta) dx_3\right] f_{k'}(y_3) dy_3 \\
&= \left(\varphi_{k'}, f_{k'}\right)_{L_2(\mathbb{R}_+)}, \quad (\varphi_{k'} \in \mathscr{D}(\mathbb{R}_+))
\end{aligned} \tag{3.5.33}$$

leads to Eq. (3.5.30), where $\mathscr{D}(\mathbb{R}_+)$ is the Schwartz space.

For the derivation of Eq. (3.5.33), we define and evaluate the following integral:

$$\begin{aligned}
I_\varepsilon &= \int_0^{y_3-\varepsilon} \left[(\mathscr{A}_{j'i'} + ip\mu\delta_{j'i'})\varphi_{i'}(x_3)\right] g_{j'k'}(x_3, y_3, \xi_r, \eta) dx_3 \\
&\quad + \int_{y_3+\varepsilon}^{\infty} \left[(\mathscr{A}_{j'i'} + ip\mu\delta_{j'i'})\varphi_{i'}(x_3)\right] g_{j'k'}(x_3, y_3, \xi_r, \eta) dx_3
\end{aligned} \tag{3.5.34}$$

where $\varepsilon > 0$. The integral by parts of I_ε becomes

$$\begin{aligned}
I_\varepsilon &= \left[\left(-\mathscr{P}_{j'i'}\varphi_{i'}(x_3)\right) g_{j'k'}(x_3, y_3, \xi_r \eta)\right]_0^{y_3-\varepsilon} \\
&\quad - \left[\varphi_{i'}(x_3)\left(-\mathscr{P}_{i'j'} g_{j'k'}(x_3, y_3, \xi_r, \eta)\right)\right]_0^{y_3-\varepsilon} \\
&\quad + \int_0^{y_3-\varepsilon} \varphi_{i'}(x_3)\left[\left(\mathscr{A}_{i'j'} + ip\mu\delta_{i'j'}\right) g_{j'k'}(x_3, y_3, \xi_r, \eta)\right] dx_3 \\
&\quad + \left[\left(-\mathscr{P}_{j'i'}\varphi_{i'}(x_3)\right) g_{j'k'}(x_3, y_3, \xi_r \eta)\right]_{y_3+\varepsilon}^{\infty} \\
&\quad - \left[\varphi_{i'}(x_3)\left(-\mathscr{P}_{i'j'} g_{j'k'}(x_3, y_3, \xi_r, \eta)\right)\right]_{y_3+\varepsilon}^{\infty} \\
&\quad + \int_{y_3+\varepsilon}^{\infty} \varphi_{i'}(x_3)\left[\left(\mathscr{A}_{i'j'} + ip\mu\delta_{i'j'}\right) g_{j'k'}(x_3, y_3, \xi_r, \eta)\right] dx_3
\end{aligned} \tag{3.5.35}$$

The application of the following properties:

$$\begin{aligned}
\left(A_{i'j'} + ip\mu\delta_{i'j'}\right) g_{j'k'}(x_3, y_3, \xi_r, \eta) &= 0, \quad (x_3 \neq y_3) \\
g_{j'k'}(y_3 + \varepsilon, y_3, \xi_r, \eta) - g_{j'k'}(y_3 - \varepsilon, y_3, \xi_r, \eta) &\to 0, \quad (\varepsilon \to 0)
\end{aligned} \tag{3.5.36}$$

to Eq. (3.5.35) yields

$$\begin{aligned} I_\varepsilon &= \varphi_{i'}(y_3+\varepsilon)\Big(-\mathscr{P}_{i'j'}g_{j'k'}(y_3+\varepsilon,y_3,\xi_r,\eta)\Big) \\ &\qquad -\varphi_{i'}(y_3-\varepsilon)\Big(-\mathscr{P}_{i'j'}g_{j'k'}(y_3-\varepsilon,y_3,\xi_r,\eta)\Big) \\ &\to \varphi_{i'}(y_3)\delta_{i'k'} = \varphi_{k'}(y_3), \quad (\varepsilon\to 0) \end{aligned} \tag{3.5.37}$$

For Eq. (3.5.37), we used

$$\mathscr{P}_{i'j'}g_{j'k'}(y_3+\varepsilon,y_3,\xi_r,\eta)-\mathscr{P}_{i'j'}g_{j'k'}(y_3-\varepsilon,y_3,\xi_r,\eta)\to -\delta_{i'k'}, \quad (\varepsilon\to 0) \tag{3.5.38}$$

which is presented in Eq. (C.2.16) in Appendix C.

We also have to show that $u_{i'}^{(+)}$ defined in Eq. (3.5.32) satisfies the free boundary conditions. We have the following:

$$\begin{aligned} \mathscr{P}_{i'j'}u_{j'}(x_3) &= \mathscr{P}_{i'j'}\int_{\mathbb{R}_+} g_{j'k'}(x_3,y_3,\xi_r,\eta)f_{k'}(y_3)dy_3 \\ &= \int_{\mathbb{R}_+} \mathscr{P}_{i'j'}g_{j'k'}(x_3,y_3,\xi_r,\eta)f_{k'}(y_3)dy_3 \\ &= 0, \quad (\text{at } x_3=0) \end{aligned} \tag{3.5.39}$$

The interchange of the order of the integral and differential operators is based on the construction of the function $g_{i'j'}$ shown in Eq. (C.3.8) in Appendix C, whereby

$$\lim_{y_3\to\infty} |y_3^n \mathscr{P}_{i'j'}g_{j'k'}(x_3,y_3,\xi_r,\eta)f_{k'}(y_3)| = 0 \tag{3.5.40}$$

for an arbitrary positive integer n and

$$\sup_{y_3\in\mathbb{R}_+} |\mathscr{P}_{i'j'}g_{j'k'}(x_3,y_3,\xi_r,\eta)f_{k'}(y_3)| < \infty \tag{3.5.41}$$

Therefore

$$|\mathscr{P}_{i'j'}g_{j'k'}(x_3,y_3,\xi_r,\eta)f_{k'}(y_3)| \in L_1(\mathbb{R}_+) \tag{3.5.42}$$

It was shown that $u_{j'}\in L_2(\mathbb{R}_+)$ in Lemma 3.2, so $u_{i'}^{(+)}\in D(\mathscr{A}_{i'j'})$. The construction of $u_{i'}^{(-)}\in D(\mathscr{A}_{i'j'})$ is also possible. As a result, the following conclusion is obtained. □

Note 3.7 Concept of resolvent and spectrum

Let A be a linear operator whose domain is denoted as $D(A)\subset\mathscr{H}$. For the case where $A-\lambda$, $(\lambda\in\mathbb{C})$ is bijective and $R(\lambda)$ defined by

$$R(\lambda) = (A-\lambda)^{-1} \tag{N3.7.1}$$

is bounded, we say that λ is in a resolvent set and $R(\lambda)$ is the resolvent operator. We denote the resolvent set of A as $\rho(A)$. We say that the complementary set of $\rho(A)$, namely $\mathbb{C}\setminus\rho(A)$, is the spectrum of A. For the case where $R(\lambda)$ has an integral representation such that

$$\Big(R(\lambda)f\Big)(x) = \int g(x,y,\lambda)f(y)dy \tag{N3.7.2}$$

we call g the *resolvent kernel*. In the main text of the book, the resolvent kernel is constructed using Green's function in the wavenumber domain.

3.5.3 RELATIONSHIP BETWEEN RESOLVENT AND EIGENFUNCTION FOR OPERATOR $\mathscr{A}_{I'J'}$

3.5.3.1 Relationship between resolvent and spectral family

Now, we can observe that the following spectral representation of the operator $\mathscr{A}_{i'j'}$ becomes possible:

$$\mathscr{A}_{i'j'} = \int_0^\infty \zeta dE_{i'j'}(\zeta) \tag{3.5.43}$$

where $E_{i'j'}$ is the spectral family, and this representation is possible because $\mathscr{A}_{i'j'}$ is both self-adjoint and non-negative. The spectral family is linked to the resolvent through the Stone theorem (Wilcox, 1976), as expressed in the equation:

$$\begin{aligned}&\Big(\big[(E_{i'j'}(b)+E_{i'j'}(b-)) - (E_{i'j'}(a)+E_{i'j'}(a-))\big]u_{j'}, v_{j'}\Big)_{L_2(\mathbb{R}_+)}\\ &= \lim_{\varepsilon\searrow 0}\frac{1}{\pi i}\int_a^b d\zeta\Big(\big[R_{i'j'}(\zeta+i\varepsilon)-R_{i'j'}(\zeta-i\varepsilon)\big]u_{j'},\, v_{i'}\Big)_{L_2(\mathbb{R}_+)}\end{aligned} \tag{3.5.44}$$

for $u_{i'},\ v_{i'} \in L_2(\mathbb{R}_+)$. Note that $R_{i'j'}$ is the resolvent of the operator $\mathscr{A}_{i'j'}$, defined as follows:

$$(R_{i'j'}(\zeta)u_{j'})(x_3) = \int_0^\infty g_{i'j'}(x_3, y_3, \xi_r, \sqrt{\zeta/\mu})u_{j'}(y_3)dy_3 \tag{3.5.45}$$

The right-hand side of Eq. (3.5.44) for the integral takes on the following form when $0 < a < \mu\eta_R^2$ and $b > \mu\xi_r^2$:

$$\begin{aligned}&\int_a^b d\zeta\Big(\big[R_{i'j'}(\zeta+i\varepsilon)-R_{i'j'}(\zeta-i\varepsilon)\big]u_{j'},\, v_{i'}\Big)_{L_2(\mathbb{R}_+)}\\ &= (-2\pi i)\operatorname*{Res}_{\eta=\eta_R}(2\eta\mu)\Big(R_{i'j'}(\zeta)u_{j'},\, v_{i'}\Big)_{L_2(\mathbb{R}_+)}\\ &\quad+\int_{\xi_r}^{\sqrt{b/\mu}} d\eta\,(2\eta\mu)\Big(\big[R_{i'j'}(\zeta+i\varepsilon)-R_{i'j'}(\zeta-i\varepsilon)\big]u_{j'},\, v_{i'}\Big)_{L_2(\mathbb{R}_+)}\end{aligned} \tag{3.5.46}$$

Here, we define $\zeta = \mu\eta^2$, and η_R is determined by the equation $F_R(\xi_r, \eta_R) = 0$. To evaluate this integral, we follow the integration path in the complex η plane, as illustrated in Fig. 3.5.1.

3.5.3.2 Eigenfunction for the point spectrum

In our current discussion, we focus on the first term on the right-hand side of Eq. (3.5.46). We will establish a relationship between this term and the eigenfunction.

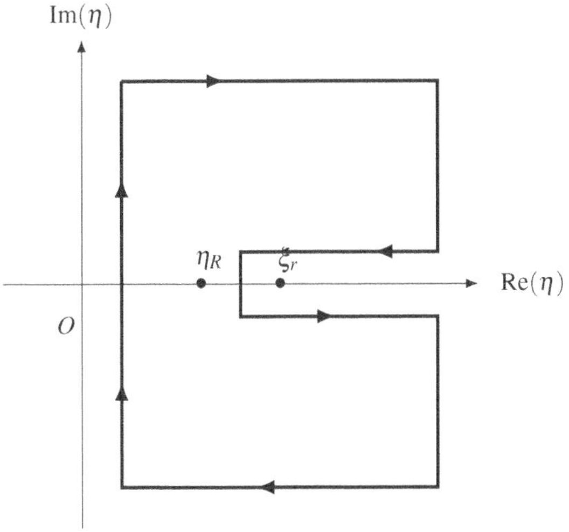

Figure 3.5.1 Path of integral in complex η plane.

To achieve this, we introduce the scalar function $W(\eta)$, which is defined as follows:

$$\begin{aligned} W(\eta) &= \left(v_{i'}, \left(\mathscr{A}_{i'j'} - \mu\eta^2\delta_{i'j'}\right)v_{j'}\right)_{L_2(\mathbb{R}_+)} \\ &\quad - \left[v_{i'}(x_3,\xi_r,\eta)\mathscr{P}_{i'j'}v_{j'}(x_3,\xi_r,\eta)\right]_{x_3=0}, \quad (\eta \in \mathbb{C}) \end{aligned} \tag{3.5.47}$$

where $v_{i'}(x_3,\xi_r,\eta)$ satisfies

$$\left(\mathscr{A}_{i'j'} - \mu\eta^2\delta_{i'j'}\right)v_{i'}(x_3,\xi_r,\eta) = 0 \tag{3.5.48}$$

Key properties of $W(\eta)$ are summarized as follows:

$$\begin{aligned} W(\eta) &= 0, \quad (\eta = \eta_R) \\ W(\eta) &\neq 0, \quad (\eta \neq \eta_R) \end{aligned} \tag{3.5.49}$$

This distinction is significant since $v_{i'}(x_3,\xi_r,\eta_R)$ represents the eigenfunction (Rayleigh wave mode) satisfying free boundary conditions, while $v_{i'}(x_3,\xi_r,\eta)$, where $\eta \neq \eta_R$, does not satisfy these conditions. Consequently, Eq. (3.5.49) is derived.

The integration by parts of Eq. (3.5.47) results in:

$$W(\eta) = I_1(\eta) - \mu\eta^2 I_2(\eta) \tag{3.5.50}$$

where

$$
\begin{aligned}
I_1(\eta) &= \int_0^\infty \Big[2\mu|\partial_3 v_{1'}|^2 + \mu|\partial_3 v_{3'}|^2 + 2\mu\xi_r^2|v_{2'}|^2 + \mu\xi_r^2|v_{3'}|^2 \\
&\quad + \lambda|\partial_3 v_{1'} - \xi_r v_{2'}|^2 + \mu|\partial_3 v_{2'} + \xi_r v_{1'}|^2\Big] dx_3 \\
I_2(\eta) &= \int_0^\infty \Big[|v_{1'}|^2 + |v_{2'}|^2 + |v_{3'}|^2\Big] dx_3
\end{aligned}
\tag{3.5.51}
$$

In the case where the eigenfunction is normalized, i.e., $I_2(\eta) = 1$, we obtain:

$$
W(\eta) = I_1(\eta) - \mu\eta^2 \tag{3.5.52}
$$

Recalling that $W(\eta_R) = 0$, this leads to the following limit:

$$
I_1(\eta) \to \mu\eta_R^2, \ \ (\eta \to \eta_R) \tag{3.5.53}
$$

Consequently, we establish the following lemma:

Lemma 3.3 *The residue of $g_{i'j'}$ at $\eta = \eta_R$ can be expressed in terms of the eigenfunction as follows:*

$$
\operatorname*{Res}_{\eta=\eta_R} g_{i'j'}(x_3, y_3, \xi_r, \eta) = -\frac{\psi_{i'm''}(x_3, \xi_r, \eta_R)\,\psi_{j'm''}(y_3, \xi_r, \eta_R)}{2\mu\eta_R} \tag{3.5.54}
$$

where $\psi_{i'm''}(x_3, \xi)$ is the eigenfunction for the point spectrum shown in Eq. (C.3.18) in Appendix C.

Remark: We employ the notation for the normalized eigenfunction by ψ. The eigenfunction for the point spectrum (Rayleigh wave mode) satisfies

$$
\begin{aligned}
\mathscr{A}_{k'l'}\psi_{l'm''}(x_3, \xi_r, \eta_R) &= \mu\eta_R^2\psi_{k'm''}(x_3, \xi_r, \eta_R) \\
\mathscr{P}_{k'l'}\psi_{l'm''}(x_3, \xi_r, \eta_R) &= 0, \ \ (\text{at } x_3 = 0)
\end{aligned}
\tag{3.5.55}
$$

We use the subscript m'' to distinguish the multiplicity of the eigenfunction. Note that for the eigenfunction for the point spectrum, $m'' = 1$.

Proof: Using Eq. (C.3.9) in Appendix C, we can construct the function $g_{i'j'}$ as follows:

$$
g_{i'j'}(x_3, y_3, \xi_r, \eta) = w_{i'j'}(x_3, y_3, \xi_r, \eta) + v_{i'm''}(x_3, \xi_r, \eta)\Delta_{j'm''}(\eta) \tag{3.5.56}
$$

where $w_{i'j'}$ is defined by

$$
\begin{aligned}
w_{i'j'}(x_3, y_3, \xi_r, \eta) &= 0, \ (x_3 > y_3) \\
\mathscr{P}_{i'j'}w_{j'k'}(x_3, y_3, \xi_r, \eta) &= +\delta_{i'k'}, \ \ (x_3 + \varepsilon = y_3, \ \varepsilon \searrow 0)
\end{aligned}
\tag{3.5.57}
$$

In addition, $\Delta_{j'm''}$ is defined to make $g_{i'j'}$ satisfy the free boundary condition:

$$\mathscr{P}_{i'j'}w_{j'k'}(x_3,y_3,\xi_r,\eta)+\mathscr{P}_{i'j'}v_{j'm''}(x_3,\xi_r,\eta)\Delta_{k'm''}(\eta)=0,\quad (\text{at } x_3=0) \tag{3.5.58}$$

The definition of $W(\eta)$ shown in Eq. (3.5.47) implies the expression:

$$W(\eta)=-v_{i'm''}(0,\xi_r,\eta)\mathscr{P}_{i'j'}v_{j'm''}(0,\xi_r,\eta) \tag{3.5.59}$$

Equations (3.5.58) and (3.5.59) yield:

$$\begin{aligned} g_{i'k'}(x_3,y_3,\xi_r,\eta) &= w_{i'k'}(x_3,y_3,\xi_r,\eta) \\ &+\frac{v_{i'm''}(x_3,\xi_r,\eta)}{W(\eta)}\left[v^*_{l'm''}(0,\xi_r,\eta)\mathscr{P}_{l'j'}w_{j'k'}(0,y_3,\xi_r,\eta)\right] \end{aligned} \tag{3.5.60}$$

Now, as η approaches η_R, due to reciprocity, we find:

$$v^*_{l'm''}(0,\xi_r,\eta_R)\mathscr{P}_{l'j'}w_{j'k'}(0,y_3,\xi_r,\eta_R)=v^*_{l'm''}(y_3,\xi_r,\eta_R)\delta_{l'k'} \tag{3.5.61}$$

Therefore

$$\begin{aligned} & v_{i'm''}(x_3,\xi_r,\eta_R)\left[v_{l'm''}(0,\xi_r,\eta_R)\mathscr{P}_{l'j'}w_{j'k'}(0,y_3,\xi_r,\eta)\right] \\ =\; & \psi_{i'm''}(x_3,\xi_r,\eta_R)\,\psi_{k'm''}(y_3,\xi_r,\eta_R) \end{aligned} \tag{3.5.62}$$

This allows us to express the residue of the resolvent kernel as:

$$\begin{aligned} & \operatorname*{Res}_{\eta=\eta_R} g_{i'j'}(x_3,y_3,\xi_r,\eta) \\ =\; & \lim_{\eta\to\eta_R}(\eta-\eta_R)\frac{\psi_{i'm''}(x_3,\xi_r,\eta_R)\psi_{j'm''}(y_3,\xi_r,\eta_R)}{W(\eta)} \\ =\; & -\frac{\psi_{i'm''}(x_3,\xi_r,\eta_R)\psi_{j'm''}(y_3,\xi_r,\eta_R)}{2\mu\eta_R} \end{aligned} \tag{3.5.63}$$

based on Eqs. (3.5.52), which concludes the proof. □

3.5.3.3 Eigenfunction for the continuous spectrum

We now analyze the second term in Eq. (3.5.46). Our aim is to establish a connection between this term and the eigenfunction. To achieve this, we examine the following expression:

$$g_{i'j'}(x_3,y_3,\xi_r,s+i\varepsilon)-g_{i'j'}(x_3,y_3,\xi_r,s-i\varepsilon) \tag{3.5.64}$$

when $s=\mathrm{Re}(\eta)>\xi_r$. The resolvent kernel $g_{i'j'}$ in the above expression is constructed as

$$\begin{aligned} & g_{i'j'}(x_3,y_3,\xi_r,s\pm i\varepsilon) \\ =\; & w_{i'j'}(x_3,y_3,\xi_r,is\pm i\varepsilon)+v_{i'm''}(x_3,\xi_r,s\pm i\varepsilon)\Delta_{j'm''}(s\pm i\varepsilon) \end{aligned} \tag{3.5.65}$$

where $v_{i'm''}$ satisfies

$$\begin{aligned}\left(\mathscr{A}_{i'j'} - \mu(s \pm i\varepsilon)^2 \delta_{i'j'}\right) v_{i'}(x_3, \xi_r, s \pm i\varepsilon) &= 0, \\ \text{(double sign correspond)} & \end{aligned} \quad (3.5.66)$$

and plays a role of the definition function as the eigenfunction for the continuous spectrum (Appendix C). The relationship between the eigenfunction for the continuous spectrum and the definition function for the continuous spectrum is expressed as

$$\psi_{i'm''}(x_3, \xi_r, s) = v_{i'm''}(x_3, \xi_r, s + i\varepsilon) - v_{i'm''}(x_3, \xi_r, s - i\varepsilon), \quad (\varepsilon \to 0) \quad (3.5.67)$$

The explicit forms of the eigenfunctions for the continuous spectrum are provided in Eqs. (C.3.28) and (C.3.36) in Appendix C. The eigenfunction for the continuous spectrum obeys

$$\mathscr{A}_{i'j'} \psi_{j'm''}(x_3, \xi_r, s) = \mu s^2 \psi_{i'm''}(x_3, \xi_r, s), \quad (s > \xi_r) \quad (3.5.68)$$

The subscript m'' is 1 or 3 in the region $\xi_r < s < c_L/c_T \xi_r$ and can be 1, 2, or 3 in the region $(c_L/c_T) \leq s$.

In line with our definition of $W(\eta)$ in Eq. (3.5.59), we introduce the following function:

$$W_{k''l''}(\xi_r, s \pm i\varepsilon) = -\psi_{i'k''}(0, \xi) \mathscr{P}_{i'j'} v_{j'l''}(0, \xi_r, s \pm i\varepsilon) \quad (3.5.69)$$

for the continuous spectrum. Substituting the explicit forms of the eigenfunction and the definition function into Eq. (3.5.69) results in the following:"

$$W_{k''l''}(\xi_r, s \pm i\varepsilon) = -\frac{\mu i}{\pi} s \delta_{k''l''}, \quad (\varepsilon \to 0) \quad (3.5.70)$$

Furthermore, it is important to note that

$$w_{i'j'}(x_3, y_3, \xi_r, s + i\varepsilon) - w_{i'j'}(x_3, y_3, \xi_r, s - i\varepsilon) \to 0, \quad (\varepsilon \to 0) \quad (3.5.71)$$

This result stems from the definition of $w_{i'k'}$ in Eq. (3.5.57). Based on Eqs. (3.5.70) and (3.5.71), we can establish the following lemma:

Lemma 3.4 *For $s > \xi_r$, the function $g_{i'j'}$ satisfies the following equation:*

$$\begin{aligned}& g_{i'j'}(x_3, y_3, \xi_r, s + i\varepsilon) - g_{i'j'}(x_3, y_3, \xi_r, s - i\varepsilon) \\ &= \pi i \frac{\psi_{i'm''}(x_3, \xi_r, s)\, \psi_{j'm''}(y_3, \xi_r, s)}{\mu s}, \quad (\varepsilon \to 0) \end{aligned} \quad (3.5.72)$$

where $\psi_{i'm''}(x_3, \xi_r, s)$ is the eigenfunction for the continuous spectrum (Appendix C).

Proof: The requirements of the free boundary condition for $g_{i'j'}$ lead to the following expression for $\Delta_{k'j}$

$$\begin{aligned}\Delta_{k'm''}(s \pm i\varepsilon) &= \left[W_{l''m''}(\xi_r, s \pm i\varepsilon)\right]^{-1} \psi_{i'l''}(0, \xi_r, s) \\ &\quad \times \mathscr{P}_{i'p'} w_{p'k'}(0, y_3, \xi_r, s \pm i\varepsilon) \end{aligned} \tag{3.5.73}$$

Incorporating the reciprocity relation:

$$\psi_{i'l''}(0, \xi_r, s) \mathscr{P}_{i'p'} w_{p'k'}(0, y_3, \xi_r, s \pm i\varepsilon) = \psi_{i'l''}(y_3, \xi_r, s) \delta_{i'k'} \tag{3.5.74}$$

into Eq. (3.5.73) yields

$$\Delta_{k''j'}(s \pm i\varepsilon) = \left[W_{l''k''}(\xi_r, s \pm i\varepsilon)\right]^{-1} \psi_{j'l''}(y_3, \xi_r, s) \tag{3.5.75}$$

Therefore, we obtain the following expression:

$$\begin{aligned} & g_{i'j'}(x_3, y_3, \xi_r, s \pm i\varepsilon) \\ &= w_{i'j'}(x_3, y_3, \xi_r, s \pm i\varepsilon) \\ &\quad + v_{i'k''}(x_3, \xi_r, s \pm i\varepsilon)\left[W_{l''k''}(\xi_r, s \pm i\varepsilon)\right]^{-1} \psi_{j'l''}(y_3, \xi_r, s) \end{aligned} \tag{3.5.76}$$

Thus, Eqs. (3.5.67), (3.5.70), and (3.5.71) conclude the proof. □

3.5.4 EIGENFUNCTION EXPANSION FOR SOLUTION FOR ELASTIC HALF-SPACE

We now denote the eigenfunctions for the point and continuous spectra as $\psi_{i'm''}(x_3 : \boldsymbol{\xi})$, where

$$\boldsymbol{\xi} = (\xi_1, \xi_2, \xi_3) \in \mathbb{R}^3_+$$

In addition, we also discuss functions denoted as $u_{j'}(\xi_r, x_3) \in L_2(\mathbb{R}_+)$ in the sense that

$$\int_0^\infty |u_{j'}(\xi_r, x_3)|^2 dx_3 < \infty \tag{3.5.77}$$

Since Eq. (3.5.46) holds for an arbitrary $u_{i'}(\xi_r, \cdot) \in L_2(\mathbb{R}_+)$, the following equation can be presented by incorporating the results for Lemmas 3.3 and 3.4:

$$\begin{aligned} & \int_a^b d\zeta \left[R_{i'j'}(\zeta + i\varepsilon) - R_{i'j'}(\zeta - i\varepsilon)\right] u_{j'} \\ &= (2\pi i) \sum_{\xi \in \sigma_p} \psi_{i'm''}(x_3 : \boldsymbol{\xi}) \left(\psi_{j'm''}(\cdot, \boldsymbol{\xi}), u_{j'}(\xi_r, \cdot)\right)_{L_2(\mathbb{R}_+)} \\ &\quad + (2\pi i) \int_{\xi_r}^{\sqrt{b/\mu}} d\xi_3 \, \psi_{i'm''}(x_3, \boldsymbol{\xi}) \left(\psi_{j'm''}(\cdot : \boldsymbol{\xi}), u_{j'}(\xi_r, \cdot)\right)_{L_2(\mathbb{R}_+)} \end{aligned} \tag{3.5.78}$$

where

$$\sigma_p = \{\boldsymbol{\xi} = (\xi_1, \xi_2, \xi_3) \in \mathbb{R}^3_+ \mid F_R(\xi_r, \xi_3) = 0\} \tag{3.5.79}$$

As mentioned earlier, $0 < a < \mu\eta_R^2$ and $b > \mu\xi_r^2$, so

$$\begin{aligned} E_{i'j'}(a) &= E_{i'j'}(a-) = 0 \\ E_{i'j'}(b) &= E_{i'j'}(b-) \end{aligned} \tag{3.5.80}$$

Therefore, Eqs. (3.5.44) and (3.5.78) yield

$$\begin{aligned} &\left(E_{i'j'}(b)u_{j'}\right)(\xi_r, x_3) \\ &= \sum_{\xi\in\sigma_p} \left(\psi_{j'm''}(\cdot, \boldsymbol{\xi}), u_{j'}(\xi_r, \cdot)\right)_{L_2(\mathbb{R}_+)} \psi_{i'm''}(x_3, \boldsymbol{\xi}) \\ &\quad + \int_{\xi_r}^{\sqrt{b/\mu}} \left(\psi_{j'm''}(\cdot, \boldsymbol{\xi}), u_{j'}(\xi_r, \cdot)\right)_{L_2(\mathbb{R}_+)} \psi_{i'm''}(x_3, \boldsymbol{\xi})\, d\xi_3 \end{aligned} \tag{3.5.81}$$

Let b in Eq. (3.5.81) approach infinity. Then, the following eigenfunction expansion form of $u_{i''}$ is obtained:

$$\begin{aligned} &\left(E_{i'j'}u_{j'}\right)(\xi_r, x_3) \\ &= \sum_{\xi\in\sigma_p} \left(\psi_{j'm''}(\cdot : \boldsymbol{\xi}), u_{j'}(\xi_r, \cdot)\right)_{L_2(\mathbb{R}_+} \psi_{i'm''}(x_3, \boldsymbol{\xi}) \\ &\quad + \int_{\xi_r}^{\infty} \left(\psi_{j'm''}(\cdot : \boldsymbol{\xi}), u_{j'}(\xi_r, \cdot)\right)_{L_2(\mathbb{R}_+)} \psi_{i'm''}(x_3, \boldsymbol{\xi})\, d\xi_3 \end{aligned} \tag{3.5.82}$$

Note that this expansion form is for $u_{i'}(\xi_r, \cdot)$ with compact support. This result can be extended to all $u_{i'}(\xi_r, \cdot) \in L_2(\mathbb{R}_+)$ by a limiting procedure; that is,

$$\left(\psi_{j'm''}(\cdot, \boldsymbol{\xi}), u_{j'}(\xi_r, \cdot)\right)_{L_2(\mathbb{R}_+)} = \underset{M\to\infty}{\text{l.i.m}} \int_0^M \psi_{j'm''}(x_3, \boldsymbol{\xi}) u_{j'}(\xi_r, x_3) dx_3 \tag{3.5.83}$$

where the convergence is in $L_2(\mathbb{R}_+)$. The transform for the function in $L_2(\mathbb{R}_+)$ obtained here can be summarized as

$$\begin{aligned} u_{m''}(\boldsymbol{\xi}) &= \left(\mathscr{F}^{(v)}_{m''j'} u_{j'}\right)(\boldsymbol{\xi}) = \int_0^{\infty} \psi_{j'm''}(x_3, \boldsymbol{\xi}) u_{j'}(\xi_r, x_3) dx_3 \\ u_{i'}(\xi_r, x_3) &= \left(\mathscr{F}^{(v)-1}_{i'm''} u_{m''}\right)(\xi_r, x_3) \\ &= \sum_{\xi\in\sigma_p} \psi_{i'm''}(x_3, \boldsymbol{\xi}) u_{m''}(\boldsymbol{\xi}) + \int_{\xi_r}^{\infty} \psi_{i'm''}(x_3, \boldsymbol{\xi}) u_{m''}(\boldsymbol{\xi}) d\xi_3 \end{aligned} \tag{3.5.84}$$

At this point, the transformation for the elastic wavefield in a half-space can be presented. Let us define the subset of the wavenumber domain:

$$\sigma_c = \{\boldsymbol{\xi} = (\xi_1, \xi_2, \xi_3) \in \mathbb{R}^3_+ \mid \xi_r > \xi_3\} \tag{3.5.85}$$

Based on Eqs. (3.5.3), (3.5.81) and the unitary matrix defined in Eq. (3.5.8), the following theorem is obtained:

Theorem 3.2 *There exists a map that satisfies the free boundary condition for the elastic half-space of the wavefield from* $L_2(\mathbb{R}^3_+)$ *to* $L_2(\sigma_p) \oplus L_2(\sigma_c)$ *defined by*

$$u_{i''}(\boldsymbol{\xi}) = \left(\mathscr{U}_{i''j}u_j\right)(\boldsymbol{\xi}) = \left(\mathscr{F}^{(v)}_{i''m'}T^*_{jm'}\mathscr{F}^{(h)}u_j\right)(\boldsymbol{\xi}) \tag{3.5.86}$$

the inverse of which is

$$u_i(\boldsymbol{x}) = \left(\mathscr{U}^{-1}_{ij''}u_{j''}\right)(\boldsymbol{x}) = \left(\mathscr{F}^{(h)-1}T_{im'}\mathscr{F}^{(v)-1}_{m'j''}u_{j''}\right)(\boldsymbol{x}) \tag{3.5.87}$$

Here, $\mathscr{U}_{i''j}u_j$ *and* $\mathscr{U}^{-1}_{ij''}u_{j''}$ *are expressed as follows:*

$$\begin{aligned}
\left(\mathscr{U}_{i''j}u_j\right)(\boldsymbol{\xi}) &= \int_{\mathbb{R}^3_+}\Lambda^*_{ji''}(\boldsymbol{\xi},\boldsymbol{x})u_j(\boldsymbol{x})d\boldsymbol{x}, \quad (\boldsymbol{\xi}\in\sigma_p\cup\sigma_c\subset\mathbb{R}^3_+)\\
\left(\mathscr{U}^{-1}_{ij''}u_{j''}\right)(\boldsymbol{x}) &= \int_{\mathbb{R}^2_+}\sum_{\boldsymbol{\xi}\in\sigma_p}\Lambda_{ij''}(\boldsymbol{\xi},\boldsymbol{x})u_{j''}(\boldsymbol{\xi})d\xi_1 d\xi_2\\
&\quad+\int_{\mathbb{R}^2_+}\int_{\xi_r}^{\infty}\Lambda_{ij''}(\boldsymbol{\xi},\boldsymbol{x})u_{j''}(\boldsymbol{\xi})d\xi_3 d\xi_1 d\xi_2, \quad (\boldsymbol{x}\in\mathbb{R}^3_+)
\end{aligned} \tag{3.5.88}$$

where

$$\Lambda_{ij''}(\boldsymbol{\xi},\boldsymbol{x}) = \frac{1}{2\pi}\exp\big(i(\xi_1x_1+\xi_2x_2)\big)\,\psi_{l'j''}(x_3,\boldsymbol{\xi})T_{il'} \tag{3.5.89}$$

3.5.5 SPECTRAL REPRESENTATION OF GREEN'S FUNCTION

We call $\Lambda_{ij''}$ defined by Eq. (3.5.89) the kernel of the generalized Fourier transform, which has the following properties as an eigenfunction:

$$L_{ij}(\partial_1,\partial_2,\partial_3)\Lambda_{jk''}(\boldsymbol{\xi},\boldsymbol{x}) = \mu\xi_3^2\Lambda_{ik''}(\boldsymbol{\xi},\boldsymbol{x}) \tag{3.5.90}$$

since

$$\begin{aligned}
&L_{ij}(\partial_1,\partial_2,\partial_3)\Lambda_{jk''}(\boldsymbol{\xi},\boldsymbol{x})\\
&= \frac{1}{2\pi}e^{i\xi_1x_1+i\xi_2x_2}\delta_{im}L_{mj}(i\xi_1,i\xi_2,\partial_3)T_{jl'}\psi_{l'k''}(x_3,\boldsymbol{\xi})\\
&= \frac{1}{2\pi}e^{i\xi_1x_1+i\xi_2x_2}T_{ip'}T^*_{mp'}L_{mj}(i\xi_1,i\xi_2,\partial_3)T_{jl'}\psi_{l'k''}(x_3,\boldsymbol{\xi})\\
&= \frac{1}{2\pi}e^{i\xi_1x_1+i\xi_2x_2}T_{ip'}A_{p'l'}\psi_{l'k''}(x_3,\boldsymbol{\xi})\\
&= -\frac{1}{2\pi}e^{i\xi_1x_1+i\xi_2x_2}T_{ip'}\mu\xi_3^2\psi_{p'k''}(x_3,\boldsymbol{\xi})\\
&= -\mu\xi_3^2\Lambda_{ik''}(\boldsymbol{\xi},\boldsymbol{x})
\end{aligned} \tag{3.5.91}$$

For Eq. (3.5.91), we used

$$\begin{aligned} T_{ip'}T^{*}_{mp'} &= \delta_{im} \\ A_{p'l'}\psi_{l'k''}(x_3,\boldsymbol{\xi}) &= \mu\xi_3^2\psi_{p'k''}(x_3,\boldsymbol{\xi}) \end{aligned}$$

which are discussed for Eqs. (3.5.10), (3.5.55), and (3.5.68).

Now, our goal is to express Green's function for an elastic half-space space in terms of the kernel of the generalized Fourier transform. Let us return to the definition of Green's function for an elastic half-space, which is as follows:

$$\left(L_{ij}(\partial_1,\partial_2,\partial_3)+\delta_{ij}\rho\omega^2\right)G_{jk}(\boldsymbol{x},\boldsymbol{y}) = -\delta_{ik}\delta(\boldsymbol{x}-\boldsymbol{y}) \tag{3.5.92}$$

$$P_{ij}(\partial_1,\partial_2,\partial_3)G_{jk}(\boldsymbol{x},\boldsymbol{y}) = 0, \text{ at } x_3=0 \tag{3.5.93}$$

Let us apply the generalized Fourier transform to the definition of Green's function. For the left-hand side of Eq. (3.5.92), we have

$$\begin{aligned} &\int_{\mathbb{R}^3_+}\Lambda^{*}_{im''}(\boldsymbol{\xi},\boldsymbol{x})\left(L_{ij}(\partial_1,\partial_2,\partial_3)+\rho\omega^2\delta_{ij}\right)G_{jk}(\boldsymbol{x},\boldsymbol{y})d\boldsymbol{x} \\ &= \int_{\mathbb{R}^3_+}\left(L_{ji}(\partial_1,\partial_2,\partial_3)+\rho\omega^2\delta_{ij}\right)\Lambda^{*}_{im''}(\boldsymbol{\xi},\boldsymbol{x})G_{jk}(\boldsymbol{x},\boldsymbol{y})d\boldsymbol{x} \\ &= (-\mu\xi_3^2+\rho\omega^2)\int_{\mathbb{R}^3_+}\Lambda^{*}_{jm''}(\boldsymbol{\xi},\boldsymbol{x})G_{jk}(\boldsymbol{x},\boldsymbol{y})d\boldsymbol{x} \\ &= (-\mu\xi_3^2+\rho\omega^2)G_{m''k}(\boldsymbol{\xi},\boldsymbol{y}) \end{aligned} \tag{3.5.94}$$

for which we use the notation

$$G_{m'',k}(\boldsymbol{\xi},\boldsymbol{y}) = \int_{\mathbb{R}^3_+}\Lambda^{*}_{jm''}(\boldsymbol{\xi},\boldsymbol{x})G_{jk}(\boldsymbol{x},\boldsymbol{y})d\boldsymbol{x} \tag{3.5.95}$$

For the right-hand side of Eq. (3.5.92), the application of the generalized Fourier transform yields

$$\int_{\mathbb{R}^3_+}\Lambda^{*}_{im''}(\boldsymbol{\xi},\boldsymbol{x})(-\delta_{ik})\delta(\boldsymbol{x}-\boldsymbol{y})d\boldsymbol{x} = -\Lambda^{*}_{km''}(\boldsymbol{\xi},\boldsymbol{y}) \tag{3.5.96}$$

Based on Eqs. (3.5.94) and (3.5.96), we have the generalized Fourier transform of Green's function such that

$$G_{m''k}(\boldsymbol{\xi},\boldsymbol{y}) = \frac{\Lambda^{*}_{km''}(\boldsymbol{\xi},\boldsymbol{y})}{\mu\xi_3^2-\rho\omega^2\mp i\varepsilon} \tag{3.5.97}$$

Therefore, Green's function for an elastic half-space in terms of the kernel of the generalized Fourier transform derived from Eqs. (3.5.88) and (3.5.97) is expressed as

$$\begin{aligned} G_{ij}(\boldsymbol{x},\boldsymbol{y}) = &\int_{\mathbb{R}^2}\sum_{\xi\in\sigma_p}\frac{\Lambda_{im''}(\boldsymbol{\xi},\boldsymbol{x})\Lambda^{*}_{jm''}(\boldsymbol{\xi},\boldsymbol{y})}{\mu\xi_3^2-\rho\omega^2\mp i\varepsilon}d\xi_1 d\xi_2 \\ &+\int_{\mathbb{R}^2}\int_{\xi_r}^{\infty}\frac{\Lambda_{im''}(\boldsymbol{\xi},\boldsymbol{x})\Lambda^{*}_{jm''}(\boldsymbol{\xi},\boldsymbol{y})}{\mu\xi_3^2-\rho\omega^2\mp i\varepsilon}d\xi_3 d\xi_1 d\xi_2 \end{aligned} \tag{3.5.98}$$

Equation (3.5.98) can be referred to as the spectral representation of Green's function for an elastic half-space. Note that the first term on the right-hand side of Eq. (3.5.98) corresponds to the residue term and the second term corresponds to the branch line integral. Both terms are unified by means of the eigenfunction $\Lambda_{im''}$. As mentioned in this section, the mathematical form of Green's function seems to be formal. In the next chapter, we apply Green's function to the analysis of scattering problems and demonstrate the usefulness of the obtained form of Green's function.

3.6 PROBLEMS

Consider a problem of sound wave propagation as depicted in Fig. 3.6.1, where two liquid half-spaces are in contact along a plane $x_3 = 0$ [4]. The mass density and the sound velocity for the region $x_3 < 0$ are denoted by ρ_1 and α_1, while those for the region $x_3 > 0$ are denoted ρ_2 and α_2, respectively. The governing equations for the wavefield are expressed as follows:

$$\begin{aligned}\left(\nabla^2+\frac{\omega^2}{\alpha_1^2}\right)p(\boldsymbol{x}) &= -q\delta(x_1)\delta(x_2)\delta(x_3+h),\ (x_3<0)\\ \left(\nabla^2+\frac{\omega^2}{\alpha_2^2}\right)p(\boldsymbol{x}) &= 0,\ (x_3>0)\end{aligned} \tag{3.6.1}$$

where p is the fluid pressure, $(0,0,-h)$ is the location of the point source and q is the strength of the point source. The boundary conditions at the interface boundary at $x_3 = 0$ are given by

$$\begin{aligned}\lim_{x_3\to+0} p(\boldsymbol{x}) &= \lim_{x_3\to-0} p(\boldsymbol{x})\\ \lim_{x_3\to+0}\frac{1}{\rho_2}\frac{\partial p(\boldsymbol{x})}{\partial x_3} &= \lim_{x_3\to-0}\frac{1}{\rho_1}\frac{\partial p(\boldsymbol{x})}{\partial x_3}\end{aligned} \tag{3.6.2}$$

The fluid pressure field for the region of $x_3 < 0$ is expressed as

$$p(\boldsymbol{x}) = p^{(Inc)}(\boldsymbol{x}) + p^{(Ref)}(\boldsymbol{x}) \tag{3.6.3}$$

where $p^{(Inc)}$ denotes the incident wavefield and $p^{(Ref)}$ represents the reflected wavefield. The fluid pressure field for the region of $x_3 > 0$ is constituted by the transmitted wave in the form of:

$$p(\boldsymbol{x}) = p^{(Tran)}(\boldsymbol{x}) \tag{3.6.4}$$

where $p^{(Tran)}$ denotes the transmitted wave.

1. Express the incident wavefield $p^{(Inc)}(\boldsymbol{x})$ by means of the Sommerfeld integral.
2. We express the reflected and transmitted wavefields as follows:

$$\begin{aligned}p^{(Ref)}(\boldsymbol{x}) &= \int_0^\infty \xi B(\xi)J_0(\xi r)\frac{\exp(-\nu_1|x_3-h|)}{\nu_1}d\xi\\ p^{(Tran)}(\boldsymbol{x}) &= \int_0^\infty \xi C(\xi)J_0(\xi r)\frac{\exp(-\nu_2 x_3-\nu_1 h)}{\nu_2}d\xi\end{aligned} \tag{3.6.5}$$

   where $B(\xi)$ and $C(\xi)$ are the functions of ξ to be determined by the boundary conditions shown in Eq. (3.6.2) and

$$\nu_j = \sqrt{\xi^2-\omega^2/\alpha_j^2},\ (j=1,2) \tag{3.6.6}$$

   Determine the functions $B(\xi)$ and $C(\xi)$ according to the boundary conditions shown in Eq. (3.6.2).
3. Approximate the reflected wavefield by the steepest descent path method, assuming that $\alpha_1 > \alpha_2$. Employ the branch cuts as well as the path of integral shown in Fig. 3.6.2 (a).
4. Assume that $\alpha_1 < \alpha_2$ and $k_2 = \omega/\alpha_2 < \xi_s$, where ξ_s is the saddle point. For this case, the branch line integral is required for the reflected wavefield. The branch cuts as well as the path of integral are illustrated in Fig. 3.6.2 (b). Evaluate the branch line integral.

---

[4] See Aki and Richards (2002), Chapter 6.2 and/or Landau and Lifshitz: fluid mechanics, § 72.

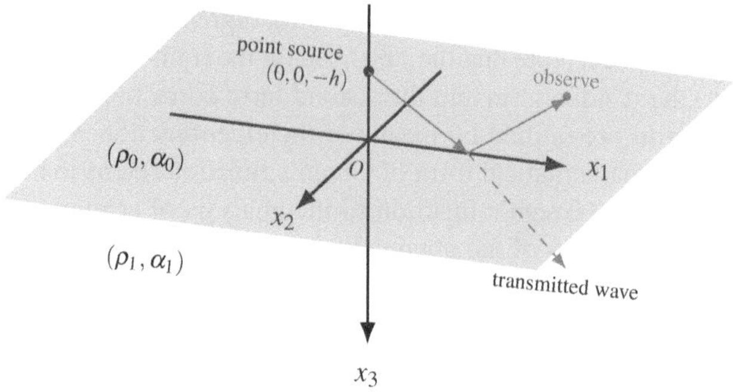

Figure 3.6.1 Sound wave propagation problem in which two liquid half spaces are in contact at $x_3 = 0$. The mass density as well as the sound velocity are denoted by ρ_j and α_j, $(j = 1, 2)$, respectively, where the subscript $j = 1$ is used for the region $x_3 < 0$ while $j = 2$ is used for $x_3 > 0$. A point source is applied on the x_3 axis of the negative region with the time factor $\exp(-i\omega t)$.

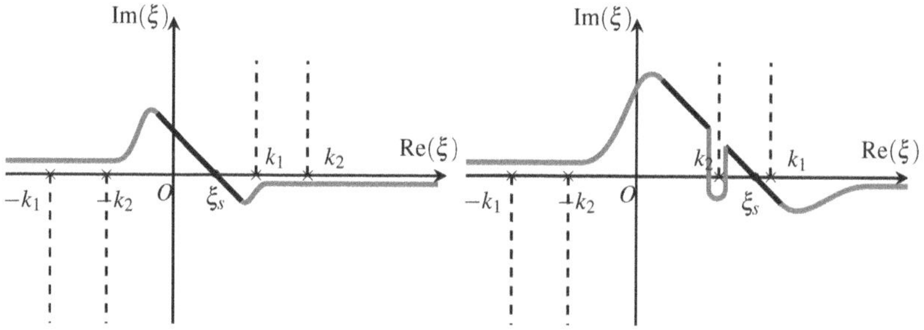

(a) Path of integral when $\alpha_1 > \alpha_2$.

(b) Path of integral when $\alpha_2 > \alpha_1$ and $\xi_s > k_2$.

Figure 3.6.2 Steepest descent paths in the complex wavenumber plane for the reflected wavefield, where ξ_s is the saddle point, $k_1 = \omega/\alpha_1$ and $k_2 = \omega/\alpha_2$, respectively. For the case of $\alpha_2 > \alpha_1$ and $\xi_s > k_2$, the branch line integral is necessary in the steepest descent path.

4 Analysis of scattering problems by means of Green's functions

4.1 APPLICATION OF REPRESENTATION THEOREM TO SOLID-FLUID INTERACTION PROBLEM

This chapter deals with computational methods in which Green's functions derived in Chapters 2 and 3 play important roles in scattering problems. We now discuss these problems in the frequency domain. We abbreviate the symbol ^ used for functions in the frequency domain, as was done in Chapter 2.

4.1.1 DEFINITION OF PROBLEM AND BASIC EQUATIONS

The representation theorem for the elastic wave equation discussed in §2.5 is a valuable tool for analyzing engineering problems. In this section, we explore the application of the representation theorem to the analysis of solid-fluid interactions. One of the significant engineering problems involving solid-fluid interaction analysis is the vibration of dam-foundation-reservoir systems. For instance, the author and Ohmachi (1993) presented a method for vibration analysis of dam-foundation-reservoir systems in the time domain, based on the representation theorem. The method combines the representation theorem with the finite element method in the time domain, using the method of weighted residuals, which yields the time stepping recurrence formula for vibration analysis of dam-foundation-reservoir systems.

However, the focus of this section is different from the analysis of dam-foundation-reservoir systems. Here, we deal with a wavefield in which a compressible fluid region is surrounded by an elastic solid in $\mathbb{R}^3$, as illustrated in Fig. 4.1.1. A sound wave, resulting from a disturbance in the fluid region, propagates towards the solid-fluid interface boundary, and we compute the fluid pressure at the boundary. While the problem shown in Fig. 4.1.1 may appear impractical, the formulation derived in this section finds practical applications in the analysis of underground energy storage facilities, which will be presented later in this section.

The fluid region, solid region, and interface boundary in Fig. 4.1.1 are denoted by Ω_f, Ω_s, and S, respectively, which are disjoint from each other and have the following property:

$$\Omega_f \cup S \cup \Omega_s = \mathbb{R}^3 \tag{4.1.1}$$

The normal vector at the solid-fluid interface boundary is denoted by $\boldsymbol{n}$, whose direction is outward from the fluid region.

DOI: 10.1201/9781003251729-4

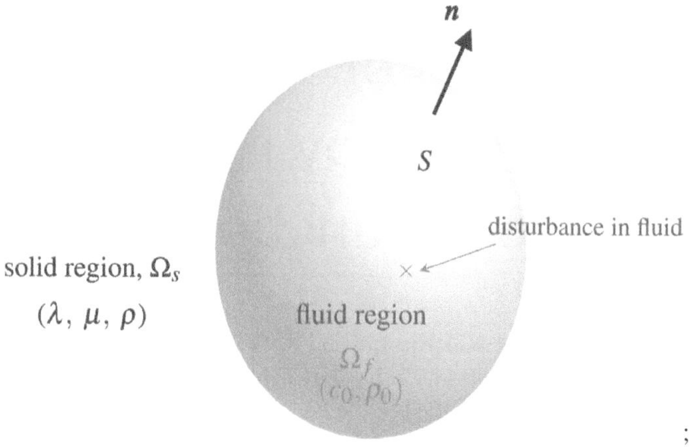

Figure 4.1.1 Concept of analysis of the wave problem in which the fluid region is surrounded by an elastic solid. A sound wave generated in the fluid region propagates toward the solid-fluid interface boundary. The sound velocity and mass density for the fluid are denoted by c_0 and ρ_0, whereas the Lamé parameters and mass density of the solid are denoted by λ, μ and ρ, respectively. In addition, the interface boundary between the solid-fluid region is denoted by S and the direction of the normal vector $\boldsymbol{n}$ at S is outward from the fluid region. The analysis is carried out in the frequency domain.

For the basic equation for the fluid region, we employ the following Helmholtz equation:

$$\nabla^2 q(\boldsymbol{x}) + \frac{\omega^2}{c_0^2} q(\boldsymbol{x}) = -f(\boldsymbol{x}), \qquad (\boldsymbol{x} \in \Omega_f) \tag{4.1.2}$$

where q is the fluid pressure, c_0 is the velocity of sound in the fluid, and f is the disturbance in the fluid region. Based on Eq. (2.4.8), the governing equation for the solid region is the following Navier-Cauchy equation:

$$(L_{ij}(\partial_1, \partial_2, \partial_3) + \rho\omega^2 \delta_{ij}) u_j(\boldsymbol{x}) = 0 \tag{4.1.3}$$

We impose the Sommerfeld-Kupradze radiation conditions shown in Eq. (2.5.21) on the elastic wavefield.

On the solid-fluid interface boundary, the sound pressure of the fluid and the vibration of the solid interact with each other according to the following equations:

$$p_i(\boldsymbol{y}) = n_i(\boldsymbol{y}) q(\boldsymbol{y}) \tag{4.1.4}$$

$$\frac{\partial q(\boldsymbol{y})}{\partial n} = -\rho_0 \omega^2 u_i(\boldsymbol{y}) n_i(\boldsymbol{y}), \quad (\boldsymbol{y} \in S) \tag{4.1.5}$$

where n_i is the component of the normal vector at the solid-fluid interface boundary, p_i is the traction of the solid, and ρ_0 is the mass density of the fluid, respectively. Equation (4.1.4) represents the equilibrium of the fluid pressure and traction of the solid at the solid-fluid interface boundary. Equation (4.1.5) is the contact condition for the solid and fluid at the solid-fluid interface boundary, where the velocity normal to the interface boundary of a fluid particle is equal to that of a solid particle.

4.1.2 REPRESENTATION THEOREM AND BOUNDARY INTEGRAL EQUATION

We begin the formulation of the problem by introducing the representation theorem for the solid and fluid regions. Based on Eq. (2.5.27), we have the following integral representation of the solution for the elastic wavefield:

$$u_i(\boldsymbol{x}) = -\int_S \Big[G_{ik}(\boldsymbol{x},\boldsymbol{y})p_k(\boldsymbol{y}) - P_{ik}(\boldsymbol{x},\boldsymbol{y})u_k(\boldsymbol{y})\Big]dS(\boldsymbol{y}), \quad (\boldsymbol{x}\in\Omega_s) \tag{4.1.6}$$

where G_{ik} and P_{ik} are Green's functions for the displacement and traction given in Eqs.(2.4.43) and (2.5.9), respectively. Note that the negative sign on the right-hand side of Eq. (4.1.6) is due to the direction of the normal vector at the solid-fluid interface boundary being inward to the solid region. In addition, we assume that the time factor of the wavefield is $\exp(i\omega t)$, therefore, G_{ik} in Eq. (4.1.6) denotes $\hat{G}_{ij}^{(-)}$ in Eq.(2.4.43).

Based on Eq. (N2.3.6), we have the following integral representation of the solution for the Helmholtz equation for Eq. (4.1.2):

$$\begin{aligned} q(\boldsymbol{x}) &= \int_S \Big[G(\boldsymbol{x},\boldsymbol{y})w(\boldsymbol{y}) - P(\boldsymbol{x},\boldsymbol{y})q(\boldsymbol{y})\Big]dS(\boldsymbol{y}) \\ &+ \int_{\Omega_f} G(\boldsymbol{x},\boldsymbol{y})f(\boldsymbol{y})dy, \quad (\boldsymbol{x}\in\Omega_f) \end{aligned} \tag{4.1.7}$$

where G is Green's function for the fluid region given by Eq. (N2.3.3) and P is the derivative of Green's function for the fluid region obtained from G such that

$$P(\boldsymbol{x},\boldsymbol{y}) = n_i(\boldsymbol{y})\frac{\partial}{\partial y_i}G(\boldsymbol{x},\boldsymbol{y}), \quad (\boldsymbol{y}\in S) \tag{4.1.8}$$

In addition, w is the gradient of the fluid pressure with respect to the normal direction of the solid-fluid interface boundary defined by

$$w(\boldsymbol{y}) = n_i(\boldsymbol{y})\frac{\partial}{\partial y_i}q(\boldsymbol{y}), \quad (\boldsymbol{y}\in S) \tag{4.1.9}$$

To couple Eqs. (4.1.6) and (4.1.7), we have to shift $\boldsymbol{x}\in\Omega_s$ for Eq. (4.1.6) and $\boldsymbol{x}\in\Omega_f$ for Eq. (4.1.7) to $\boldsymbol{x}\in S$ by taking the limit. To realize $\boldsymbol{x}\in S$ for both equations, we encounter the problem of the singularity of Green's function where $\boldsymbol{y}\in S$ matches $\boldsymbol{x}$.

For the limit of Eq. (4.1.6), we define $\tilde{S}$ and $\tilde{\Omega}_s$ from S and Ω_s, respectively, such that

$$\begin{aligned} \tilde{S} &= S_\varepsilon + S\setminus S_\varepsilon \\ \tilde{\Omega}_s &= V_\varepsilon + \Omega_s\setminus V_\varepsilon \end{aligned} \tag{4.1.10}$$

where V_ε and S_ε are a hemisphere of radius $\varepsilon > 0$ and its surface, respectively, and $\Omega_s \subset \tilde{\Omega}_s$, as shown in Fig. 4.1.2 (a). We take $\boldsymbol{x} \in S$ as the center of V_ε. Then, we see that $\boldsymbol{x} \in \tilde{\Omega}_s$ and the limit of $\boldsymbol{x} \in \Omega_s \to \boldsymbol{x} \in S$ is expressed as

$$\begin{aligned}
&\lim_{\varepsilon\to 0}\int_{\tilde{S}}\Big[G_{ik}(\boldsymbol{x},\boldsymbol{y})p_k(\boldsymbol{y}) - P_{ik}(\boldsymbol{x},\boldsymbol{y})u_k(\boldsymbol{y})\Big]dS(\boldsymbol{y}) \\
&= \lim_{\varepsilon\to 0}\int_{S_\varepsilon}\Big[G_{ik}(\boldsymbol{x},\boldsymbol{y})p_k(\boldsymbol{y}) - P_{ik}(\boldsymbol{x},\boldsymbol{y})u_k(\boldsymbol{y})\Big]dS(\boldsymbol{y}) \\
&\quad + \lim_{\varepsilon\to 0}\int_{S\setminus S_\varepsilon}\Big[G_{ik}(\boldsymbol{x},\boldsymbol{y})p_k(\boldsymbol{y}) - P_{ik}(\boldsymbol{x},\boldsymbol{y})u_k(\boldsymbol{y})\Big]dS(\boldsymbol{y}) \\
&= -(1/2)u_i(\boldsymbol{x}) + \text{P.V.}\int_S\Big[G_{ik}(\boldsymbol{x},\boldsymbol{y})p_k(\boldsymbol{y}) - P_{ik}(\boldsymbol{x},\boldsymbol{y})u_k(\boldsymbol{y})\Big]dS(\boldsymbol{y}) \quad (4.1.11)
\end{aligned}$$

For the derivation of the right-hand side of Eq. (4.1.11), we use the following singularities of Green's functions:

$$\begin{aligned}
G_{ij}(\boldsymbol{x},\boldsymbol{y}) &= O(r^{-1}) \\
P_{ij}(\boldsymbol{x},\boldsymbol{y}) &= \frac{1}{4\pi r^2}\Big[(c_T/c_L)^2\Big(-\delta_{ij}\frac{\partial r}{\partial n} + \frac{\partial r}{\partial y_i}n_j - \frac{\partial r}{\partial y_j}n_i\Big) \\
&\quad -3(1-(c_T/c_L)^2)\frac{\partial r}{\partial y_i}\frac{\partial r}{\partial y_j}\frac{\partial r}{\partial n}\Big] + O(r^{-1}) \qquad (4.1.12)
\end{aligned}$$

for the case $r = |\boldsymbol{x}-\boldsymbol{y}| \to 0$. The details of the derivation of Eq. (4.1.11) are shown in Note 4.1.

For the investigation of the limit of $\boldsymbol{x} \in \Omega_f \to \boldsymbol{x} \in S$ for Eq. (4.1.7), we also define $\tilde{S}'$ and $\tilde{\Omega}_f$ such that

$$\begin{aligned}
\tilde{S}' &= S'_\varepsilon + S\setminus S'_\varepsilon \\
\tilde{\Omega}_f &= V'_\varepsilon + \Omega_f\setminus V'_\varepsilon \qquad (4.1.13)
\end{aligned}$$

where $V_{\varepsilon'}$ and S'_ε are a hemisphere of radius $\varepsilon > 0$ and its surface, respectively, and $\Omega_f \subset \tilde{\Omega}_f$, as shown in Fig. 4.1.2 (b). We take $\boldsymbol{x} \in S$ as the center of V'_ε. Then, we see that $\boldsymbol{x} \in \tilde{\Omega}_f$ and the limit of $\boldsymbol{x} \in \Omega_f \to \boldsymbol{x} \in S$ is expressed as

$$\begin{aligned}
&\lim_{\varepsilon\to 0}\int_{\tilde{S}'}\Big[P(\boldsymbol{x},\boldsymbol{y})q(\boldsymbol{y}) - G(\boldsymbol{x},\boldsymbol{y})w(\boldsymbol{y})\Big]dS(\boldsymbol{y}) \\
&= \lim_{\varepsilon\to 0}\int_{S'_\varepsilon}\Big[P(\boldsymbol{x},\boldsymbol{y})q(\boldsymbol{y}) - G(\boldsymbol{x},\boldsymbol{y})w(\boldsymbol{y})\Big]dS(\boldsymbol{y}) \\
&\quad + \lim_{\varepsilon\to 0}\int_{S\setminus S'_\varepsilon}\Big[P(\boldsymbol{x},\boldsymbol{y})q(\boldsymbol{y}) - G(\boldsymbol{x},\boldsymbol{y})w(\boldsymbol{y})\Big]dS(\boldsymbol{y}) \\
&= -(1/2)q(\boldsymbol{x}) + \text{P.V.}\int_S\Big[P(\boldsymbol{x},\boldsymbol{y})q(\boldsymbol{y}) - G(\boldsymbol{x},\boldsymbol{y})w_k(\boldsymbol{y})\Big]dS(\boldsymbol{y}) \quad (4.1.14)
\end{aligned}$$

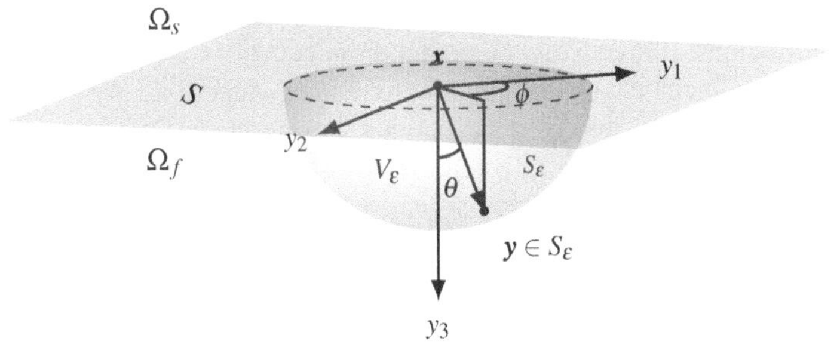

(a) Hemisphere for Eq. (4.1.11).

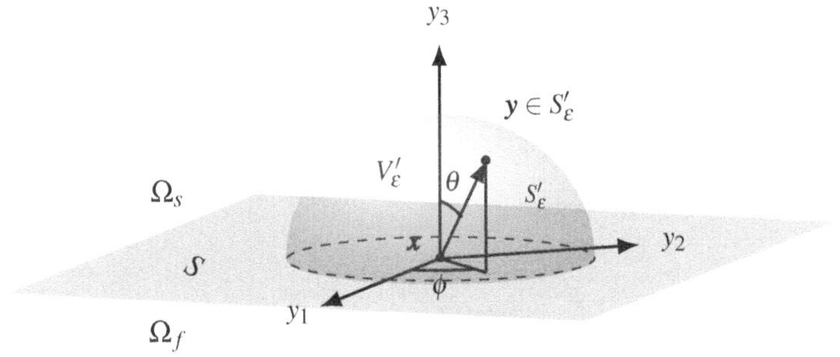

(b) Hemisphere for Eq. (4.1.14).

Figure 4.1.2 Hemispheres used to evaluate singularities of Green's functions in boundary integral equation and the coordinate (θ, ϕ) for the surface of the hemisphere.

For the derivation of the right-hand side of Eq. (4.1.14), we use the following singularities Green's functions:

$$\begin{aligned} G(\boldsymbol{x},\boldsymbol{y}) &= O(r^{-1}) \\ P(\boldsymbol{x},\boldsymbol{y}) &= -\frac{1}{4\pi r^2}\frac{\partial r}{\partial n} + O(r^{-1}) \end{aligned} \tag{4.1.15}$$

for the case $r = |\boldsymbol{x} - \boldsymbol{y}| \to 0$. The details of the derivation of Eq. (4.1.14) are shown in Note 4.1.

Now, as a result of taking the limit of Eqs. (4.1.11) and (4.1.14), we have the following integral equations for the solid and fluid regions, respectively:

$$(1/2)u_i(\boldsymbol{x}) = -\text{P.V.} \int_S \Big[G_{ik}(\boldsymbol{x},\boldsymbol{y})p_k(\boldsymbol{y}) - P_{ik}(\boldsymbol{x},\boldsymbol{y})u_k(\boldsymbol{y}) \Big] dS(\boldsymbol{y}) \tag{4.1.16}$$

$$\begin{aligned} (1/2)q(\boldsymbol{x}) &= \text{P.V.} \int_S \Big[G(\boldsymbol{x},\boldsymbol{y})w(\boldsymbol{y}) - P(\boldsymbol{x},\boldsymbol{y})q(\boldsymbol{y}) \Big] dS(\boldsymbol{y}) \\ &\quad + \int_{\Omega_f} G(\boldsymbol{x},\boldsymbol{y})f(\boldsymbol{y})d\boldsymbol{y}, \quad (\boldsymbol{x} \in S) \end{aligned} \tag{4.1.17}$$

We refer to Eqs. (4.1.16) and (4.1.17) as the *boundary integral equations* for the solid and fluid regions, respectively. Note that these equations are singular integral equations since they include the singular point for Green's functions. We couple Eqs. (4.1.16) and (4.1.17) via the interaction equations shown in Eqs. (4.1.4) and (4.1.5).

Note 4.1 Derivation of Eqs. (4.1.11) and (4.1.14)

Based on the singularities of Green's functions shown in Eq. (4.1.12) and the coordinate system (θ, ϕ) shown in Fig. 4.1.2 (a), we have

$$\left|\int_{S_\varepsilon} G_{ki}(\boldsymbol{x},\boldsymbol{y})p_i(\boldsymbol{y})dS(\boldsymbol{y})\right| \leq \varepsilon^2 O(\varepsilon^{-1}) \sup_{\boldsymbol{y}\in S_\varepsilon} |p_i(\boldsymbol{y})| \to 0, \ (\varepsilon \to 0) \tag{N4.1.1}$$

$$\int_{S_\varepsilon} P_{ki}(\boldsymbol{x},\boldsymbol{y})u_i(\boldsymbol{y})dS(\boldsymbol{y}) = \int_{S_\varepsilon} \left(\frac{1}{4\pi\varepsilon^2}P^{(s)}_{ki}(\boldsymbol{x},\boldsymbol{y}) + O(\varepsilon^{-1})\right)u_i(\boldsymbol{y})dS(\boldsymbol{y}) \tag{N4.1.2}$$

$$\begin{aligned}\int_{S_\varepsilon} \frac{1}{4\pi\varepsilon^2}P^{(s)}_{ki}(\boldsymbol{x},\boldsymbol{y})u_i(\boldsymbol{y})dS(\boldsymbol{y}) &= \frac{1}{4\pi}\int_0^{\pi/2} \sin\theta d\theta \int_0^{2\pi} d\phi\, P^{(s)}_{ki}(\boldsymbol{x},\boldsymbol{y})u_k(\boldsymbol{x}) \\ &+ \frac{1}{4\pi}\int_0^{\pi/2} \sin\theta d\theta \int_0^{2\pi} d\phi\, P^{(s)}_{ki}(\boldsymbol{x},\boldsymbol{y})\left(u_k(\boldsymbol{y}) - u_k(\boldsymbol{x})\right) \\ &\to -(1/2)u_k(\boldsymbol{x}), \ (\varepsilon \to 0)\end{aligned} \tag{N4.1.3}$$

for the evaluation of the left-hand side of Eq. (4.1.11). We assume that

$$|u_i(\boldsymbol{y}) - u_i(\boldsymbol{x})| = O(\varepsilon) \tag{N4.1.4}$$

for Eq. (N4.1.3). In Eq. (N4.1.3), $P^{(s)}_{ij}$ is expressed as

$$\begin{aligned}P^{(s)}_{ij}(\boldsymbol{x},\boldsymbol{y}) &= (c_T/c_L)^2\left(-\delta_{ij}\frac{\partial r}{\partial n} + \frac{\partial r}{\partial y_i}n_j - \frac{\partial r}{\partial y_j}n_i\right) \\ &\quad -3(1-(c_T/c_L)^2)\frac{\partial r}{\partial y_i}\frac{\partial r}{\partial y_j}\frac{\partial r}{\partial n}\end{aligned} \tag{N4.1.5}$$

where

$$\begin{aligned}\frac{\partial r}{\partial y_1} &= \sin\theta\cos\phi \ (= -n_1) \\ \frac{\partial r}{\partial y_2} &= \sin\theta\sin\phi \ (= -n_2) \\ \frac{\partial r}{\partial y_3} &= \cos\theta \ (= -n_3)\end{aligned} \tag{N4.1.6}$$

Note 4.1 Derivation of Eqs. (4.1.11) and (4.1.14) (continued)

Therefore, we have the right-hand side of Eq. (4.1.11) in the following form:

$$\begin{aligned}&\lim_{\varepsilon\to 0}\int_{S_\varepsilon+S\setminus S_\varepsilon}\left[G_{ik}(\boldsymbol{x},\boldsymbol{y})p_k(\boldsymbol{y})-P_{ik}(\boldsymbol{x},\boldsymbol{y})u_k(\boldsymbol{y})\right]dS(\boldsymbol{y})\\&=\;-(1/2)u_i(\boldsymbol{x})+\text{P.V.}\int_S\left[G_{ik}(\boldsymbol{x},\boldsymbol{y})p_k(\boldsymbol{y})-P_{ik}(\boldsymbol{x},\boldsymbol{y})u_k(\boldsymbol{y})\right]dS(\boldsymbol{y})\end{aligned}\tag{N4.1.7}$$

Likewise, based on the singularities of Green's functions shown in Eq. (4.1.15) and the coordinate system (θ,ϕ) shown in Fig. 4.1.2 (b), we have

$$\begin{aligned}&\lim_{\varepsilon\to 0}\int_{S_\varepsilon+S\setminus S_\varepsilon}\left[G(\boldsymbol{x},\boldsymbol{y})w(\boldsymbol{y})-P(\boldsymbol{x},\boldsymbol{y})q(\boldsymbol{y})\right]dS(\boldsymbol{y})\\&=\;(1/2)q(\boldsymbol{x})+\text{P.V.}\int_S\left[G(\boldsymbol{x},\boldsymbol{y})w(\boldsymbol{y})-P(\boldsymbol{x},\boldsymbol{y})q(\boldsymbol{y})\right]dS(\boldsymbol{y})\end{aligned}\tag{N4.1.8}$$

4.1.3 COUPLING OF BOUNDARY INTEGRAL EQUATIONS VIA DISCRETIZATION METHOD

Our task at present is to couple the boundary integral equations shown in Eqs. (4.1.16) and (4.1.17) via the interaction equations, namely Eqs. (4.1.4) and (4.1.5), after discretizing all these equations. For the discretization procedure, we approximate the boundary S using the sum of segments, which are referred to as boundary elements. The boundary S is approximated by

$$S\simeq\bigcup_{\alpha=1}S_\alpha\tag{4.1.18}$$

where S_α is a boundary element. $\{S_\alpha\}_{\alpha=1}^N$ is the set of boundary elements that are disjoint from each other, where N is the number of boundary elements. We assume that the boundary values are constant for a given boundary element. For this assumption, the boundary integral equations can be approximated by

$$\begin{aligned}&(1/2)u_k(\boldsymbol{y}_\beta)-\sum_{\alpha=1}^N\text{P.V.}\int_{S_\alpha}P_{ki}(\boldsymbol{y}_\beta,\boldsymbol{y})dS_\alpha(\boldsymbol{y})\,u_i(\boldsymbol{y}_\alpha)\\&=\;-\sum_{\alpha=1}^N\int_{S_\alpha}G_{ki}(\boldsymbol{y}_\beta,\boldsymbol{y})dS_\alpha(\boldsymbol{y})\,p_i(\boldsymbol{y}_\alpha)\end{aligned}\tag{4.1.19}$$

$$\begin{aligned}&(1/2)q(\boldsymbol{y}_\beta)+\sum_{\alpha=1}^N\text{P.V.}\int_{S_\alpha}P(\boldsymbol{y}_\beta,\boldsymbol{y})dS_\alpha(\boldsymbol{y})\,q(\boldsymbol{y}_\alpha)\\&=\;\sum_{\alpha=1}^N\int_{S_\alpha}G(\boldsymbol{y}_\beta,\boldsymbol{y})dS_\alpha(\boldsymbol{y})\,w(\boldsymbol{y}_\alpha)+\int_{V_f}G(\boldsymbol{y}_\beta,\boldsymbol{y})f(\boldsymbol{y})d\boldsymbol{y}\end{aligned}\tag{4.1.20}$$

Note that $\boldsymbol{y}_\alpha$ and $\boldsymbol{y}_\beta$ are defined by $\boldsymbol{y}_\alpha\in S_\alpha$ and $\boldsymbol{y}_\beta\in S_\beta$, respectively, which are referred to as the *nodal points* of the boundary elements. Since we assume that the boundary values are constant for a given boundary element in the formulation in this section, we set the nodal point at the centroid of the boundary element. Numerical integration for each boundary element in Eqs. (4.1.19) and (4.1.20) yields the following algebraic equations:

$$\sum_{\alpha=1}^{N}\left[P_{ki}^{(\beta\alpha)}\right]\left(u_i^{(\alpha)}\right) = -\sum_{\alpha=1}^{N}\left[G_{ki}^{(\beta\alpha)}\right]\left(p_i^{(\alpha)}\right) \quad (4.1.21)$$

$$\sum_{\alpha=1}^{N} P^{(\beta\alpha)} q^{(\alpha)} = \sum_{\alpha=1}^{N} G^{(\beta\alpha)} w^{(\alpha)} + f^{(\beta)} \quad (4.1.22)$$

where $P_{ki}^{(\beta\alpha)}$, $G_{ki}^{(\beta\alpha)}$, $P^{(\beta\alpha)}$, $G^{(\beta\alpha)}$, and $f^{(\alpha)}$ are respectively obtained as

$$\begin{aligned}
P_{ki}^{(\beta\alpha)} &= (1/2)\delta_{ki}\delta_{\beta\alpha} - \text{P.V.}\int_{S_\alpha} P_{ki}(\boldsymbol{y}_\beta,\boldsymbol{y})dS_\alpha(\boldsymbol{y}) \\
G_{ki}^{(\beta\alpha)} &= \int_{S_\alpha} G_{ki}(\boldsymbol{y}_\beta,\boldsymbol{y})dS_\alpha(\boldsymbol{y}) \\
P^{(\beta\alpha)} &= (1/2)\delta_{\beta\alpha} + \text{P.V.}\int_{S_\alpha} P(\boldsymbol{y}_\beta,\boldsymbol{y})dS_\alpha(\boldsymbol{y}) \\
G^{(\beta\alpha)} &= \int_{S_\alpha} G(\boldsymbol{y}_\beta,\boldsymbol{y})dS_\alpha(\boldsymbol{y}) \\
f^{(\beta)} &= \int_{\Omega_f} G(\boldsymbol{y}_\beta,\boldsymbol{y})f(\boldsymbol{y})d\boldsymbol{y}
\end{aligned} \quad (4.1.23)$$

In addition, $u_i^{(\alpha)}$, $p_i^{(\alpha)}$, $q^{(\alpha)}$, and $w^{(\alpha)}$ are respectively defined by

$$\begin{aligned}
u_i^{(\alpha)} &= u_i(\boldsymbol{y}_\alpha) \\
p_i^{(\alpha)} &= p_i(\boldsymbol{y}_\alpha) \\
q^{(\alpha)} &= q(\boldsymbol{y}_\alpha) \\
w^{(\alpha)} &= w(\boldsymbol{y}_\alpha)
\end{aligned} \quad (4.1.24)$$

Note that $\left[P_{ki}^{(\beta\alpha)}\right]$ and $\left[G_{ki}^{(\beta\alpha)}\right]$ in Eq. (4.1.21) are 3×3 matrices, and $\left(p_i^{(\alpha)}\right)$ and $\left(u_i^{(\alpha)}\right)$ are column vectors with three components.

The solid-interaction equations shown in Eqs. (4.1.4) and (4.1.5) can also be discretized as

$$\begin{aligned}
p_i^{(\alpha)} &= n_i^{(\alpha)} q^{(\alpha)} \\
w^{(\alpha)} &= -\rho_0\omega^2 n_i^{(\alpha)} u_i^{(\alpha)}
\end{aligned} \quad (4.1.25)$$

where $n_i^{(\alpha)} = n_i(y_\alpha)$. Incorporating Eq. (4.1.25) into Eqs. (4.1.21) and (4.1.22) yields

$$\sum_{\alpha=1}^{N} \begin{bmatrix} P_{11}^{(\beta\alpha)} & P_{12}^{(\beta\alpha)} & P_{13}^{(\beta\alpha)} & G_{1i}^{(\beta\alpha)} n_i^{(\alpha)} \\ P_{21}^{(\beta\alpha)} & P_{22}^{(\beta\alpha)} & P_{23}^{(\beta\alpha)} & G_{2i}^{(\beta\alpha)} n_i^{(\alpha)} \\ P_{31}^{(\beta\alpha)} & P_{32}^{(\beta\alpha)} & P_{33}^{(\beta\alpha)} & G_{3i}^{(\beta\alpha)} n_i^{(\alpha)} \\ m_1^{(\alpha)} G^{(\beta\alpha)} & m_2^{(\alpha)} G^{(\beta\alpha)} & m_3^{(\alpha)} G^{(\beta\alpha)} & P^{(\beta\alpha)} \end{bmatrix} \begin{bmatrix} u_1^{(\alpha)} \\ u_2^{(\alpha)} \\ u_3^{(\alpha)} \\ q^{(\alpha)} \end{bmatrix} = \begin{bmatrix} 0 \\ 0 \\ 0 \\ f^{(\beta)} \end{bmatrix} \tag{4.1.26}$$

for $\beta = 1, \ldots, N$, where

$$m_i^{(\alpha)} = \rho_0 \omega^2 n_i^{(\alpha)} \tag{4.1.27}$$

A linear algebraic equation with a size of $4N \times 4N$ with respect to the solid displacement and fluid pressure at the solid-fluid interface boundary can be composed from Eq. (4.1.26). We can then compute the fluid pressure at the solid-fluid interface boundary due to an internal disturbance in the fluid region.

4.1.4 ANALYSIS OF SOUND PROPAGATION IN VIRTUAL LPG TANK

We present a numerical example of a solid-fluid interaction analysis that originates from a practical engineering problem, namely the analysis of sound propagation in a virtual LPG tank. Figure 4.1.3 shows a model of the virtual LPG tank [1]. It is assumed that the tank was constructed by excavation through hard bedrock. The tank is filled with LPG and its spatial extent is 630 m × 200 m × 50 m.

We discretize the surface of the LPG tank into boundary elements. The mesh division of the surface of the LPG tank is shown in Fig. 4.1.4. Most of the mesh cells are squares with dimensions of 3 m×3 m. Some mesh cells are triangles with a single side length of approximately 3 m. The number of boundary elements for the entire surface of the LPG tank is 21,480. Therefore, we have to solve a simultaneous equation with a size of 85920 × 85920.

The properties of the materials used in the analysis are summarized in Table 4.1. As shown, the velocity of sound in LPG is 0.750 km s^{-1}, the mass density of LPG is 0.507 g cm^{-3}, and the P and S wave velocities of the surrounding bedrock are 5.60 and 2.80 km s^{-1}, respectively. In addition, the mass density of the surrounding bedrock is 2.67 g cm^{-3}. We impose a Q value of 100 to represent the attenuation effects of wave propagation in the surrounding bedrock. By means of the Q value, we define the wavenumbers for the P and S waves as complex values such that

$$\begin{aligned} k_T &= \frac{\omega}{c_T}\left(1 - \frac{i}{2Q}\right) \\ k_L &= \frac{\omega}{c_L}\left(1 - \frac{i}{2Q}\right) \end{aligned} \tag{4.1.28}$$

[1]This analysis model uses an LPG underground rock storage tank constructed by JOGMEC as a commissioned project from the Agency for Natural Resources and Energy. We would like to thank JOGMEC for their cooperation in this analysis.

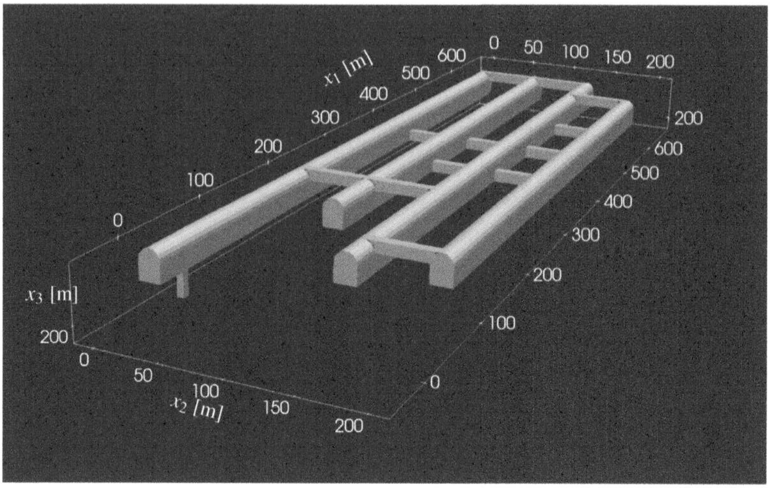

Figure 4.1.3 Model of virtual LPG tank surrounded by elastic solid.

Figures 4.1.5 (a)-(f) show the distribution of the fluid pressure at the free surface due to the disturbance of the point force, the location of which is shown in Fig. 4.1.4. The disturbance is expressed as

$$f(\boldsymbol{x}) = f_0 \delta(\boldsymbol{x} - \boldsymbol{x}_s) \tag{4.1.29}$$

for Eq. (4.1.2), where $\boldsymbol{x}_s$ and f_0 are the location and the amplitude the time harmonic point force, respectively. The analysis of the frequency range is 40 to 140 Hz and the amplitude of the force f_0 is 1 N. Figure 4.1.5 (a) shows that the frequency response of the fluid pressure at 40 Hz is very small. As the analysis frequency increases to 140 Hz, the fluid pressure also increases. The sound waves propagate to the adjacent tunnel when the analysis frequency exceeds 80 Hz, which shows that the sound waves can propagate to the connecting tunnels when they exceed a certain frequency. When the analysis frequency is 120 Hz, the fluid pressure takes the maximum value.

Table 4.1
Properties of materials used for analysis.

	Wave velocity (km s^{-1})	Mass density(g cm^{-3})	Q value
Surrounding bedrock	$c_L = 5.60,\ c_T = 2.80$	2.67	100
LPG	0.750	0.507	–

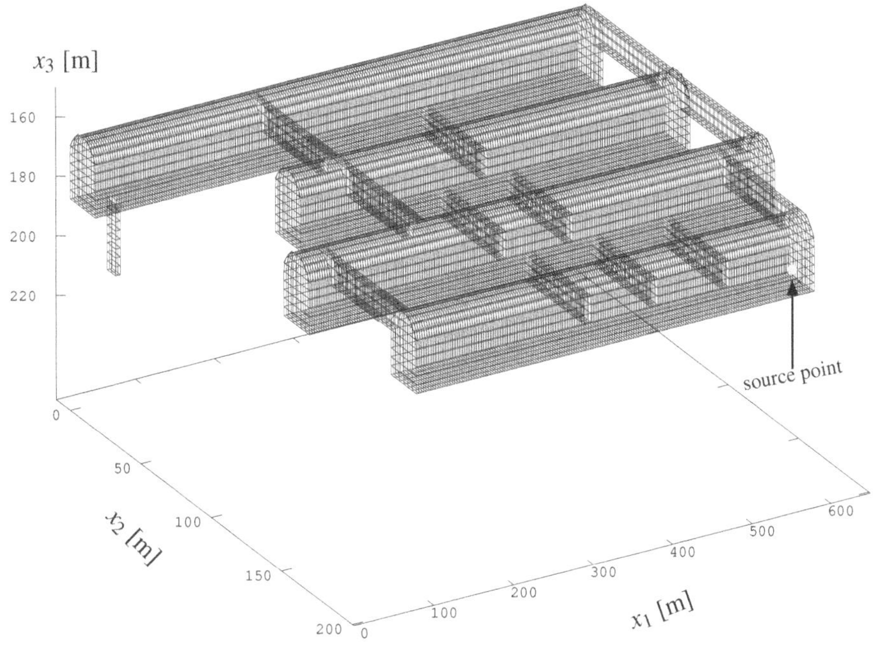

Figure 4.1.4 Boundary element discretization of virtual LPG tank and the location of the source point. Boundary elements are set at the surface wall of the tank.

When the analysis frequency is 140 Hz, the fluid pressure decreases compared to that for 120 Hz. These results verify the existence of a frequency in which the sound pressure in tunnels is significantly amplified.

4.2 APPLICATION OF THE SPECTRAL REPRESENTATION OF GREEN'S FUNCTION TO SCATTERING PROBLEM

4.2.1 OVERVIEW OF PROBLEM

In this section, we show an example of the application of the spectral representation of Green's function to a scattering problem. The spectral representation of Green's function and the generalized Fourier transform developed in § 3.5 are in rather abstract forms. Here, we show that the spectral representation and the generalized Fourier transform can be used as a computational method for a scattering problem[2]. Figure 4.2.1 shows the concept of the considered scattering problem, in

[2]This section incorporates the findings from a published article in 'A fast volume integral equation method for elastic wave propagation in a half space' by Touhei, T. , published in the International Journal of Solids and Structures (2011, Vol. 48, pages 3194-3208, Copyright Elsevier). Several modifications have been made to the content of this published article.

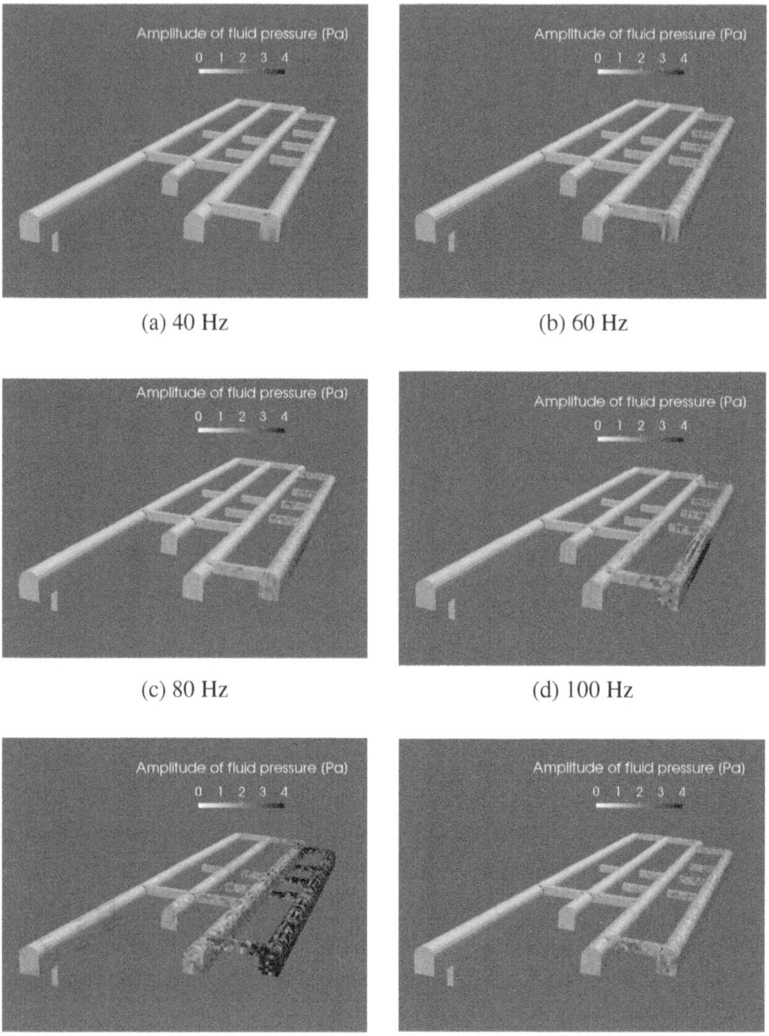

(a) 40 Hz (b) 60 Hz

(c) 80 Hz (d) 100 Hz

(e) 120 Hz (f) 140 Hz

Figure 4.1.5 Frequency response of pressure at surface of virtual LPG tank due to the disturbance of the unit point force.

which a point source is applied to a 3-D elastic half-space and scattered waves are generated by the interaction between the waves from the point source and the fluctuation of the wavefield. The effects of the fluctuation of the wavefield on the scattered

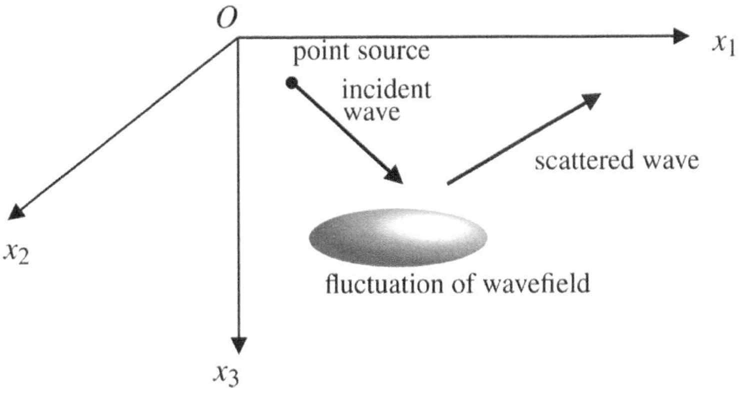

Figure 4.2.1 Concept of scattering problem for application of spectral form of Green's function in elastic half-space.

wavefield in an elastic half-space are described by the Lippmann-Schwinger equation (volume integral equation). The purpose of the discussion in this section is to develop a numerical method for the Lippmann-Schwinger equation by means of the generalized Fourier transform and to show the usefulness of this transform for a certain class of integral equations.

4.2.2 DERIVATION OF LIPPMANN-SCHWINGER EQUATION FOR SCATTERING PROBLEM

The fluctuation of the wavefield is expressed by the Lamé constants and the mass density of the background structure of the wavefield such that

$$\begin{aligned} \lambda(\boldsymbol{x}) &= \lambda_0 + \tilde{\lambda}(\boldsymbol{x}) \\ \mu(\boldsymbol{x}) &= \mu_0 + \tilde{\mu}(\boldsymbol{x}) \\ \rho(\boldsymbol{x}) &= \rho_0 + \tilde{\rho}(\boldsymbol{x}), \quad (\boldsymbol{x} \in \mathbb{R}^3_+) \end{aligned} \tag{4.2.1}$$

where λ_0 and μ_0 are the background Lamé constants, ρ_0 is the mass density of the background, and $\tilde{\lambda}$, $\tilde{\mu}$, and $\tilde{\rho}$ are their fluctuations, respectively. We assume that the region of the fluctuations are disjoint from the free surface of an elastic half-space. The governing equation for the wave problem is derived from the equilibrium equation

$$\partial_j \sigma_{ij} + \delta_{ij}\rho(\boldsymbol{x})u_j(\boldsymbol{x}) = -q_i\delta(\boldsymbol{x} - \boldsymbol{x}_s) \tag{4.2.2}$$

and the constitutive equation

$$\sigma_{ij} = \lambda(\boldsymbol{x})\delta_{ij}\partial_k u_k(\boldsymbol{x}) + \mu(\boldsymbol{x})\left(\partial_i u_j(\boldsymbol{x}) + \partial_j u_i(\boldsymbol{x})\right) \tag{4.2.3}$$

where $\boldsymbol{x}_s$ is the spatial point at which the point force is applied, q_i is the amplitude of the point source, σ_{ij} is the stress tensor, and u_i is the displacement field. The coupling of Eqs. (4.2.2) and (4.2.3) yields the following governing equation:

$$\begin{aligned}\left(L_{ij}(\partial_1,\partial_2,\partial_3)+\delta_{ij}\rho_0\omega^2\right)u_j(\boldsymbol{x})\\ = N_{ij}(\partial_1,\partial_2,\partial_3,\boldsymbol{x})u_j(\boldsymbol{x})-q_i\delta(\boldsymbol{x}-\boldsymbol{x}_s)\end{aligned} \tag{4.2.4}$$

where the operator L_{ij} was previously shown in Eq. (3.2.44) in the following form:

$$L_{ij}(\partial_1,\partial_2,\partial_3)=(\lambda_0+\mu_0)\partial_i\partial_j+\delta_{ij}\mu_0\partial_k\partial_k \tag{4.2.5}$$

and the operator N_{ij}, which is included due to the presence of the fluctuation of the wavefield, is expressed as

$$\begin{aligned}N_{ij}(\partial_1,\partial_2,\partial_3,\boldsymbol{x}) &= -\left(\tilde{\lambda}(\boldsymbol{x})+\tilde{\mu}(\boldsymbol{x})\right)\partial_i\partial_j-\delta_{ij}\tilde{\mu}(\boldsymbol{x})\partial_k\partial_k-\partial_i\tilde{\lambda}(\boldsymbol{x})\partial_j\\ &\quad -\delta_{ij}\partial_k\tilde{\mu}(\boldsymbol{x})\partial_k-\partial_j\tilde{\mu}(\boldsymbol{x})\partial_i-\delta_{ij}\tilde{\rho}(\boldsymbol{x})\omega^2\end{aligned} \tag{4.2.6}$$

The boundary condition for the present problem is given as

$$P_{ij}(\partial_1,\partial_2,\partial_3)u_j(\boldsymbol{x})=0, \quad (\text{at } x_3=0) \tag{4.2.7}$$

where the operator P_{ij} is given by Eq. (3.2.45).

Assume that the right-hand side of Eq. (4.2.4) is the inhomogeneous term of the equation. Then, the solution for Eq. (4.2.4), together with the boundary condition shown in Eq. (4.2.7), is expressed by the following Lippmann-Schwinger equation:

$$u_i(\boldsymbol{x})=f_i(\boldsymbol{x},\boldsymbol{x}_s)-\int_{\mathbb{R}^3_+}G_{ij}(\boldsymbol{x},\boldsymbol{y})N_{jk}(\partial_1,\partial_2,\partial_3,\boldsymbol{y})u_k(\boldsymbol{y})d\boldsymbol{y} \tag{4.2.8}$$

where f_i is the incident wavefield and G_{ij} is the Green's function that satisfies the free boundary condition.

By means of Green's function, the incident wavefield is expressed as

$$f_i(\boldsymbol{x},\boldsymbol{x}_s)=G_{ij}(\boldsymbol{x},,\boldsymbol{x}_s)q_j \tag{4.2.9}$$

It is convenient to express the Lippmann-Schwinger equation in terms of the scattered wavefield.

$$v_i(\boldsymbol{x})=u_i(\boldsymbol{x})-f_i(\boldsymbol{x},\boldsymbol{x}_s) \tag{4.2.10}$$

which becomes

$$\begin{aligned}v_i(\boldsymbol{x}) &= -\int_{\mathbb{R}^3_+}G_{ij}(\boldsymbol{x},\boldsymbol{y})N_{jk}(\partial_1,\partial_2,\partial_3,\boldsymbol{y})f_k(\boldsymbol{y},\boldsymbol{x}_s)d\boldsymbol{y}\\ &\quad -\int_{\mathbb{R}^3_+}G_{ij}(\boldsymbol{x},\boldsymbol{y})N_{jk}(\partial_1,\partial_2,\partial_3,\boldsymbol{y})v_k(\boldsymbol{y})d\boldsymbol{y}\end{aligned} \tag{4.2.11}$$

4.2.3 APPLICATION OF GENERALIZED FOURIER TRANSFORM TO LIPPMANN-SCHWINGER EQUATION

The generalized Fourier transform and its inverse are presented in Eqs. (3.5.86) and (3.5.87), respectively, in Chapter 3. These transforms are respectively expressed as

$$\begin{aligned}
\left(\mathscr{U}_{i''j}u_j\right)(\boldsymbol{\xi}) &= \left(\mathscr{F}^{(v)}_{i''m'}T^*_{jm'}\mathscr{F}^{(h)}u_j\right)(\boldsymbol{\xi}) \\
&= \int_{\mathbb{R}^3_+}\Lambda^*_{ji''}(\boldsymbol{\xi},\boldsymbol{x})u_j(\boldsymbol{x})d\boldsymbol{x}, \quad (\boldsymbol{\xi}\in\sigma_p\cup\sigma_c\subset\mathbb{R}^3_+) \\
\left(\mathscr{U}^{-1}_{ij''}u_{j''}\right)(\boldsymbol{x}) &= \left(\mathscr{F}^{(h)-1}T_{im'}\mathscr{F}^{(v)-1}_{m'j''}u_{j''}\right)(\boldsymbol{x}) \\
&= \int_{\mathbb{R}^2}\sum_{\boldsymbol{\xi}\in\sigma_p}\Lambda_{ij''}(\boldsymbol{\xi},\boldsymbol{x})u_{j''}(\boldsymbol{\xi})d\xi_1 d\xi_2 \\
&\quad +\int_{\mathbb{R}^2}\int_{\xi_r}^{\infty}\Lambda_{ij''}(\boldsymbol{\xi},\boldsymbol{x})u_{j''}(\boldsymbol{\xi})d\xi_3 d\xi_1 d\xi_2, \quad (\boldsymbol{x}\in\mathbb{R}^3_+)
\end{aligned} \tag{4.2.12}$$

The notation for the generalized Fourier transform is given in § 3.5. The Fourier transform and its inverse with respect to the horizontal coordinate system are denoted by $\mathscr{F}^{(h)}$ and $\mathscr{F}^{(h)-1}$, respectively, given in Eq. (3.5.3), those with respect to the vertical coordinate system are denoted by $\mathscr{F}^{(v)}$ and $\mathscr{F}^{(v)-1}$, respectively, given in Eq. (3.5.84), and $T_{im'}$ is the unitary matrix given in Eq. (3.5.8). In addition, the subsets of the wavenumber domain σ_p and σ_c are presented in Eqs. (3.5.79) and (3.5.85).

Now, the problem is to apply the generalized Fourier transform to the following function:

$$u_i(\boldsymbol{x}) = \int_{\mathbb{R}^3_+} G_{ij}(\boldsymbol{x},\boldsymbol{y})f_j(\boldsymbol{y})d\boldsymbol{y} \tag{4.2.13}$$

which is expressed as the integral of the product of Green's function and a given function f, which is assumed to be in $L_2(\mathbb{R}^3_+)$. The application of the transform to Eq. (4.2.13) yields

$$\begin{aligned}
\left(\mathscr{U}_{m''i}u_i\right)(\boldsymbol{\xi}) &= \int_{\mathbb{R}^3_+}\Lambda^*_{im''}(\boldsymbol{\xi},\boldsymbol{x})\left[\int_{\mathbb{R}^3_+}G_{ij}(\boldsymbol{x},\boldsymbol{y})f_j(\boldsymbol{y})d\boldsymbol{y}\right]d\boldsymbol{x} \\
&= \int_{\mathbb{R}^3_+}\left[\int_{\mathbb{R}^3_+}\Lambda^*_{im''}(\boldsymbol{\xi},\boldsymbol{x})G_{ij}(\boldsymbol{x},\boldsymbol{y})d\boldsymbol{x}\right]f_j(\boldsymbol{y})d\boldsymbol{y}
\end{aligned} \tag{4.2.14}$$

From Eqs. (3.5.95) and (3.5.97), we have

$$\int_{\mathbb{R}^3_+}\Lambda^*_{im''}(\boldsymbol{\xi},\boldsymbol{x})G_{ij}(\boldsymbol{x},\boldsymbol{y})d\boldsymbol{x} = G(\boldsymbol{\xi})\Lambda^*_{jm''}(\boldsymbol{\xi},\boldsymbol{y}) \tag{4.2.15}$$

where

$$G(\boldsymbol{\xi}) = \frac{1}{\mu_0 \xi_3^2 - \rho_0 \omega^2 + i\varepsilon} \tag{4.2.16}$$

Note that the time factor of the wavefield is assumed to be $\exp(i\omega t)$ for the wavefield for Eq. (4.2.16).

Equation (4.2.14) becomes

$$\begin{aligned}
&\left(\mathscr{U}_{m''i} \int_{\mathbb{R}^3_+} G_{ij}(\cdot, \mathbf{y}) f_j(\mathbf{y}) d\mathbf{y}\right)(\boldsymbol{\xi}) \\
&= \int_{\mathbb{R}^3_+} G(\boldsymbol{\xi}) \Lambda^*_{jm''}(\boldsymbol{\xi}, \mathbf{y}) f_j(\mathbf{y}) d\mathbf{y} \\
&= G(\boldsymbol{\xi}) \left(\mathscr{U}_{m''j} f_j\right)(\boldsymbol{\xi})
\end{aligned} \tag{4.2.17}$$

We find from Eq. (4.2.17) that a function with the integral representation shown in Eq. (4.2.13) can be factorized into the product of $G(\boldsymbol{\xi})$ and the generalized Fourier transform of f_j.

By means of the properties of Eq. (4.2.17), let us apply the generalized Fourier transform to Eq. (4.2.11). Then, we have the following result:

$$\begin{aligned}
v_{m''}(\boldsymbol{\xi}) &= -G(\boldsymbol{\xi}) \left(\mathscr{U}_{m''j}(N_{jk} f_k)\right)(\boldsymbol{\xi}) \\
&\quad -G(\boldsymbol{\xi}) \left(\mathscr{U}_{m''j}(N_{jk}(\mathscr{U}^{-1}_{kl''} v_{l''}))\right)(\boldsymbol{\xi}), \quad (\boldsymbol{\xi} \in \sigma_p \cup \sigma_c)
\end{aligned} \tag{4.2.18}$$

For simplicity, let

$$\begin{aligned}
b_{m''}(\boldsymbol{\xi}) &= -G(\boldsymbol{\xi}) \left(\mathscr{U}_{m''j}(N_{jk} f_k)\right)(\boldsymbol{\xi}) \\
\mathscr{A}_{m''l''} &= -G(\boldsymbol{\xi})\, \mathscr{U}_{m''j} N_{jk} \mathscr{U}^{-1}_{kl''}
\end{aligned} \tag{4.2.19}$$

Then, Eq. (4.2.18) is modified as

$$v_{m''}(\boldsymbol{\xi}) = b_{m''}(\boldsymbol{\xi}) + \mathscr{A}_{m''l''} v_{l''}(\boldsymbol{\xi}), \quad (\boldsymbol{\xi} \in \sigma_p \cup \sigma_c) \tag{4.2.20}$$

Note that $\mathscr{A}_{m''l''}$ is the linear operator. Equation (4.2.20) can be characterized as the Lippmann-Schwinger equation in the wavenumber domain. Now, an alternative approach is available for Eq. (4.2.11). Rather than solving Eq. (4.2.11) directly, we can address the scattering problem by employing Eq. (4.2.20) by developing a fast algorithm for generalized transforms. In view of the above, we will formulate a numerical method for Eq. (4.2.20) and show its efficiency.

4.2.4 NUMERICAL METHOD FOR LIPPMANN-SCHWINGER EQUATION IN WAVENUMBER DOMAIN

4.2.4.1 Overview of structure of discretized form of generalized Fourier transform

We need to discretize the generalized Fourier transform to develop a numerical method for Eq. (4.2.20). By means of Eq. (4.2.12) (which is based on Eqs. (3.5.86) and (3.5.87)), the discretized form of the structure of the generalized Fourier transform can be expressed as:

$$\begin{aligned} u_{i''}(\boldsymbol{\xi}) &= \left(\mathscr{U}_{(D)i''j}u_j\right)(\boldsymbol{\xi}) \\ &= \left(\mathscr{F}^{(v)}_{(D)i''m'}\,T^*_{(D)jm'}\,\mathscr{F}^{(h)}_{(D)}u_j\right)(\boldsymbol{\xi}),\quad (\boldsymbol{\xi}\in D_\Xi) \\ u_i(\boldsymbol{x}) &= \left(\mathscr{U}^{-1}_{ij''}u_{j''}\right)(\boldsymbol{x}) \\ &= \left(\mathscr{F}^{(h)-1}_{(D)}\,T_{(D)im'}\,\mathscr{F}^{(v)-1}_{(D)m'j''}\,u_{j''}\right)(\boldsymbol{x}),\quad (\boldsymbol{x}\in D_X) \end{aligned} \tag{4.2.21}$$

where D_X and D_Ξ are the sets of the finite number of grid points for the space and wavenumber domains, respectively. They are respectively defined by

$$\begin{aligned} D_X &= \{(n_1\Delta x_1, n_2\Delta x_2, n_3\Delta x_3)\,|\,n_1\in\mathbb{N}_1, n_2\in\mathbb{N}_2, n_3\in\mathbb{N}_3\} \\ D_\Xi &= D_{\Xi_p}\cup D_{\Xi_c} \\ &= \{(r_1\Delta\xi_1, r_2\Delta\xi_2, \eta_R)\,|\,r_1\in\mathbb{N}_1, r_2\in\mathbb{N}_2, F_R(\xi_r,\eta_R)=0\} \\ &\quad \cup\left\{\left(r_1\Delta\xi_1, r_2\Delta\xi_2, (r_3^2\Delta\xi_3^2+\xi_r^2)^{1/2}\right)\,|\,r_1\in\mathbb{N}_1, r_2\in\mathbb{N}_2, r_3\in\mathbb{N}_3\right\} \end{aligned} \tag{4.2.22}$$

Note that the subscript (D) for the operators denotes that the operators are in the discretized form. For the definition of D_X and D_Ξ, we used $\mathbb{N}_1$, $\mathbb{N}_2$, and $\mathbb{N}_3$, which are sets of integers given by

$$\begin{aligned} \mathbb{N}_1 &= \{n\,|\,-N_1/2\le n<N_1/2\} \\ \mathbb{N}_2 &= \{n\,|\,-N_2/2\le n<N_2/2\} \\ \mathbb{N}_3 &= \{n\,|\,0\le n<N_3\} \end{aligned} \tag{4.2.23}$$

Δx_j $(j=1,2,3)$ are the intervals of the grid in the space domain, $\Delta\xi_j$ $(j=1,2)$ are the horizontal intervals of the grid in the wavenumber domain, and $\Delta\xi_3$ is the interval for discretizing $\mathscr{F}^{(v)}$ and $\mathscr{F}^{(v)-1}$. In addition, note that (N_1, N_2, N_3) defines the total number of grid points for the discretized space and wavenumber domains.

We also need to prepare the sets of the following finite number of grid points to discretize the operators $\mathscr{F}^{(v)}$ and $\mathscr{F}^{(v)-1}$:

$$\begin{aligned} D_{\hat{X}} &= \left\{(r_1\Delta\xi_1, r_2\Delta\xi_2, n_3\Delta x_3)\,|\,r_1\in\mathbb{N}_1, r_2\in\mathbb{N}_2, n_3\in\mathbb{N}_3\right\} \\ D_{\Xi_{c'}} &= \left\{\left(r_1\Delta\xi_1, r_2\Delta\xi_2, (c_L/c_T)(r_3'^2\Delta\xi_3^2+\xi_r^2)^{1/2}\right)\,|\,r_1\in\mathbb{N}_1, r_2\in\mathbb{N}_2, r_3'\in\mathbb{N}_3\right\} \end{aligned} \tag{4.2.24}$$

Now, we have to obtain the discretized forms of the operators $\mathscr{F}^{(h)}_{(D)}$, $\mathscr{F}^{(h)-1}_{(D)}$, $\mathscr{F}^{(v)}_{(D)}$, and $\mathscr{F}^{(v)-1}_{(D)}$. The explicit forms of $T_{(D)im'}$ and $T^*_{(D)im'}$ are directly derived from Eq. (3.5.8), which is almost self-evident. Therefore, the discussion of $T_{(D)im'}$ and $T^*_{(D)im'}$ is omitted in the remainder of this section.

4.2.4.2 Explicit forms of $\mathscr{F}_{(D)}^{(h)}$ and $\mathscr{F}_{(D)}^{(h)-1}$

Equation (3.5.3) shows that $\mathscr{F}_{(D)}^{(h)}$ and $\mathscr{F}_{(D)}^{(h)-1}$ in Eq. (4.2.21) are expressed by the standard discrete Fourier transforms in the following form:

$$
\begin{aligned}
&\left(\mathscr{F}_D^{(h)}[u]\right)(\hat{\boldsymbol{x}}^{(r_1,r_2,n_3)}) \\
&= \frac{\Delta x_1 \Delta x_2}{2\pi} \sum_{(n_1,n_2)\in \mathbb{N}_1\times\mathbb{N}_2} u(\boldsymbol{x}^{(n_1,n_2,n_3)}) \exp\left(-i\boldsymbol{x}^{(n_1,n_2,n_3)}\cdot\hat{\boldsymbol{x}}^{(r_1,r_2,n_3)}\right) \\
&\left(\mathscr{F}_D^{(h)-1}[u]\right)(\boldsymbol{x}^{(n_1,n_2,n_3)}) \\
&= \frac{\Delta \xi_1 \Delta \xi_2}{2\pi} \sum_{(r_1,r_2)\in \mathbb{N}_1\times\mathbb{N}_2} u(\hat{\boldsymbol{x}}^{(r_1,r_2,n_3)}) \exp\left(i\boldsymbol{x}^{(n_1,n_2,n_3)}\cdot\hat{\boldsymbol{x}}^{(r_1,r_2,n_3)}\right)
\end{aligned}
\tag{4.2.25}
$$

where

$$
\begin{aligned}
\boldsymbol{x}^{(n_1,n_2,n_3)} &= (\ n_1\Delta x_1,\ \ n_2\Delta x_2,\ \ n_3\Delta x_3\ ) \\
\hat{\boldsymbol{x}}^{(r_1,r_2,n_3)} &= (\ r_1\Delta \xi_1,\ \ r_2\Delta \xi_2,\ \ n_3\Delta x_3\ )
\end{aligned}
\tag{4.2.26}
$$

and

$$
\boldsymbol{x}^{(n_1,n_2,n_3)}\cdot\hat{\boldsymbol{x}}^{(r_1,r_2,n_3)} = n_1 r_1 \Delta x_1 \Delta \xi_1 + n_2 r_2 \Delta x_2 \Delta \xi_2 \tag{4.2.27}
$$

4.2.4.3 Decomposition of $\mathscr{F}_{(D)i''m'}^{(v)}$ into discrete Laplace and Fourier transforms

The operators $\mathscr{F}_{(D)i''m'}^{(v)}$ and $\mathscr{F}_{(D)m'j''}^{(v)-1}$ have rather complex forms. They can, however, be decomposed into the discrete Laplace, Fourier sine, and Fourier cosine transforms. For the case where the operator $\mathscr{F}_{(D)i''m'}^{(v)}$ is a map from $u_{m'}(\hat{\boldsymbol{x}})$, $(\hat{\boldsymbol{x}} \in D_{\hat{X}})$ to $u_{i''}(\boldsymbol{\xi})$, $(\boldsymbol{\xi} \in D_{\Xi_p})$, the operator is expressed as

$$
\begin{aligned}
&\left(\mathscr{F}_{(D)i''m'}^{(v)} u_{m'}\right)(\boldsymbol{\xi}^{(r_1,r_2,\eta_R)}) \\
&= \left(a_{i''m'}^{L\gamma}\mathscr{L}_{(D)\gamma}[u_{m'}] + a_{i''m'}^{L\nu}\mathscr{L}_{(D)\nu}[u_{m'}]\right)(\boldsymbol{\xi}^{(r_1,r_2,\eta_R)})
\end{aligned}
\tag{4.2.28}
$$

For the case where the operator $\mathscr{F}_{(D)i''m'}^{(v)}$ is a map from $u_{m'}(\hat{\boldsymbol{x}})$, $(\hat{\boldsymbol{x}} \in D_{\hat{X}})$ to $u_{i''}(\boldsymbol{\xi})$, $(\boldsymbol{\xi} \in D_{\Xi_c})$, the operator is expressed as

$$
\begin{aligned}
\left(\mathscr{F}_{(D)i''m'}^{(v)} u_{m'}\right)(\boldsymbol{\xi}^{(r_1,r_2,r_3)}) &= \Big(a_{i''m'}^{L\gamma}\mathscr{L}_{(D)\gamma}[u_{m'}] + a_{i''m'}^{Fc\bar{\gamma}}\, l_{\Xi_c\leftarrow\Xi_c'}\,\mathscr{F}_{(D)c\bar{\gamma}}[u_{m'}] \\
&\quad + a_{i''m'}^{Fs\bar{\gamma}}\, l_{\Xi_c\leftarrow\Xi_c'}\,\mathscr{F}_{s\bar{\gamma}}[u_{m'}] + a_{i''m'}^{Fc\bar{\nu}}\mathscr{F}_{(D)c\bar{\nu}}[u_{m'}] \\
&\quad + a_{i''m'}^{Fs\bar{\nu}}\mathscr{F}_{(D)s\bar{\nu}}[u_{m'}]\Big)(\boldsymbol{\xi}^{(r_1,r_2,r_3)})
\end{aligned}
\tag{4.2.29}
$$

where $\mathscr{L}_{(D)\gamma}$ and $\mathscr{L}_{(D)\nu}$ are the operators for the discrete Laplace transforms and $\mathscr{F}_{(D)c\bar{\gamma}}$, $\mathscr{F}_{(D)c\bar{\nu}}$, $\mathscr{F}_{(D)s\bar{\gamma}}$, and $\mathscr{F}_{(D)s\bar{\nu}}$ are the operators for the discrete Fourier cosine and sine transforms, respectively, whose explicit forms are summarized in Notes 4.2 and 4.3. In addition, $a^{L\gamma}_{i''m'}$, $a^{L\nu}_{i''m'}$, $a^{Fc\bar{\gamma}}_{i''m'}$, $a^{Fc\bar{\nu}}_{i''m'}$, $a^{Fs\bar{\gamma}}_{i''m'}$, and $a^{Fs\bar{\nu}}_{i''m'}$ are the functions of the wavenumbers, which are derived from eigenfunctions for the point and continuous spectra. The explicit forms of the functions for the wavenumber are summarized at the end of this section in Note 4.6. In addition, $l_{\Xi_c \leftarrow \Xi_{c'}}$ is a map from a function defined on $D_{\Xi_{c'}}$ to that defined on D_{Ξ_c} to adjust the grid interval between D_{Ξ_c} and $D_{\Xi_{c'}}$.

4.2.4.4 Decomposition of $\mathscr{F}^{(\nu)-1}_{(D)m'j''}$ into discrete Laplace and Fourier transforms

The operator $\mathscr{F}^{(\nu)-1}_{(D)m'j''}$ is a map from a function defined on D_Ξ to that on $D_{\hat{X}}$ and is expressed as

$$\begin{aligned}\left(\mathscr{F}^{(\nu)-1}_{(D)m'j''}u_{j''}\right)\left(\hat{\boldsymbol{x}}^{(r_1,r_2,n_3)}\right) &= \Big(\mathscr{L}^*_{(D)\gamma(p)}[a^{L*\gamma(p))}_{m'j''}u_{j''}] + \mathscr{L}^*_{(D)\nu(p)}[a^{L*\nu(p))}_{m'j''}\hat{u}_{j''}] \\ &\quad + \mathscr{L}^*_{(D)\gamma}[a^{L^*\gamma}_{m'j''}u_{j''}] + \mathscr{F}^*_{(D)c\bar{\gamma}}[a^{F^*c\bar{\gamma}}_{m'j''}l_{\Xi_c\to\Xi_{c'}}u_{j''}] \\ &\quad + \mathscr{F}^*_{(D)s\bar{\gamma}}[a^{F^*s\bar{\gamma}}_{m'j''}l_{\Xi_c\to\Xi_{c'}}u_{j''}] + \mathscr{F}^*_{(D)c\bar{\nu}}[a^{F^*c\bar{\nu}}_{m'j''}u_{j''}] \\ &\quad + \mathscr{F}^*_{(D)s\bar{\nu}}[a^{F^*s\bar{\nu}}_{m'j''}u_{j''}]\Big)\left(\hat{\boldsymbol{x}}^{(r_1,r_2,n_3)}\right) \qquad (4.2.30)\end{aligned}$$

where $\mathscr{L}^*_{(D)\gamma(p)}$, $\mathscr{L}^*_{(D)\nu(p)}$, and $\mathscr{L}^*_{(D)\gamma}$ are the operators for the discrete Laplace transform and $\mathscr{F}^*_{(D)c\bar{\nu}}$, $\mathscr{F}^*_{(D)c\bar{\gamma}}$, $\mathscr{F}^*_{(D)s\bar{\nu}}$, and $\mathscr{F}^*_{(D)s\bar{\gamma}}$ are the operators for the discrete Fourier cosine and sine transforms, which are shown in Notes 4.4 and 4.5. The functions of the wavenumber, namely $a^{L*\gamma(p))}_{m'j''}$, $a^{L*\nu(p))}_{m'j''}$, $a^{L*\gamma}_{m'j''}$, $a^{F*c\bar{\gamma}}_{m'j''}$, $a^{F*s\bar{\gamma}}_{m'j''}$, $a^{F*c\bar{\nu}}_{m'j''}$, and $a^{F*s\bar{\nu}}_{m'j''}$, are directly derived from the explicit forms of the eigenfunctions, which are summarized in Note 4.6 at the end of this section.

Note 4.2 Operators for discrete Laplace transforms in $\mathscr{F}^{(\nu)}_{(D)m''j'}$

The explicit forms of the discrete Laplace transforms presented in Eqs. (4.2.28) and (4.2.29) are

$$\begin{aligned}\left(\mathscr{L}_{(D)\gamma}[u]\right)\left(\boldsymbol{\xi}^{(r_1,r_2,\eta_R)}\right) &= \Delta x_3 \sum_{n_3=0}^{N-1} u\left(\hat{\boldsymbol{x}}^{(r_1,r_2,n_3)}\right)\exp(-\gamma^{(\eta_R)}x_3^{(n_3)}) \\ \left(\mathscr{L}_{(D)\gamma}[u]\right)\left(\boldsymbol{\xi}^{(r_1,r_2,r_3)}\right) &= \Delta x_3 \sum_{n_3=0}^{N-1} u\left(\hat{\boldsymbol{x}}^{(r_1,r_2,n_3)}\right)\exp(-\gamma^{(r_3)}x_3^{(n_3)}) \\ \left(\mathscr{L}_{(D)\nu}[u]\right)\left(\boldsymbol{\xi}^{(r_1,r_2,\eta_R)}\right) &= \Delta x_3 \sum_{n_3=0}^{N-1} u\left(\hat{\boldsymbol{x}}^{(r_1,r_2,n_3)}\right)\exp(-\nu^{(\eta_R)}x_3^{(n_3)}) \qquad (N4.2.1)\end{aligned}$$

where

$$\begin{aligned}\gamma^{(\eta_R)} &= \sqrt{\xi_r^2 - (c_T/c_L)^2\eta_R^2} \\ \gamma^{(r_3)} &= \sqrt{\xi_r^2 - (c_T/c_L)^2(\xi_r^2 + r_3^2\Delta\xi_3^2)} \\ \nu^{(\eta_R)} &= \sqrt{\xi_r^2 - \eta_R^2} \qquad (N4.2.2)\end{aligned}$$

Note 4.3 Operators for discrete Fourier transforms in $\mathscr{F}^{(v)}_{(D)m''j'}$

The explicit forms of the discrete Fourier transforms presented in Eq. (4.2.29) are

$$
\begin{aligned}
\left(\mathscr{F}_{(D)c\bar{\gamma}}[u]\right)\left(\boldsymbol{\xi}^{(r_1,r_2,r_3')}\right) &= \Delta x_3 \sum_{n_3=0}^{N-1} u\left(\hat{\boldsymbol{x}}^{(r_1,r_2,n_3)}\right) \cos\left(\bar{\gamma}^{(r_3')} x_3^{(n_3)}\right) \\
\left(\mathscr{F}_{(D)s\bar{\gamma}}[u]\right)\left(\boldsymbol{\xi}^{(r_1,r_2,r_3')}\right) &= \Delta x_3 \sum_{n_3=0}^{N-1} u\left(\hat{\boldsymbol{x}}^{(r_1,r_2,n_3)}\right) \sin\left(\bar{\nu}^{(r_3')} x_3^{(n_3)}\right) \\
\left(\mathscr{F}_{(D)c\bar{\nu}}[u]\right)\left(\boldsymbol{\xi}^{(r_1,r_2,r_3)}\right) &= \Delta x_3 \sum_{n_3=0}^{N-1} u\left(\hat{\boldsymbol{x}}^{(r_1,r_2,n_3)}\right) \cos\left(\bar{\nu}^{(r_3)} x_3^{(n_3)}\right) \\
\left(\mathscr{F}_{(D)s\bar{\gamma}}[u]\right)\left(\boldsymbol{\xi}^{(r_1,r_2,r_3)}\right) &= \Delta x_3 \sum_{n_3=0}^{N-1} u\left(\hat{\boldsymbol{x}}^{(r_1,r_2,n_3)}\right) \sin\left(\bar{\gamma}^{(r_3)} x_3^{(n_3)}\right) \qquad \text{(N4.3.1)}
\end{aligned}
$$

where

$$
\begin{aligned}
\bar{\nu}^{(r_3)} &= r_3 \Delta\xi_3 \\
\bar{\gamma}^{(r_3')} &= r_3' \Delta\xi_3 \qquad \text{(N4.3.2)}
\end{aligned}
$$

Note 4.4 Operators for discrete Laplace transforms in $\mathscr{F}^{(v)-1}_{(D)m''j'}$

The explicit forms of the discrete Laplace transforms presented in Eq. (4.2.30) are

$$
\begin{aligned}
\left(\mathscr{L}^*_{(D)\gamma(p)}[u]\right)\left(\hat{\boldsymbol{x}}^{(r_1,r_2,n_3)}\right) &= u\left(\boldsymbol{\xi}^{(r_1,r_2,\eta_R)}\right) \exp\left(-\gamma^{(\eta_R)} x_3^{(n_3)}\right) \\
\left(\mathscr{L}^*_{(D)\nu(p)}[u]\right)\left(\hat{\boldsymbol{x}}^{(r_1,r_2,n_3)}\right) &= u\left(\boldsymbol{\xi}^{(r_1,r_2,\eta_R)}\right) \exp\left(-\nu^{(\eta_R)} x_3^{(n_3)}\right) \\
\left(\mathscr{L}^*_{(D)\gamma}[u]\right)\left(\hat{\boldsymbol{x}}^{(r_1,r_2,r_3)}\right) &= \Delta\xi_3 \sum_{r_3=0}^{N-1} u\left(\hat{\boldsymbol{x}}^{(r_1,r_2,r_3)}\right) \exp\left(-\gamma^{(r_3)} x_3^{(n_3)}\right) \qquad \text{(N4.4.1)}
\end{aligned}
$$

Note 4.5 Operators for discrete Fourier transforms in $\mathscr{F}^{(v)-1}_{(D)m''j'}$

The explicit forms of the discrete Fourier transforms presented in Eq. (4.2.30) are

$$
\begin{aligned}
\left(\mathscr{F}^*_{(D)c\bar{\gamma}}[u]\right)\left(\hat{\boldsymbol{x}}^{(r_1,r_2,n_3)}\right) &= \Delta\xi_3 \sum_{r_3=0}^{N-1} u\left(\boldsymbol{\xi}^{(r_1,r_2,r_3')}\right) \cos\left(\bar{\gamma}^{(r_3')} x_3^{(n_3)}\right) \\
\left(\mathscr{F}^*_{(D)s\bar{\gamma}}[u]\right)\left(\hat{\boldsymbol{x}}^{(r_1,r_2,n_3)}\right) &= \Delta\xi_3 \sum_{r_3=0}^{N-1} u\left(\boldsymbol{\xi}^{(r_1,r_2,r_3')}\right) \sin\left(\bar{\gamma}^{(r_3')} x_3^{(n_3)}\right) \\
\left(\mathscr{F}^*_{(D)c\bar{\nu}}[u]\right)\left(\hat{\boldsymbol{x}}^{(r_1,r_2,n_3)}\right) &= \Delta\xi_3 \sum_{n_3=0}^{N-1} u\left(\boldsymbol{\xi}^{(r_1,r_2,r_3)}\right) \cos\left(\bar{\nu}^{(r_3)} x_3^{(n_3)}\right) \\
\left(\mathscr{F}^*_{(D)s\bar{\nu}}[u]\right)\left(\hat{\boldsymbol{x}}^{(r_1,r_2,n_3)}\right) &= \Delta\xi_3 \sum_{r_3=0}^{N-1} u\left(\boldsymbol{\xi}^{(r_1,r_2,r_3)}\right) \sin\left(\bar{\nu}^{(r_3)} x_3^{(n_3)}\right) \qquad \text{(N4.5.1)}
\end{aligned}
$$

4.2.4.5 Fast algorithm for generalized Fourier transform

We can apply the fast Fourier transform to $\mathscr{F}^{(h)}_{(D)}$ and $\mathscr{F}^{(h)-1}_{(D)}$. We can also apply the fast Fourier transform and the fast Laplace transform (Strain, 1992) to $\mathscr{F}^{(v)}_{(D)}$ and $\mathscr{F}^{(v)-1}_{(D)}$. As a result, a fast algorithm for the generalized Fourier transform becomes possible, which enables us to develop a fast method for the Lippmann-Schwinger equation in the wavenumber domain shown in Eq. (4.2.20).

The fast algorithm for the discretized generalized Fourier transforms is shown in Fig. 4.2.2. For Fig. 4.2.2, the function $u_j(ip,kk)$ defined at the grid points in the space domain is transformed into the functions and $u_{j''}(ip,kk)$ defined at the grid points in the wavenumber domain, where ip $(ip = 1,\cdots,N_1N_2)$ and kk $(kk \times 1,\cdots,N_3)$ are parameters that define the horizontal and vertical components, respectively, of the grid points. Based on Eq. (4.2.21), the first step of the generalized Fourier transform is the application of $\mathscr{F}^{(h)}_{(D)}$ and the second step is the multiplication of $T^*_{(D)jm'}$. The application of $\mathscr{F}^{(h)}_{(D)}$, which is denoted as FFT2D in Fig. 4.2.2, must be repeated N_3 times and the multiplication of $T^*_{(D)jm'}$ must be repeated $N_1 \times N_2 \times N_3$ times. The final step of the transform, namely $\mathscr{F}^{(v)}_{(D)i''m'}$, which is denoted as FFTV1D in Fig. 4.2.2, must be repeated $N_1 \times N_2$ times. The computational complexity of FFT2D is $O\Big(N_1N_2\log(N_1N_2)\Big)$ and that of FFT1DV is $O(N_3 \log N_3)$ due to the 1D fast Fourier transform. Note that the computational complexity of the fast Laplace transform included in FFT1DV is $O(N_3)$, which is smaller than that of the fast Fourier transform. As a result, the computational complexity of the generalized Fourier transform is $O(N \log N)$, where $N = N_1N_2N_3$, which is the number of grid points for the elastic half-space.

4.2.4.6 Fast method for Lippmann-Schwinger equation in wavenumber domain

Now, let us return to the Lippmann-Schwinger equation in the wavenumber domain given in Eq. (4.2.20). We express the equation in discrete form as

$$v_{m''}(\boldsymbol{\xi}) = b_{m''}(\boldsymbol{\xi}) + \mathscr{A}_{(D)m''l''}\, v_{l''}(\boldsymbol{\xi}), \quad (\boldsymbol{\xi} \in D_{\Xi}) \tag{4.2.31}$$

where $\mathscr{A}_{(D)m''l''}$ is the discretized operator of $\mathscr{A}_{m''l''}$. Based on Eqs. (4.2.19) and (4.2.20), we have

$$\begin{aligned}\mathscr{A}_{(D)m''l''} &= -G_{(D)}(\boldsymbol{\xi})\, \mathscr{U}_{(D)m''j} N_{(D)jk} \mathscr{U}^{-1}_{(D)kl''} \\ &= -G_{(D)}(\boldsymbol{\xi})\, \mathscr{F}^{(v)}_{(D)m''p'} T^*_{(D)jp'} \mathscr{F}^{(h)}_{(D)} N_{(D)jk} \mathscr{F}^{(h)-1}_{(D)} T_{(D)kq'} \mathscr{F}^{(v)-1}_{(D)q'l''}\end{aligned} \tag{4.2.32}$$

```
do kk = 1,··· ,N3
    do ip = 1,··· ,N1 * N2
        w_j(ip) ← u_j(ip,kk),    (j = 1,2,3)
    end do
     apply FFT2D to w_j → y_j, (j = 1,2,3)
    do ip = 1,··· ,N1 * N2
        u_m'(ip,kk) ← T*_jm' y_j(ip), (i,j = 1,2,3)
    end do
end do

do ip = 1,··· ,N1 * N2
    do kk = 1,··· ,N3
        w_m'(kk) ← u_m'(ip,kk), (m' = 1,2,3)
    end do
     apply FFT1DV to w_m' → z_i'', (i'' = 1,2,3)
    do kk = 1,··· ,N3
     z_i''(kk) → û_i''(ip,kk), (i'' = 1,2,3)
    end do
end do
```

Figure 4.2.2 Fast algorithm for discretized form of generalized Fourier transform. Note that w_m, y_m, $u_{m'}$, and $z_{j''}$ are temporary work areas.

The fast algorithm is applicable to the above equation. To solve Eq. (4.2.20), the Krylov subspace iteration method can be used since it was designed for simultaneous equations in the following form:

$$A\vec{x} = \vec{b} \tag{4.2.33}$$

where A is a matrix, $\vec{x}$ is an unknown vector, and $\vec{b}$ is the given vector. The Krylov subspace is defined by

$$K_m = \text{span}\,\{\vec{b}, A\vec{b}, A^2\vec{b}, \ldots, A^m\vec{b}\} \tag{4.2.34}$$

where m is the number of iterations. In the Krylov subspace iteration method, the coefficients for the recurrence formula used to approximate the solution are determined during the construction of the orthonormal basis from K_m. It is possible to construct the Krylov subspace for Eq. (4.2.20) because $\mathscr{A}_{(D)m''l''}$ is a linear operator for a finite-dimensional vector space. Note that it is not necessary to specify all of the elements of $\mathscr{A}_{(D)m''l''}$ as the matrix; the specification for the linear transform is sufficient for the construction of the Krylov subspace. During the iterative process, the generalized Fourier transform and the inverse Fourier transform are used repeatedly.

Table 4.2
Properties of material used for background structure.

λ_0 [GPa]	μ_0 [GPa]	ρ_0 [g/cm^3]
4.0	2.0	2.0

4.2.5 NUMERICAL EXAMPLES

4.2.5.1 Definition of background structure of wavefield and its grid intervals

For the numerical computations, we provide a summary of the background Lamé constants and mass density of an elastic half-space in Table 4.2. Using this information, we can determine the P and S wave velocities to be 2 km/s and 1 km/s, respectively, whereas the Rayleigh wave velocity is approximately 0.93 km/s.

To facilitate the numerical computations, we need to define D_X and D_Ξ as presented in Eq. (4.2.22). For D_X, we set $N_1 = N_2 = 256$, $N_3 = 128$, and $\Delta x_1 = \Delta x_2 = \Delta x_3 = 0.25$ km. As a result, the values of $\Delta\xi_1$ and $\Delta\xi_2$ for D_Ξ are calculated as follows:

$$\Delta\xi_j = \frac{2\pi}{N_j \Delta x_j} = 0.09817 \text{ km}^{-1}, \ (j = 1, 2) \tag{4.2.35}$$

Similarly, the value of $\Delta\xi_3$ for D_Ξ is determined using the formula:

$$\Delta\xi_3 = \frac{\pi}{N_3 \Delta x_3} = 0.09817 \text{ km}^{-1} \tag{4.2.36}$$

It is nearly self-evident that sets $D_{\hat{X}}$ and $D_{\Xi_{c'}}$ defined by Eq. (4.2.24) can be uniquely determined by conditions mentioned above. We use the sets D_X, D_Ξ, $D_{\hat{X}}$ and $D_{\Xi_{c'}}$ for the analysis at a frequency of 1 Hz.

We have to be aware that the accuracy of Green's function composed by the generalized Fourier transform is significant in the analysis of scattered wavefield. In Appendix D, we evaluate the accuracy of Green's function based on the conditions defined earlier. We compare Green's function by the generalized Fourier transform with with the results of the Fourier-Hankel transform as well as the steepest descent path method in Appendix D. Our findings indicate that the accuracy of Green's function composed by the generalized Fourier transform exhibits a satisfactory level of precision.

4.2.5.2 Comparison of converged solution and Born approximation of Lippmann-Schwinger equation

We investigate the solution of Eq. (4.2.31) by examining the convergence properties of the solution as well as comparison with the Born approximation. For the analysis,

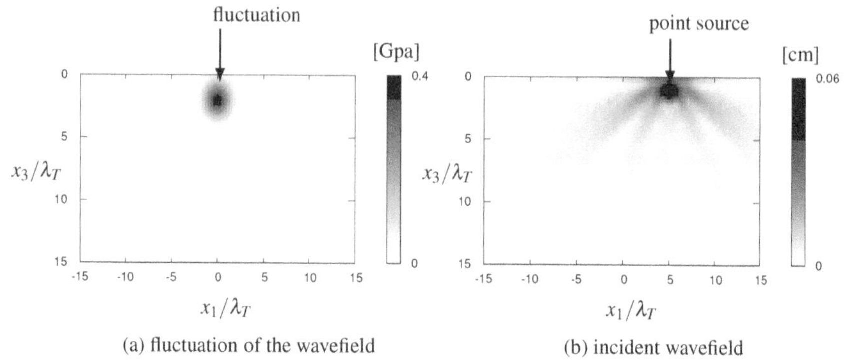

Figure 4.2.3 Fluctuations and incident wavefield.

we set the location of the point source to define the incident wavefield as:

$$\boldsymbol{x}_s = (5,0,1) \ \text{[km]} \tag{4.2.37}$$

The amplitude of the point source is 1×10^7 kN, acting in the vertical direction of excitation. The fluctuations of the wavefield are define as:

$$\begin{aligned} \tilde{\lambda}(\boldsymbol{x}) &= A_\lambda \exp(-\eta_\lambda |\boldsymbol{x} - \boldsymbol{x}_c|^2) \\ \tilde{\mu}(\boldsymbol{x}) &= A_\mu \exp(-\eta_\mu |\boldsymbol{x} - \boldsymbol{x}_c|^2) \end{aligned} \tag{4.2.38}$$

where $\boldsymbol{x}_c$ is the center of the fluctuation region, which is

$$\boldsymbol{x}_c = (0,0,2) \ \text{[km]} \tag{4.2.39}$$

The parameters A_λ, A_μ, η_λ, and η_μ for Eq. (4.2.38) describe the amplitudes and spreads of the fluctuations. They are given as

$$A_\lambda = A_\mu = 0.4 \ \text{[GPa]} \tag{4.2.40}$$

$$\eta_\lambda = \eta_\mu = 0.5 \ [\text{km}^{-2}] \tag{4.2.41}$$

The fluctuation model as well as the incident wavefield at the $x_2 = 0$ km plane are shown in Figs. 4.2.3 (a) and (b), which clarify how an incident wave propagates toward the fluctuation region. It's important to note that in Figs. 4.2.3 (a) and (b), the x_1 and x_3 axes are expressed in dimensionless form as x_1/λ_T and x_3/λ_T, respectively. Here, λ_T represents the wavelength of the S wave, which is 1 km in this specific case.

As mentioned, we solve Eq. (4.2.31) by means of the Krylov subspace iteration method. For the present numerical examples, we employ the Bi-CGSTAB method.

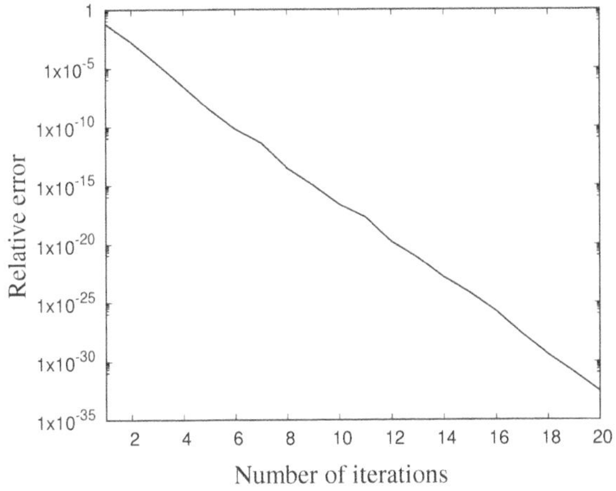

Figure 4.2.4 Convergence of solution obtained using Bi-CGSTAB method.

Figure 4.2.4 shows the convergence properties of the solution for Eq. (4.2.31) during the iterative process. Based on Eq. (4.2.31), the relative error is defined by

$$\varepsilon_r = \frac{\|v_{i''} - \mathscr{A}_{i''j''} v_{j''} - b_{i''}\|}{\|b_{i''}\|} \tag{4.2.42}$$

where $\|\cdot\|$ is the norm of the function in the wavenumber domain defined by

$$\|f_{i''}(\boldsymbol{\xi})\|^2 = \sum_{\boldsymbol{\xi}\in D_\Xi} \left[|f_{1''}(\boldsymbol{\xi})|^2 + |f_{2''}(\boldsymbol{\xi})|^2 + |f_{3''}(\boldsymbol{\xi})|^2\right] \tag{4.2.43}$$

Figure 4.2.4 shows that the relative error rapidly decreases; it is less than 1.0×10^{-5} after five iterations.

Figure 4.2.5 shows the effect of the number of iterations on the solution for Eq. (4.2.31). In the figure, the solution is presented along the x_1 axis at the free surface. The converged solution (result after 20 iterations) is compared with the solution obtained using an iterative procedure.

Figure 4.2.5(a) compares the converged solution and the Born approximation (no iterations). The approximate solution of Eq. (4.2.31) by using the Born approximation is given by

$$v_{m''}(\boldsymbol{\xi}) \sim b_{m''}(\boldsymbol{\xi}), \quad (\boldsymbol{\xi} \in D_\Xi) \tag{4.2.44}$$

There are differences in the displacement amplitude in the region just above the fluctuation region and a forward region. That is, the displacement amplitude just above the fluctuation region obtained using the Born approximation is larger than that obtained using the proposed method. On the other hand, the displacement amplitude in a forward region obtained using the Born approximation is smaller than the converged solution.

Figure 4.2.5(b) presents a comparison between the converged solution and the solution obtained from a single iteration. The discrepancy in the displacement amplitude just above the fluctuation region as well as in a forward region begin to resolve after a single iteration.

Figure 4.2.5 (c) presents a comparison between the converged solution and the solution obtained after two iterations. The difference between these solutions has almost disappeared. Two iterations are thus sufficient for investigating the surface response for the present numerical model.

Figures 4.2.6(a) and 4.2.6(b) show the Born approximation and the converged solution, respectively, for Eq. (4.2.31) for the distributions of displacement amplitudes at the free surface. The plots can be characterized by high-displacement-amplitude regions just above the fluctuation region as well as a forward region. The displacement amplitudes just above the fluctuation region obtained using the Born approximation are slightly stronger than those for the converged solution. On the other hand, the displacement amplitudes in the forward region just above the fluctuation region obtained using the Born approximation are weaker than those for the converged solution.

Figures 4.2.7 (a) and 4.2.7 (b) show the Born approximation and the converged solution for the distributions of displacement amplitudes in a vertical plane. The differences between these two solutions are very small along the free surface. On the other hand, the spread of the high-displacement-amplitude region around the fluctuation of the wavefield is different between these two solutions. That is, the spread of the high-displacement-amplitude region for the converged solution is narrower than that for the Born approximation.

The above numerical examples show the differences between the Born approximation and the converged solution. The efficiency of the proposed method and the required CPU time, which should be proportional to $N\log(N)$, where N is the number of grid points, are verified in previous studies (Touhei, 2009, 2011).

Note that the concept of the method used for the Lippmann-Schwinger equation is applicable not only to an elastic half-space. It should be also applicable to the analysis of partial differential equations expressed by a self-adjoint operator. We can express the solution for the equation using the Lippmann-Schwinger equation and introduce the spectral representation of Green's function into the Lippmann-Schwinger equation.

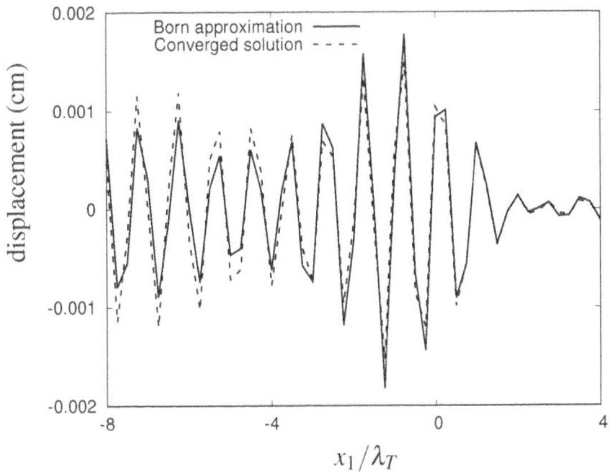

(a) Comparison of converged solution and Born approximation

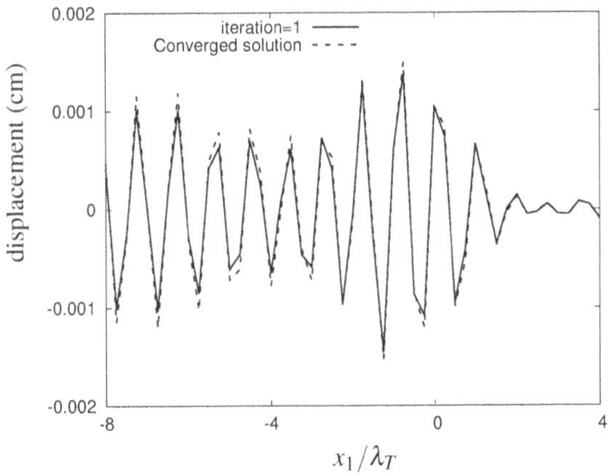

(b) Comparison of converged solution and solution obtained after one iteration

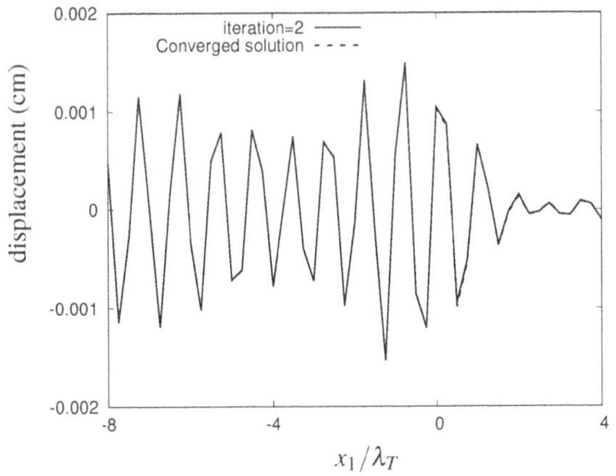

(c) Comparison of converged solution and solution obtained after two iterations

Figure 4.2.5 Comparison of displacement at free surface with respect to number of iterations.

Note 4.6 Explicit forms of $a_{m'j''}$

The functions of the wavenumber, namely $a_{m'j''}^{L\gamma}$, $a_{m'j''}^{L\nu}$, $a_{m'j''}^{Fc\bar{\gamma}}$, $a_{m'j''}^{Fs\bar{\gamma}}$, $a_{m'j''}^{Fc\bar{\nu}}$, $a_{m'j''}^{Fs\bar{\nu}}$, $a_{m'j''}^{L*\gamma(p))}$, $a_{m'j''}^{L*\nu(p))}$, $a_{m'j''}^{F*c\bar{\gamma}}$, $a_{m'j''}^{F*s\bar{\gamma}}$, $a_{m'j''}^{F*c\bar{\nu}}$ and $a_{m'j''}^{F*s\bar{\nu}}$ are directly derived from the explicit forms of the eigenfunctions $\psi_{i'j''}$ given in Appendix C.

For the case where $\xi \in \sigma_p$, these functions are

$$a_{1'1''}^{L*\gamma(p)} = -\gamma\Delta_1 \quad a_{1'1''}^{L*\nu(p)} = \xi_r^2\Delta_3 \quad a_{2'1''}^{L*\gamma(p)} = \xi_r\Delta_1 \quad a_{2'1''}^{L*\nu(p)} = -\xi_r\nu\Delta_3$$
$$a_{1'1''}^{L\gamma} = -\gamma\Delta_1 \quad a_{1'1''}^{L\nu} = \xi_r^2\Delta_3 \quad a_{1'2''}^{L\gamma} = \xi_r\Delta_1 \quad a_{1'2''}^{L\nu} = -\xi_r\nu\Delta_3 \qquad \text{(N4.6.1)}$$

where the coefficients Δ_1 and Δ_3 are for Eq. (C.3.15). For the case where $\xi \in \sigma_c$, where $\xi_r < \xi_3 < (c_L/c_T)\xi_r$, these functions are

$$a_{1'1''}^{L\gamma} = -\gamma\Delta_1, \qquad a_{1'1''}^{Fc\bar{\nu}} = \xi_r^2\Delta_3, \qquad a_{1'1''}^{Fs\bar{\nu}} = \xi_r\Delta_4,$$
$$a_{1'2''}^{L\gamma} = \xi_r\Delta_1, \qquad a_{1'2''}^{Fs\bar{\nu}} = -\xi_r\bar{\nu}\Delta_3, \qquad a_{1'2''}^{Fc\bar{\nu}} = \xi_r\bar{\nu}\Delta_4,$$
$$a_{1'1''}^{L^*\gamma} = -\frac{\gamma^2c_L^2}{\xi_3c_T^2}C_3, \qquad a_{1'1''}^{F^*c\bar{\nu}} = \frac{\bar{\nu}\xi_r^2}{\xi_3}\Delta_3 \qquad a_{1'1''}^{F^*s\bar{\nu}} = \frac{\bar{\nu}\xi_r^2}{\xi_3}\Delta_4 \qquad \text{(N4.6.2)}$$
$$a_{2'1''}^{L^*\gamma} = \frac{\xi_rc_L^2}{\xi_3c_T^2}\Delta_3, \qquad a_{2'1''}^{F^*s\bar{\nu}} = -\frac{\bar{\nu}^2\xi_r}{\xi_3}\Delta_3 \qquad a_{2'1''}^{F^*c\bar{\nu}} = \frac{\xi_r\bar{\nu}^2}{\xi_3}\Delta_4$$

where the coefficients Δ_1, Δ_3 and Δ_4 for Eq. (C.3.20). In addition, for the case where $\xi \in \sigma_c$, where $(c_L/c_T)\xi_r < \xi_3$, these functions are

$$a_{1''1'}^{Fc\bar{\gamma}} = \gamma\Delta_1, \qquad a_{1''1'}^{Fc\bar{\nu}} = \bar{\alpha}(\xi)\xi_r^2\Delta_1, \qquad a_{1''2'}^{Fs\bar{\gamma}} = \xi_r\Delta_1, \qquad a_{1''2'}^{Fs\bar{\nu}} = -\bar{\alpha}(\xi)\xi_r\bar{\nu}\Delta_1$$
$$a_{2''1'}^{Fs\bar{\gamma}} = -\gamma\Delta_2, \qquad a_{2''2'}^{Fs\bar{\nu}} = \beta(\xi)\xi_r^2\Delta_2, \qquad a_{2''2'}^{Fc\bar{\gamma}} = \xi_r\Delta_2, \qquad a_{2''2'}^{Fc\bar{\nu}} = \beta(\xi)\xi_r\bar{\nu}\Delta_2,$$
$$a_{1'1''}^{F^*c\bar{\gamma}} = \frac{\gamma^2c_L^2}{\xi_3c_T^2}\Delta_1, \qquad a_{1'1''}^{F^*c\bar{\nu}} = \frac{\bar{\nu}\bar{\alpha}\xi_r^2}{\xi_3}\Delta_1 \qquad a_{2'1''}^{F^*s\bar{\gamma}} = \frac{\bar{\gamma}\xi_rc_L^2}{\xi_3c_T^2}\Delta_1 \qquad a_{2'1''}^{F^*s\bar{\nu}} = -\frac{\bar{\nu}^2\xi_r\bar{\alpha}}{\xi_3}\Delta_1$$
$$a_{1'2''}^{F^*s\bar{\gamma}} = -\frac{\bar{\gamma}^2c_L^2}{\xi_3c_T^2}\Delta_2, \qquad a_{2'1''}^{F^*s\bar{\nu}} = \frac{\bar{\nu}\xi_r^2\beta}{\xi_3}\Delta_2 \qquad a_{2'2''}^{F^*c\bar{\gamma}} = \frac{\xi_r\bar{\gamma}c_L^2}{\xi_3c_T^2}\Delta_2 \qquad a_{2'2''}^{F^*c\bar{\nu}} = \frac{\beta\xi_r\bar{\nu}^2}{\xi_3}\Delta_2$$
$$\text{(N4.6.3)}$$

where the coefficients Δ_1 and Δ_2 are given by Eq. (C.3.35). Finally, $a_{3'3''}^{Fc\bar{\nu}}$ and $a_{3'3''}^{F^*c\bar{\nu}}$ have the following form for the region $\xi \in \sigma_c$:

$$a_{3'3''}^{Fc\bar{\nu}} = \xi_r\Delta_5, \quad a_{3'3''}^{F^*c\bar{\nu}} = \frac{\bar{\nu}\xi_r}{\xi_3}\Delta_5 \qquad \text{(N4.6.4)}$$

where Δ_5 is given by Eq. (C.3.26).

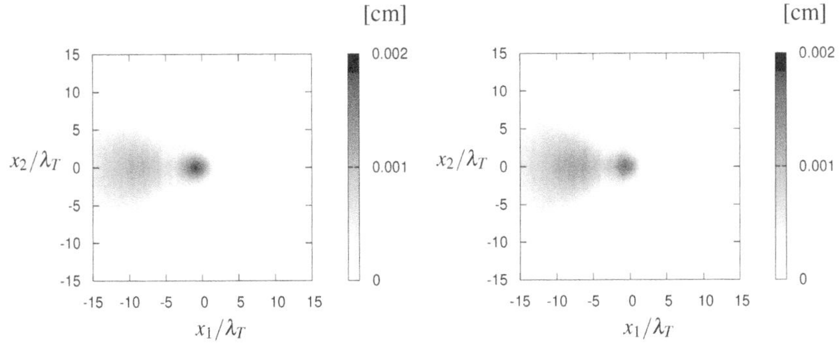

(a) Born approximation (b) Converged solution

Figure 4.2.6 Comparison of displacement amplitude at free surface between Born approximation and converged solution.

4.3 INVERSION OF POINT-LIKE SCATTERERS IN ELASTIC HALF-SPACE

4.3.1 DEFINITION OF PROBLEM AND BASIC NOTATION

So far, we have dealt with forward scattering problems. We have computed the scattered wavefield using the information of the media in which the wave propagates. In addition to forward scattering problems, we also have to be aware of the importance of inverse scattering problems, in which we determine the location, shape, and/or

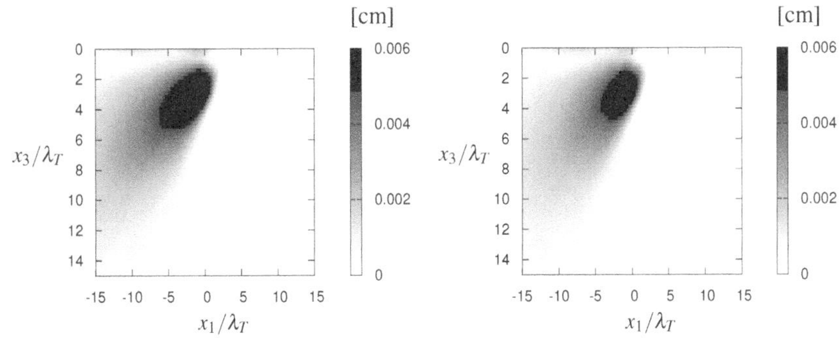

(a) Born approximation (b) Converged solution

Figure 4.2.7 Comparison of displacement amplitude at vertical plane between Born approximation and converged solution.

properties of the scatterers based on the observed scattered waves. The theories and applications of inverse scattering analysis are growing rapidly; because there is a vast literature, their overview is omitted here. In this section, our problem is to identify the locations of point-like scatterers in an elastic half-space using the assumption of the Born approximation [3]. We show the usefulness of Green's function for an elastic half-space approximated by the steepest descent path method as presented in §3.4.

Figure 4.3.1 shows the outline of the wave problem considered in this section. The wavefield is a 3D elastic half-space, in which the incident waves from point sources at the free surface propagate toward point-like scatterers embedded in the half-space. We observe the scattered waves that propagate back to the free surface. The problem considered in this section is to develop a method for identifying the locations of the point-like scatterers characterized by fluctuations of the Lamé parameters and mass density from the background structure of the wavefield. The developed method couples the MUSIC algorithm (see Appendix E) and a novel pseudo-projection method.

The concept of point-like scatterers is that the spatial scale of scatterers is very small compared to that of the wavelength of waves that propagate in the background structure of the wavefield. In addition, the spatial spread of the point-like scatterers is described by the Dirac impulse. In this context, the Lamé parameters and the mass density are expressed as

$$
\begin{aligned}
\lambda(\boldsymbol{x}) &= \lambda_0 + \sum_{\boldsymbol{y}_m \in E} \tilde{\lambda}_m \delta(\boldsymbol{x} - \boldsymbol{y}_m) \\
\mu(\boldsymbol{x}) &= \mu_0 + \sum_{\boldsymbol{y}_m \in E} \tilde{\mu}_m \delta(\boldsymbol{x} - \boldsymbol{y}_m) \\
\rho(\boldsymbol{x}) &= \rho_0 + \sum_{\boldsymbol{y}_m \in E} \tilde{\rho}_m \delta(\boldsymbol{x} - \boldsymbol{y}_m), \quad (\boldsymbol{x} \in \mathbb{R}^3_+)
\end{aligned} \tag{4.3.1}
$$

where λ and μ are the Lamé parameters and ρ is the mass density with background values λ_0, μ_0, and ρ_0, respectively. $\tilde{\lambda}_m$, $\tilde{\mu}_m$, and $\tilde{\rho}_m$ are the amplitudes of the fluctuations and $\boldsymbol{y}_m$ is the position of the point-like scatterers. Note that the set of the locations of the point-like scatterers is denoted as E. In the following, we sometimes express the fluctuations of the Lamé parameters and mass density as

$$
\begin{aligned}
\tilde{\lambda}(\boldsymbol{x}) &= \sum_{\boldsymbol{y}_m \in E} \tilde{\lambda}_m \delta(\boldsymbol{x} - \boldsymbol{y}_m) \\
\tilde{\mu}(\boldsymbol{x}) &= \sum_{\boldsymbol{y}_m \in E} \tilde{\mu}_m \delta(\boldsymbol{x} - \boldsymbol{y}_m) \\
\tilde{\rho}(\boldsymbol{x}) &= \sum_{\boldsymbol{y}_m \in E} \tilde{\rho}_m \delta(\boldsymbol{x} - \boldsymbol{y}_m), \quad (\boldsymbol{x} \in \mathbb{R}^3_+)
\end{aligned} \tag{4.3.2}
$$

[3]This section incorporates the findings from a published article in 'Pseudo-projection approach to reconstruct locations of point-like scatterers characterized by Lamé parameters and mass densities in an elastic half-space' by Touhei, T., published in the International Journal of Solids and Structures (2019, Vol.169, pages 187-204, Copyright Elsevier). Several modifications have been made to the content of this published article.

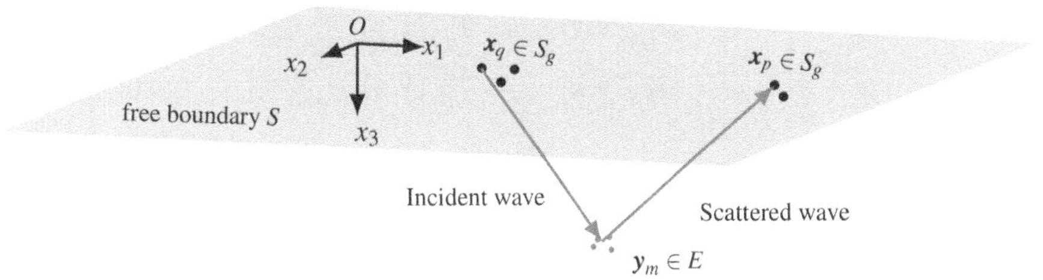

Figure 4.3.1 Schematic diagram of scattering problem. An incident wave generated from a point source propagates to point-like scatterers. We observe the scattered waves that propagate back to the free surface. Note that S is the free surface of an elastic half-space, whereas S_g is the set of position vectors for the source and receiver sensors defined in S such that $S_g \subset S$.

As notation for Green's functions, $G^{\nwarrow}_{ij}(\boldsymbol{x},\boldsymbol{y})$ and $G^{\swarrow}_{ij}(\boldsymbol{x},\boldsymbol{y})$ are used in this section to clarify the direction of the waves of a Green's function and the structure of the factorization of the far-field operator derived from the pseudo-projections. The definitions for this notation are

$$\begin{aligned} G^{\nwarrow}_{ij}(\boldsymbol{x},\boldsymbol{y}) &\stackrel{\text{def}}{=} G_{ij}(\boldsymbol{x},\boldsymbol{y}), \text{ for } \boldsymbol{x} \in S \text{ and } \boldsymbol{y} \in \mathbb{R}^3_+ \setminus S \\ G^{\swarrow}_{ij}(\boldsymbol{x},\boldsymbol{y}) &\stackrel{\text{def}}{=} G_{ij}(\boldsymbol{x},\boldsymbol{y}), \text{ for } \boldsymbol{x} \in \mathbb{R}^3_+ \setminus S \text{ and } \boldsymbol{y} \in S \end{aligned} \tag{4.3.3}$$

where S is the free surface of the elastic half-space, as shown in Fig. 4.3.1. Based on the reciprocity of Green's functions, the following relationship can be established:

$$G^{\nwarrow}_{ij}(\boldsymbol{x},\boldsymbol{y}) = G^{\swarrow}_{ji}(\boldsymbol{y},\boldsymbol{x}), \quad \boldsymbol{x} \in S, \ \boldsymbol{y} \in \mathbb{R}^3_+ \setminus S \tag{4.3.4}$$

The derivatives of Green's functions are also necessary in the following. The notation for the derivatives of Green's function $G^{\nwarrow}_{ij}(\boldsymbol{x},\boldsymbol{y})$ is

$$G^{\nwarrow}_{ij,k}(\boldsymbol{x},\boldsymbol{y}) = \partial_{y_k} G^{\nwarrow}_{ij}(\boldsymbol{x},\boldsymbol{y}), \qquad (\boldsymbol{x} \in S, \ \boldsymbol{y} \in \mathbb{R}^3_+ \setminus S, \ k = 1,2,3) \tag{4.3.5}$$

and that for the derivatives of Green's function $G^{\swarrow}_{ji}(\boldsymbol{y},\boldsymbol{x})$ is

$$G^{\swarrow}_{ji,k}(\boldsymbol{y},\boldsymbol{x}) = \partial_{y_k} G^{\swarrow}_{ji}(\boldsymbol{y},\boldsymbol{x}), \qquad (\boldsymbol{x} \in S, \boldsymbol{y} \in \mathbb{R}^3_+ \setminus S, \ k = 1,2,3). \tag{4.3.6}$$

The tensors formed by the derivatives of Green's functions are also important in the formulation of the proposed method (see below). They are

$$T^{\nwarrow}_{ijk}(\boldsymbol{x},\boldsymbol{y}) = (1/2)\Big(G^{\nwarrow}_{ij,k}(\boldsymbol{x},\boldsymbol{y}) + G^{\nwarrow}_{ik,j}(\boldsymbol{x},\boldsymbol{y})\Big) \tag{4.3.7}$$

$$T^{\swarrow}_{kji}(\boldsymbol{y},\boldsymbol{x}) = (1/2)\Big(G^{\swarrow}_{ji,k}(\boldsymbol{y},\boldsymbol{x}) + G^{\swarrow}_{ki,j}(\boldsymbol{y},\boldsymbol{x})\Big) \tag{4.3.8}$$

From the reciprocity and symmetry of Green's functions, the following relationships are established:

$$T^{\swarrow}_{kji}(\boldsymbol{y},\boldsymbol{x}) = T^{\nwarrow}_{ikj}(\boldsymbol{x},\boldsymbol{y}) \tag{4.3.9}$$

$$
\begin{aligned}
T^{\nwarrow}_{ijk}(\boldsymbol{x},\boldsymbol{y}) &= T^{\nwarrow}_{ikj}(\boldsymbol{x},\boldsymbol{y}) \\
T^{\swarrow}_{kji}(\boldsymbol{y},\boldsymbol{x}) &= T^{\swarrow}_{jki}(\boldsymbol{y},\boldsymbol{x})
\end{aligned}
\tag{4.3.10}
$$

4.3.2 METHOD FOR THE IDENTIFICATION OF THE LOCATION OF POINT-LIKE SCATTERERS

4.3.2.1 Representation of scattered wavefield

Let u_i and ε_{ij} be respectively the displacement field and strain tensor corresponding to the total field for the wave problem. We decompose the wavefield in the following form:

$$
\begin{aligned}
u_i(\boldsymbol{x}) &= u_i^{(0)}(\boldsymbol{x}) + u_i^{(s)}(\boldsymbol{x}) \\
\varepsilon_{ij}(\boldsymbol{x}) &= \varepsilon_{ij}^{(0)}(\boldsymbol{x}) + \varepsilon_{ij}^{(s)}(\boldsymbol{x})
\end{aligned}
\tag{4.3.11}
$$

where the superscripts (0) and (s) denote the incident background wavefield and the scattered wavefield, respectively. For the Born approximation, the decomposition of the stress tensor can be expressed as

$$
\begin{aligned}
\sigma_{ij}(\boldsymbol{x}) &= \lambda_0 \delta_{ij} \varepsilon_{kk}^{(0)}(\boldsymbol{x}) + 2\mu_0 \varepsilon_{ij}^{(0)}(\boldsymbol{x}) \\
&\quad + \lambda_0 \delta_{ij} \varepsilon_{kk}^{(s)}(\boldsymbol{x}) + 2\mu_0 \varepsilon_{ij}^{(s)}(\boldsymbol{x}) \\
&\quad + \tilde{\lambda}(\boldsymbol{x}) \delta_{ij} \varepsilon_{kk}^{(0)}(\boldsymbol{x}) + 2\tilde{\mu}(\boldsymbol{x}) \varepsilon_{ij}^{(0)}(\boldsymbol{x}).
\end{aligned}
\tag{4.3.12}
$$

Based on Eq. (4.2.11), along with the assumption of the Born approximation, the scattered wavefield can be represented by the following integral:

$$
u_i^{(s)}(\boldsymbol{x}) = -\int_\Omega G_{ij}(\boldsymbol{x},\boldsymbol{y}) N_{jk}(\partial_1,\partial_2,\partial_3,\boldsymbol{y}) u_k^{(0)}(\boldsymbol{y}) d\boldsymbol{y} \tag{4.3.13}
$$

where Ω is an arbitrary domain that includes the region for the point-like scatterers E and N_{jk} is the operator for the fluctuation of the wavefield. The operator N_{jk} has been already defined by Eq. (4.2.6) and is expressed as

$$
\begin{aligned}
N_{jk}(\partial_1,\partial_2,\partial_3,\boldsymbol{y}) &= -(\tilde{\lambda}(\boldsymbol{y}) + \tilde{\mu}(\boldsymbol{y}))\partial_j\partial_k - \delta_{jk}(\tilde{\mu}(\boldsymbol{y})\partial_l\partial_l + \tilde{\rho}(\boldsymbol{y})\omega^2) \\
&\quad -\partial_j\tilde{\lambda}(\boldsymbol{y})\,\partial_k - \partial_k\tilde{\mu}(\boldsymbol{y})\,\partial_j - \delta_{jk}\partial_l\tilde{\mu}(\boldsymbol{y})\,\partial_l.
\end{aligned}
\tag{4.3.14}
$$

As shown in Fig. 4.3.2, the domain Ω is surrounded by the boundaries Γ and S, where $\Gamma \cap S = \emptyset$. Based on the decomposition of the wavefield presented in Eq. (4.3.12), Eq. (4.3.13) can be modified as

$$
\begin{aligned}
u_i^{(s)}(\boldsymbol{x}) &= \int_\Omega G^{\nwarrow}_{ij}(\boldsymbol{x},\boldsymbol{y}) \Big(\partial_{y_k} (\lambda(\boldsymbol{y})\delta_{jk}\varepsilon_{ll}(\boldsymbol{y}) + 2\mu(\boldsymbol{y})\varepsilon_{jk}(\boldsymbol{y})) + \rho(\boldsymbol{y})\omega^2 u_j(\boldsymbol{y}) \Big) d\boldsymbol{y} \\
&\quad - \int_\Omega G^{\nwarrow}_{ij}(\boldsymbol{x},\boldsymbol{y}) \Big(\partial_{y_k} (\lambda_0\delta_{jk}\varepsilon_{ll}(\boldsymbol{y}) + 2\mu_0\varepsilon_{jk}(\boldsymbol{y})) + \rho_0\omega^2 u_j(\boldsymbol{y}) \Big) d\boldsymbol{y}
\end{aligned}
\tag{4.3.15}
$$

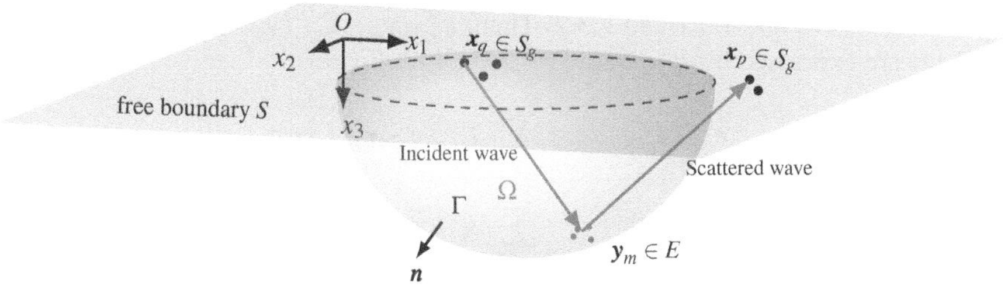

Figure 4.3.2 Region Ω that contains point-like scatterers. Region Ω is surrounded by the boundaries Γ and S, where $\Gamma \cap S = \emptyset$. Note that applying integration by parts to Eq. (4.3.17) does not yield the boundary terms for Γ and S.

Also in the context of the Born approximation, we used

$$\rho(\boldsymbol{x})u_i(\boldsymbol{x}) = \rho_0 u_i(\boldsymbol{x}) + \tilde{\rho}(\boldsymbol{x})u_i^{(0)}(\boldsymbol{x}) \tag{4.3.16}$$

for Eq. (4.3.15).

Integration by parts of the first term on the right-hand side of Eq. (4.3.15) yields

$$\begin{aligned}
&\int_\Omega G_{ij}^{\nwarrow}(\boldsymbol{x},\boldsymbol{y})\Big(\partial_{y_k}\big(\lambda(\boldsymbol{y})\delta_{jk}\varepsilon_{ll}(\boldsymbol{y}) + 2\mu(\boldsymbol{y})\varepsilon_{jk}(\boldsymbol{y})\big) + \rho(\boldsymbol{y})\omega^2 u_j(\boldsymbol{y})\Big)d\boldsymbol{y} \\
&= \int_\Omega \Big[-\lambda(\boldsymbol{y})T_{ikk}^{\nwarrow}(\boldsymbol{x},\boldsymbol{y})\varepsilon_{ll}(\boldsymbol{y}) - 2\mu(\boldsymbol{y})T_{ikl}^{\nwarrow}(\boldsymbol{x},\boldsymbol{y})\varepsilon_{kl}(\boldsymbol{y}) + \rho(\boldsymbol{y})\omega^2 G_{ij}^{\nwarrow}(\boldsymbol{x},\boldsymbol{y})u_j(\boldsymbol{y})\Big]d\boldsymbol{y} \\
&\quad + \int_\Gamma G_{ij}^{\nwarrow}(\boldsymbol{x},\boldsymbol{y})\, n_k(\boldsymbol{y})\,\big[\lambda_0\delta_{jk}\varepsilon_{ll}(\boldsymbol{y}) + 2\mu_0\varepsilon_{jk}(\boldsymbol{y})\big]d\Gamma(\boldsymbol{y})
\end{aligned} \tag{4.3.17}$$

where n_k is the component of the normal vector defined at the boundary Γ whose direction is outward from the region Ω. From Eq. (4.3.17), it is not difficult to see that the boundary terms generated by applying integration by parts to the first and second terms on the right-hand side of Eq. (4.3.15) cancel each other out. Therefore, introducing Eq. (4.3.1) into the results of the integration by parts of Eq. (4.3.15) yields

$$\begin{aligned}
u_i^{(s)}(\boldsymbol{x}) &= -\sum_{y_m \in E} \tilde{\lambda}_m T_{ikk}^{\nwarrow}(\boldsymbol{x},\boldsymbol{y}_m)\varepsilon_{ll}^{(0)}(\boldsymbol{y}_m) \\
&\quad - \sum_{y_m \in E} 2\tilde{\mu}_m T_{ikl}^{\nwarrow}(\boldsymbol{x},\boldsymbol{y}_m)\varepsilon_{kl}^{(0)}(\boldsymbol{y}_m) \\
&\quad + \sum_{y_m \in E} \tilde{\rho}_m\omega^2 G_{ij}^{\nwarrow}(\boldsymbol{x},\boldsymbol{y}_m)u_k^{(0)}(\boldsymbol{y}_m)
\end{aligned} \tag{4.3.18}$$

Consider the background wavefield resulting from point force $\{f_j(\boldsymbol{x}_q)\}, \boldsymbol{x}_q \in S_g$, where S_g represents the set of position vectors for the source and receiver sensors, as illustrated in Figs. 4.3.1 and 4.3.2. Then, $\varepsilon_{ll}^{(0)}(\boldsymbol{y}_m)$, $\varepsilon_{kl}^{(0)}(\boldsymbol{y}_m)$ and $u_k^{(0)}(\boldsymbol{y}_m)$ in Eq. (4.3.18) are expressed by

$$\begin{aligned}
\varepsilon_{ll}^{(0)}(\boldsymbol{y}_m) &= \sum_{\boldsymbol{x}_q \in S_g} T_{llj}^{\swarrow}(\boldsymbol{y}_m, \boldsymbol{x}_q) f_j(\boldsymbol{x}_q) \\
\varepsilon_{kl}^{(0)}(\boldsymbol{y}_m) &= \sum_{\boldsymbol{x}_q \in S_g} T_{klj}^{\swarrow}(\boldsymbol{y}_m, \boldsymbol{x}_q) f_j(\boldsymbol{x}_q) \\
u_k^{(0)}(\boldsymbol{y}_m) &= \sum_{\boldsymbol{x}_q \in S_g} G_{kj}^{\swarrow}(\boldsymbol{y}_m, \boldsymbol{x}_q) f_j(\boldsymbol{x}_q)
\end{aligned} \tag{4.3.19}$$

The substitution of Eq. (4.3.19) into Eq. (4.3.18) results in the following expression for scattered wave field at $\boldsymbol{x}_p \in S_g$:

$$\begin{aligned}
u_i^{(s)}(\boldsymbol{x}_p) &= \sum_{q=1}^{N} \mathscr{N}_{ij}(\boldsymbol{x}_p, \boldsymbol{x}_q) f_j(\boldsymbol{x}_q) \\
&= \sum_{q=1}^{N} \sum_{\boldsymbol{y}_m \in E} \Big(-\tilde{\lambda}_m T_{ikk}^{\nwarrow}(\boldsymbol{x}_p, \boldsymbol{y}_m) T_{llj}^{\swarrow}(\boldsymbol{y}_m, \boldsymbol{x}_q) \\
&\quad -2\tilde{\mu}_m T_{ikl}^{\nwarrow}(\boldsymbol{x}_p, \boldsymbol{y}_m) T_{klj}^{\swarrow}(\boldsymbol{y}_m, \boldsymbol{x}_q) \\
&\quad +\tilde{\rho}_m \omega^2 G_{ik}^{\nwarrow}(\boldsymbol{x}_p, \boldsymbol{y}_m) G_{kj}^{\swarrow}(\boldsymbol{y}_m, \boldsymbol{x}_q) \Big) f_j(\boldsymbol{x}_q), \\
&\quad (\boldsymbol{x}_p, \boldsymbol{x}_q \in S_g)
\end{aligned} \tag{4.3.20}$$

where $\mathscr{N}_{ij}(\cdot,\cdot)$ is referred to as the kernel of the near-field operator and N is the number of source and receiver sensors at S_g. Occasionally, Eq. (4.3.20) is also represented as

$$u_i^{(s)}(\boldsymbol{x}_p) = \Big(\mathscr{N}_{ij} f_j\Big)(\boldsymbol{x}_p), \ (\boldsymbol{x}_q \in S_g) \tag{4.3.21}$$

In the context of Eq. (4.3.21), $\mathscr{N}_{ij}$ can be interpreted as the near-field operator.

4.3.2.2 Pseudo-projections derived from far-field properties of Green's functions

We know that Green's function for an elastic half-space includes the contributions of P, SV, and SH waves, along with the Rayleigh wave, as discussed in §3.4.3. However, for the purposes of the current discussions, we will exclude the Rayleigh wave effects from Green's function, since our primary focus lies in the treatment of the near-field observation. Within these constraints, we introduce an approximation of Green's function derived from the steepest descent path method obtained in §3.4.3. Our main objective is to develop operators to isolate one type of waves such as P, SV and/or SH waves from the near field operator. We refer to these operators developed in our current discussions as *pseudo-projections*.

We express Eq. (3.4.36) discussed in §.3.4.3 in the following form:

$$G^{\nwarrow}_{ij}(\boldsymbol{x},\boldsymbol{y}) = \sum_{\#\in K} \frac{\exp(ik_{(\#)}|\boldsymbol{x}-\boldsymbol{y}|)}{4\pi|\boldsymbol{x}-\boldsymbol{y}|} D^{(\#)}_{ij}(\theta,\varphi) + O(|\boldsymbol{x}-\boldsymbol{y}|^{-2}) \quad (4.3.22)$$

where

$$K = \{T_V, T_H, L\} \quad (4.3.23)$$

and

$$k_{(T_V)} = k_{(T_H)} = k_T \quad (4.3.24)$$

In addition, $D^{(T_V)}$, $D^{(T_H)}$, and $D^{(L)}$ are the directivity tensors, which can be factorize in the following form:

$$D^{(\#)}_{ij}(\theta,\varphi) = A^{(\#)}(\theta) W^{(\#)}_i(\theta,\varphi) V^{(\#)}_j(\theta,\varphi), \quad (\# \in K) \quad (4.3.25)$$

where $A^{(\#)}$, $W^{(\#)}_i$, and $V^{(\#)}_j$ are given in Eqs. (3.4.38) to (3.4.41). The derivative of Green's function $G^{\nwarrow}_{ij,k}$ is derived from the dipole source. After an elaborate derivation process, the derivative of Green's functions is expressed as

$$G^{\nwarrow}_{ij,k}(\boldsymbol{x},\boldsymbol{y}) = \sum_{\#\in K} \frac{\exp(ik_{(\#)}|\boldsymbol{x}-\boldsymbol{y}|)}{4\pi|\boldsymbol{x}-\boldsymbol{y}|} D^{\#}_{ijk}(\theta,\varphi) + O(|\boldsymbol{x}-\boldsymbol{y}|^{-2}) \quad (4.3.26)$$

where the directivity tensor $D^{(\#)}_{ijk}$ can be factorized as

$$\begin{aligned} & D^{(\#)}_{ijk}(\theta,\varphi) \\ & = -ik_{(\#)} A^{(\#)}(\theta) W^{(\#)}_i(\theta,\varphi) V^{(\#)}_j(\theta,\varphi) V^{(L)}_k(\theta,\varphi), \quad (\# \in K) \end{aligned} \quad (4.3.27)$$

The angles θ and φ define the direction of the vector $\boldsymbol{x}-\boldsymbol{y}$, as shown in Fig. 4.3.3.

We now construct $W^{(\#\star)}_i$ that satisfies the following properties:

$$W^{(\#)}_i(\theta,\varphi) W^{(\#'\star)}_i(\theta,\varphi) = \delta_{\#\#'} W^{(\#)}_i(\theta,\varphi) W^{(\#'\star)}_i(\theta,\varphi) \quad (4.3.28)$$

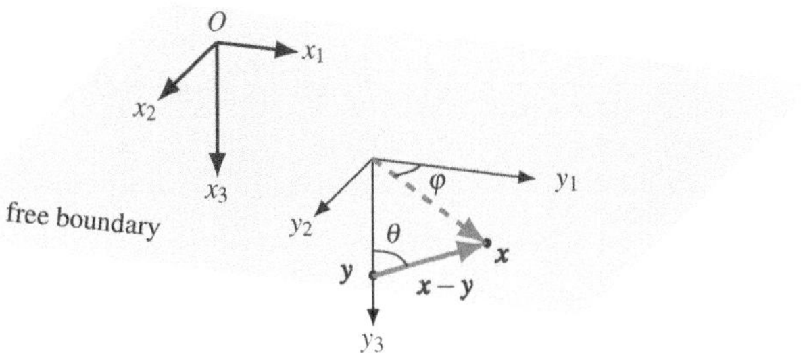

Figure 4.3.3 Definitions of angles θ and φ, which determine direction of $\boldsymbol{x}-\boldsymbol{y}$.

for $\#, \#' \in K$. The explicit forms of $W_i^{\#\star}$ are expressed as

$$\begin{aligned}
\left(W_i^{(L\star)}(\theta,\varphi)\right) &= \left(\cos\varphi\sin\theta \quad \sin\varphi\sin\theta \quad -\frac{\cos\theta\sin^2\theta}{\kappa^{(T_V)}(\theta)} \right) \\
\left(W_i^{(T_V\star)}(\theta,\varphi)\right) &= \left(\cos\varphi\sin\theta \quad \sin\varphi\sin\theta \quad -\frac{\sin^2\theta}{\kappa^{(L)}(\theta)} \right) \\
\left(W_i^{(T_H\star)}(\theta,\varphi)\right) &= \left(W_i^{(T_H)}(\theta,\varphi)\right)
\end{aligned} \tag{4.3.29}$$

where $\kappa^{(L)}$ and $\kappa^{(T_V)}$ are given in Eq. (3.4.40). We also introduce an operator $Q_{ij}^{(\#)}$ using the structure of the directivity tensors $D_{ij}^{(\#)}$ and $D_{ijk}^{(\#)}$ shown in Eqs. (4.3.25) and (4.3.27), along with $W_i^{(\#\star)}$ as follows:

$$Q_{ij}^{(\#)}(\theta,\varphi) = \frac{W_i^{(\#)}(\theta,\varphi)\,W_j^{(\#\star)}(\theta,\varphi)}{W_l^{(\#)}(\theta,\varphi)\,W_l^{(\#\star)}(\theta,\varphi)}, \quad (\# \in K) \tag{4.3.30}$$

The operator defined by Eq. (4.3.30) satisfies the following identity:

$$Q_{ij}^{(\#)}(\theta,\varphi)Q_{jk}^{(\#')}(\theta,\varphi) = \delta_{\#\#'}Q_{ik}^{(\#)}(\theta,\varphi) \tag{4.3.31}$$

As a result, through straightforward calculations derived from Eqs. (4.3.22) and (4.3.25)-(4.3.27), in conjunction with the properties outlined for $W_i^{(\#\star)}$ in Eq. (4.3.28), we can establish the following theorem:

Theorem 4.1 *The operator defined by Eq. (4.3.30) isolate specific types of wave from Green's function as follows:*

$$\begin{aligned}
&Q_{ij}^{(\#)}(\theta,\varphi)\,G_{jk}^{\nwarrow}(\boldsymbol{x},\boldsymbol{y}) \\
&\quad = \frac{\exp(ik_{(\#)}|\boldsymbol{x}-\boldsymbol{y}|)}{4\pi|\boldsymbol{x}-\boldsymbol{y}|}W_i^{(\#)}(\theta,\varphi)U_k^{(\#)}(\theta,\varphi) + O(|\boldsymbol{x}-\boldsymbol{y}|^{-2}) \\
&Q_{ij}^{(\#)}(\theta,\varphi)\,T_{jkl}^{\nwarrow}(\boldsymbol{x},\boldsymbol{y}) \\
&\quad = \frac{\exp(ik_{(\#)}|\boldsymbol{x}-\boldsymbol{y}|)}{4\pi|\boldsymbol{x}-\boldsymbol{y}|}W_i^{(\#)}(\theta,\varphi)U_{kl}^{(\#)}(\theta,\varphi) + O(|\boldsymbol{x}-\boldsymbol{y}|^{-2})
\end{aligned} \tag{4.3.32}$$

where

$$U_k^{(\#)}(\theta,\varphi) = A^{(\#)}(\theta)V_k^{(\#)}(\theta,\varphi) \tag{4.3.33}$$

$$\begin{aligned}
U_{kl}^{(\#)}(\theta,\varphi) &= -(1/2)ik_{(\#)}A^{(\#)}(\theta)\times \\
&\quad \left(V_k^{(\#)}(\theta,\varphi)V_l^{(L)}(\theta,\varphi) + V_l^{(\#)}(\theta,\varphi)V_k^{(L)}(\theta,\varphi)\right)
\end{aligned} \tag{4.3.34}$$

[Remark]
In general, when a set of linear operators $\{P^{(\#)}\}_{\#}$ exhibits the following properties

$$\begin{aligned} P^{(\#)}P^{(\#')} &= \delta_{\#\#'}P^{(\#)} \\ P^{(\#)} &= \left(P^{(\#)}\right)^{H} \end{aligned}$$

where $\left(P^{(\#)}\right)^{H}$ denotes the Hermitian adjoint of $P^{(\#)}$, we classify $P^{(\#)}$ as a projection operator and $\{P^{(\#)}\}_{\#}$ forms a family of projections.

It should be noted that $Q_{ij}^{(\#)}(\theta,\varphi)$ defined by Eq. (4.3.30) as newly introduced for the purpose of this discussion. While it does not possess Hermitian properties, it does exhibit the characteristics outlined in Eq. (4.3.31). In this context, $Q_{ij}^{(\#)}(\theta,\varphi)$ is referred to as the *pseudo-projection*. The significance of the operator $Q_{ij}^{(\#)}(\theta,\varphi)$ lies in its ability to isolate specific types of waves, which will be discussed later.

The actions of the pseudo-projections on $G_{ij}^{\swarrow}$ as well as $T_{ijk}^{\swarrow}$ are also important. Based on the reciprocity of Green's functions, these actions are expressed as follows:

$$\begin{aligned} & Q_{ij}^{(\#)}(\theta,\varphi)G_{kj}^{\swarrow}(\boldsymbol{y},\boldsymbol{x}) \\ = & \frac{\exp(ik_{(\#)}|\boldsymbol{x}-\boldsymbol{y}|)}{4\pi|\boldsymbol{x}-\boldsymbol{y}|}W_i^{(\#)}(\theta,\varphi)U_k^{(\#)}(\theta,\varphi)+O(|\boldsymbol{x}-\boldsymbol{y}|^{-2}) \\ & Q_{ij}^{(\#)}(\theta,\varphi)T_{klj}^{\swarrow}(\boldsymbol{y},\boldsymbol{x}) \\ = & \frac{\exp(ik_{(\#)}|\boldsymbol{x}-\boldsymbol{y}|)}{4\pi|\boldsymbol{x}-\boldsymbol{y}|}W_i^{(\#)}(\theta,\varphi)U_{kl}^{(\#)}(\theta,\varphi)+O(|\boldsymbol{x}-\boldsymbol{y}|^{-2}). \end{aligned} \tag{4.3.35}$$

Note that the angles (θ,φ) also describe the direction of the vector $\boldsymbol{x}-\boldsymbol{y}$, even for $G_{kj}^{\swarrow}(\boldsymbol{y},\boldsymbol{x})$ and $T_{klj}^{\swarrow}(\boldsymbol{y},\boldsymbol{x})$.

4.3.2.3 Introduction of the pseudo-projections to the near-field operator

Our focus in the current discussions is on introducing pseudo-projections into the near-field operator as a means to develop a method for identifying the locations of point-like scatterers. We delve into the problem formulation, specifically examining cases where particular types of waves are isolated from the near-field operator.

Let us define the operator $\mathscr{P}_{ij}^{(\#)}$ by using the pseudo-projections as follows:

$$\mathscr{P}_{ij}^{(\#)}(\boldsymbol{x}_p,\boldsymbol{z}_s)=l(k_{(\#)},\boldsymbol{x}_p,\boldsymbol{z}_s)Q_{ij}^{(\#)}(\theta_{ps},\varphi_{ps}) \tag{4.3.36}$$

where

$$l(k_{(\#)},\boldsymbol{x}_p,\boldsymbol{z}_s)=4\pi|\boldsymbol{x}_p-\boldsymbol{z}_s|\exp(-ik_{(\#)}|\boldsymbol{x}_p-\boldsymbol{z}_s|) \tag{4.3.37}$$

The angles $(\theta_{ps},\varphi_{ps})$ define the direction of the vector $\boldsymbol{x}_p-\boldsymbol{z}_s$. In the following, we also use the notation for angles $(\theta_{pm},\varphi_{pm})$ to define the direction of the vector

$\boldsymbol{x}_p - \boldsymbol{y}_m$. Additionally, we designate $\boldsymbol{z}_s$ as a probe point for determining the presence of point-like scatterers through the use of an indicator function, which will be formulated later. The role of $l(k_{(\#)}, \boldsymbol{x}_p, \boldsymbol{z}_s)$ on $\mathscr{P}_{ij}^{(\#)}$ is to eliminate the effects of the geometrical decay and the phase of the waves from the probe point.

We describe the actions of the operator $\mathscr{P}_{ij}^{(\#)}$ on Green's function as follows:

$$
\begin{aligned}
\mathscr{P}_{il}^{(\#)}(\boldsymbol{x}_p, \boldsymbol{z}_s) T_{ljk}^{\nwarrow}(\boldsymbol{x}_p, \boldsymbol{y}_m) &= H_{ijk}^{(\#)}(\boldsymbol{x}_p, \boldsymbol{z}_s, \boldsymbol{y}_m) + O\left(|\boldsymbol{x}_p - \boldsymbol{z}_s|\,|\boldsymbol{x}_p - \boldsymbol{y}_m|^{-2}\right) \\
\mathscr{P}_{il}^{(\#)}(\boldsymbol{x}_p, \boldsymbol{z}_s) G_{lk}^{\nwarrow}(\boldsymbol{x}_p, \boldsymbol{y}_m) &= H_{ik}^{(\#)}(\boldsymbol{x}_p, \boldsymbol{z}_s, \boldsymbol{y}_m) + O\left(|\boldsymbol{x}_p - \boldsymbol{z}_s|\,|\boldsymbol{x}_p - \boldsymbol{y}_m|^{-2}\right) \\
T_{knl}^{\swarrow}(\boldsymbol{y}_m, \boldsymbol{x}_q) \mathscr{P}_{jl}^{(\#)}(\boldsymbol{x}_q, \boldsymbol{z}_s) &= H_{jkn}^{(\#)}(\boldsymbol{x}_q, \boldsymbol{z}_s, \boldsymbol{y}_m) + O\left(|\boldsymbol{x}_q - \boldsymbol{z}_s|\,|\boldsymbol{x}_q - \boldsymbol{y}_m|^{-2}\right) \\
G_{kl}^{\swarrow}(\boldsymbol{x}_q, \boldsymbol{y}_m) \mathscr{P}_{jl}^{(\#)}(\boldsymbol{x}_q, \boldsymbol{z}_s) &= H_{jk}^{(\#)}(\boldsymbol{x}_q, \boldsymbol{z}_s, \boldsymbol{y}_m) + O\left(|\boldsymbol{x}_q - \boldsymbol{z}_s|\,|\boldsymbol{x}_q - \boldsymbol{y}_m|^{-2}\right)
\end{aligned}
\tag{4.3.38}
$$

Based on Eqs. (4.3.32), (4.3.36), and (4.3.38), the tensors $H_{ikj}^{(\#)}$ and $H_{ik}^{(\#)}$ can be respectively expressed as

$$
\begin{aligned}
H_{ikj}^{(\#)}(\boldsymbol{x}_p, \boldsymbol{z}_s, \boldsymbol{y}_m) &= W_i^{(\#)}(\theta_{ps}, \varphi_{ps}) \sum_{\#'=1}^{3} B^{(\#\#')}(\boldsymbol{x}_p, \boldsymbol{z}_s, \boldsymbol{y}_m) U_{kj}^{(\#')}(\theta_{pm}, \varphi_{pm}) \\
H_{ik}^{(\#)}(\boldsymbol{x}_p, \boldsymbol{z}_s, \boldsymbol{y}_m) &= W_i^{(\#)}(\theta_{ps}, \varphi_{ps}) \sum_{\#'=1}^{3} B^{(\#\#')}(\boldsymbol{x}_p, \boldsymbol{z}_s, \boldsymbol{y}_m) U_{k}^{(\#')}(\theta_{pm}, \varphi_{pm})
\end{aligned}
\tag{4.3.39}
$$

where

$$
\begin{aligned}
& B^{(\#\#')}(\boldsymbol{x}_p, \boldsymbol{z}_s, \boldsymbol{y}_m) \\
&= \frac{|\boldsymbol{x}_p - \boldsymbol{z}_s|}{|\boldsymbol{x}_p - \boldsymbol{y}_m|} \frac{\exp\left(ik_{(\#')}|\boldsymbol{x}_p - \boldsymbol{y}_m|\right)}{\exp\left(ik_{(\#)}|\boldsymbol{x}_p - \boldsymbol{z}_s|\right)} \\
&\quad \times \frac{W_k^{(\#')}(\theta_{pm}, \varphi_{pm}) W_k^{(\#\star)}(\theta_{ps}, \varphi_{ps})}{W_l^{(\#)}(\theta_{ps}, \varphi_{ps}) W_l^{(\#\star)}(\theta_{ps}, \varphi_{ps})}.
\end{aligned}
\tag{4.3.40}
$$

The isolation of specific types of waves from the near-field operator, as defined by Eqs. (4.3.20) and (4.3.21), is achieved through the following procedure:

$$
\begin{aligned}
& \mathscr{P}_{ik}^{(\#)}(\boldsymbol{x}_p, \boldsymbol{z}_s) \mathscr{N}_{kl}(\boldsymbol{x}_p, \boldsymbol{x}_q) \mathscr{P}_{jl}^{(\#)}(\boldsymbol{x}_q, \boldsymbol{z}_s) \\
&= \mathscr{A}_{ij}^{\infty(\#)}(\boldsymbol{x}_p, \boldsymbol{x}_q, \boldsymbol{z}_s) + O\left(|\boldsymbol{x}_p - \boldsymbol{z}_s|\,|\boldsymbol{x}_p - \boldsymbol{y}_m|^{-2}\right) \\
&\quad + O\left(|\boldsymbol{x}_q - \boldsymbol{z}_s|\,|\boldsymbol{x}_q - \boldsymbol{y}_m|^{-2}\right)
\end{aligned}
\tag{4.3.41}
$$

Multiplying the operator $\mathscr{P}^{(\#)}$ with the near-field operator not only isolates specific types of waves but also extracts their far-field properties. In this context, we introduce the symbol '$\mathscr{A}_{ij}^{\infty(\#)}$' in Eq. (4.3.41), representing the kernel of the far-field operator for # type of waves. The explicit form of $\mathscr{A}_{ij}^{\infty(\#)}$ is obtained by substituting Eq. (4.3.20) into Eq. (4.3.41), resulting in:

$$\begin{aligned}
&\mathscr{A}_{ij}^{\infty(\#)}(\boldsymbol{x}_p,\boldsymbol{x}_q,\boldsymbol{z}_s)\\
&= \sum_{\boldsymbol{y}_m\in E}\Big(-\tilde{\lambda}_m H_{ikk}^{(\#)}(\boldsymbol{x}_p,\boldsymbol{z}_s,\boldsymbol{y}_m)H_{jll}^{(\#)}(\boldsymbol{x}_q,\boldsymbol{z}_s,\boldsymbol{y}_m)\\
&\qquad -2\tilde{\mu}_m H_{ikl}^{(\#)}(\boldsymbol{x}_p,\boldsymbol{z}_s,\boldsymbol{y}_m)H_{jkl}^{(\#)}(\boldsymbol{x}_q,\boldsymbol{z}_s,\boldsymbol{y}_m)\\
&\qquad +\tilde{\rho}_m\omega^2 H_{ik}^{(\#)}(\boldsymbol{x}_p,\boldsymbol{z}_s,\boldsymbol{y}_m)H_{jk}^{(\#)}(\boldsymbol{x}_q,\boldsymbol{z}_s,\boldsymbol{y}_m)\Big)\\
&\qquad\qquad (\#\in K)
\end{aligned} \tag{4.3.42}$$

Occasionally, we also express the action of $\mathscr{A}_{ij}^{\infty(\#)}$ on function $\{f_j(\boldsymbol{x}_q)\}_{\boldsymbol{x}_q\in S_g}$ as:

$$\left(\mathscr{A}_{ij}^{\infty(\#)}(\boldsymbol{z}_s)f_j\right)(\boldsymbol{x}_p)=\sum_{q=1}^{N}\mathscr{A}_{ij}^{\infty(\#)}(\boldsymbol{x}_p,\boldsymbol{x}_q,\boldsymbol{z}_s)f_j(\boldsymbol{x}_q) \tag{4.3.43}$$

In this context, the $\mathscr{A}_{ij}^{\infty(\#)}$ on the left-hand side of Eq. (4.3.43) can be considered as the far-field operator for # type of wave. In the following, we sometimes use the notation

$$\Big((f_j(\boldsymbol{x}_q))\Big)=\Big((f_j(\boldsymbol{x}_q))_{j=1,2,3}\Big)_{q=1,2...,N}\in\mathbb{C}^{3N} \tag{4.3.44}$$

4.3.2.4 Indicator function to identify the location of point-like scatterers

Our focus on the current discussion is for characterizing the far-field operator and developing an indicator function for identifying the location of the point-like scatterers. The formulation of this indicator function is based on the results obtained from our characterization of the far-field operator.

We see from Eq. (4.3.43) that the far-field operator depends on a probe point $\boldsymbol{z}_s\in E$, where E is the set of the position of the point-like scatterers. The characterization of the far-field operator is conducted, whether $\boldsymbol{z}_s$ belongs to E or not.

We require two preparations for the characterization. The first preparation is given by

$$B^{(\#\#')}(\boldsymbol{x}_p,\boldsymbol{z}_s,\boldsymbol{y}_m)\longrightarrow\delta_{\#\#'},\ (\boldsymbol{z}_s\to\boldsymbol{y}_s\in E) \tag{4.3.45}$$

as described by in Eqs. (4.3.28) and (4.3.40). Consequently, we obtain

$$\begin{aligned}
H_{ij}^{(\#)}(\boldsymbol{x}_p,\boldsymbol{z}_s,\boldsymbol{y}_m) &\longrightarrow Wi^{(\#)}(\theta_{pm},\varphi_{pm})U_j^{(\#)}(\theta_{pm},\varphi_{pm})\\
&= D_{ij}^{(\#)}(\theta_{pm},\varphi_{pm})\\
H_{ijk}^{(\#)}(\boldsymbol{x}_p,\boldsymbol{z}_s,\boldsymbol{y}_m) &\longrightarrow Wi^{(\#)}(\theta_{pm},\varphi_{pm})U_{jk}^{(\#)}(\theta_{pm},\varphi_{pm})\\
&= (1/2)\Big(D_{ijk}^{(\#)}(\theta_{pm},\varphi_{pm})+D_{ikj}^{(\#)}(\theta_{pm},\varphi_{pm})\Big)
\end{aligned} \tag{4.3.46}$$

when $\boldsymbol{z}_s\to\boldsymbol{y}_m$. Note that we used Eqs. (4.3.25), (4.3.27), (4.3.33), and (4.3.34) to obtain Eq. (4.3.46).

The next preparation is to define the subspace in $\mathbb{C}^{3N}$, denoted as $\mathscr{M}^{(\#)}(\boldsymbol{z}_s)$, such that

$$\mathscr{M}^{(\#)}(\boldsymbol{z}_s) = \operatorname*{span}_{1\leq k,l\leq 3}\left\{\left(D_{ik}^{(\#)}(\theta_{ps},\varphi_{ps})\right),\left(D_{ikl}^{(\#)}(\theta_{ps},\varphi_{ps})\right)\right\} \tag{4.3.47}$$

Note that

$$\begin{aligned}\left(D_{ik}^{(\#)}(\theta_{ps},\varphi_{ps})\right) &= \left(\left(D_{ik}^{(\#)}(\theta_{ps},\varphi_{ps})\right)_{i=1,2,3}\right)_{p=1,2\dots,N} \in \mathbb{C}^N \\ \left(D_{ikl}^{(\#)}(\theta_{ps},\varphi_{ps})\right) &= \left(\left(D_{ikl}^{(\#)}(\theta_{ps},\varphi_{ps})\right)_{i=1,2,3}\right)_{p=1,2\dots,N} \in \mathbb{C}^N\end{aligned} \tag{4.3.48}$$

for fixed k and l. The definition of $\mathscr{M}^{(\#)}(z_S)$ by Eq. (4.3.47) states that it represents the set of all linear combinations of $\left(D_{ik}^{(\#)}(\theta_{ps},\varphi_{ps})\right)$ and $\left(D_{ikl}^{(\#)}(\theta_{ps},\varphi_{ps})\right)$.

Based on the above preparations, we can state the following theorem:

Theorem 4.2 *The range of the far-field operator $\mathscr{A}_{ij}^{(\#)}$ is characterized as follows:*

$$\exists \boldsymbol{y}_{m^*} \in E \text{ such that } \boldsymbol{z}_s = \boldsymbol{y}_s \Longleftrightarrow \mathscr{M}^{(\#)}(\boldsymbol{z}_s) \subset ran\mathscr{A}_{ij}^{(\#)}(\boldsymbol{z}_s) \tag{4.3.49}$$

[Remark]
The notation 'ran' refers to the range of the operator, as discussed in Note 3.7.

[Proof]
Now, assuming $\exists \boldsymbol{y}_{m^*} \in E$ such that

$$\boldsymbol{y}_{m^*} = \boldsymbol{z}_s \tag{4.3.50}$$

we can express the far-field operator as follows:

$$\begin{aligned}&\left(\mathscr{A}_{ij}^{\infty(\#)}(\boldsymbol{z}_s)f_j\right)(\boldsymbol{x}_p) \\ &= \left(\mathscr{A}_{ij}^{\infty(\#)m^*}(\boldsymbol{z}_s)f_j\right)(\boldsymbol{x}_p) + \left(\overline{\mathscr{A}_{ij}^{\infty(\#)m^*}}(\boldsymbol{z}_s)f_j\right)(\boldsymbol{x}_p)\end{aligned} \tag{4.3.51}$$

where

$$\begin{aligned}&\left(\mathscr{A}_{ij}^{\infty(\#)m^*}(\boldsymbol{z}_s)f_j\right)(\boldsymbol{x}_p) \\ &= D_{ikk}^{(\#)}(\theta_{ps},\varphi_{ps}) \\ &\qquad \times \sum_{q=1}^{N} -\tilde{\lambda}_{m^*} W_j^{(\#)}(\theta_{qs},\varphi_{qs})U_{ll}^{(\#)}(\theta_{qs},\varphi_{qs})f_j(\boldsymbol{x}_q) \\ &\quad +(1/2)\left(D_{ikl}^{(\#)}(\theta_{ps},\varphi_{ps})+D_{ilk}^{(\#)}(\theta_{ps},\varphi_{ps})\right) \\ &\qquad \times \sum_{q=1}^{N} -2\tilde{\mu}_{m^*} W_j^{(\#)}(\theta_{qs},\varphi_{qs})U_{kl}^{(\#)}(\theta_{qs},\varphi_{qs})f_j(\boldsymbol{x}_q) \\ &\quad +D_{ik}^{(\#)}(\theta_{ps},\varphi_{ps}) \\ &\qquad \times \sum_{q=1}^{N} \tilde{\rho}_{m^*}\,\omega^2 W_j^{(\#)}(\theta_{qs},\varphi_{qs})U_{k}^{(\#)}(\theta_{qs},\varphi_{qs})f_j(\boldsymbol{x}_q)\end{aligned} \tag{4.3.52}$$

and

$$
\begin{aligned}
&\left(\overline{\mathscr{A}_{ij}^{\infty(\#)m^*}}(z_s)f_j\right)(x_p)\\
&= \sum_{q=1}^{N}\sum_{y_m\in E\setminus\{y_{m^*}\}}\Big(-\tilde{\lambda}_m H_{ikk}^{(\#)}(x_p,z_s,y_m)H_{jll}^{(\#)}(x_q,z_s,y_m)\\
&\qquad -2\tilde{\mu}_m H_{ikl}^{(\#)}(x_p,z_s,y_m)H_{jkl}^{(\#)}(x_q,z_s,y_m)\\
&\qquad +\tilde{\rho}_m H_{ik}^{(\#)}(x_p,z_s,y_m)H_{jk}^{(\#)}(x_q,z_s,y_m)\Big) f_j(x_q)
\end{aligned}
\tag{4.3.53}
$$

Therefore, for this case, we have the following characterization of the far-field operator:

$$
\mathscr{M}^{(\#)}(z_s) \quad\subset\quad \operatorname{ran}\mathscr{A}_{ij}^{\infty(\#)m^*}(z_s)\subset \operatorname{ran}\mathscr{A}_{ij}^{\infty(\#)}(z_s) \tag{4.3.54}
$$

Conversely, let $z_s \notin E$. Using Eq. (4.3.42), the expression for the far-field operator is

$$
\begin{aligned}
&\left(\mathscr{A}_{ij}^{\infty(\#)}(z_s)f_j\right)(x_p)\\
&= \sum_{y_m\in E}\Big[W_i^{(\#)}(\theta_{ps},\varphi_{ps})\sum_{\#'}B^{(\#\#')}(x_p,z_s,y_m)\,U_{kl}^{(\#')}(\theta_{pm},\varphi_{pm})\\
&\qquad\times\sum_{q=1}^{N}C_{(1)jkl}^{(\#)}(x_q,z_s,y_m)f_j(x_q)\\
&\quad +W_i^{(\#)}(\theta_{ps},\varphi_{ps})\sum_{\#'}B^{(\#\#')}(x_p,z_s,y_m)\,U_{k}^{(\#')}(\theta_{pm},\varphi_{pm})\\
&\qquad\times\sum_{q=1}^{N}C_{(2)jk}^{(\#)}(x_q,z_s,y_m)f_j(x_q)\Big]
\end{aligned}
\tag{4.3.55}
$$

where

$$
\begin{aligned}
C_{(1)jkl}^{(\#)}(x_q,z_s,y_m) &= -\tilde{\lambda}_m\delta_{kl}H_{jnn}^{(\#)}(x_q,z_s,y_m)-2\tilde{\mu}_m H_{jkl}^{(\#)}(x_q,z_s,y_m)\\
C_{(2)jk}^{(\#)}(x_q,z_s,y_m) &= \tilde{\rho}_m\omega^2 H_{jk}^{(\#)}(x_q,z_s,y_m)
\end{aligned}
\tag{4.3.56}
$$

Equation (4.3.55) is a factorization of the operator and we see that

$$
\begin{aligned}
&\left(D_{ijk}^{(\#)}(\theta_{ps},\varphi_{ps})\right)=\left((-ik_{(\#)}A^{(\#)}(\theta_{ps})W_i^{(\#)}(\theta_{ps},\varphi_{ps})V_j^{(\#)}(\theta_{ps},\varphi_{ps})V_k^{(L)}(\theta_{ps},\varphi_{ps})\right)\\
&\notin \operatorname*{span}_{\#'\in K}\Big\{\Big(W_i^{(\#)}(\theta_{ps},\varphi_{ps})\sum_{\#'}B^{(\#\#')}(x_p,z_s,y_m)A^{(\#')}(\theta_{pm})\\
&\qquad\times V_j^{(\#')}(\theta_{pm},\varphi_{pm})V_k^{(L)}(\theta_{pm},\varphi_{pm})\Big)\Big\}\\
&\left(D_{ij}^{(\#)}(\theta_{ps},\varphi_{ps})\right)=\left(A^{(\#)}(\theta_{ps})W_i^{(\#)}(\theta_{ps},\varphi_{ps})V_j^{(\#)}(\theta_{ps},\varphi_{ps})\right)\\
&\notin \operatorname*{span}_{\#'\in K}\Big\{\Big(W_i^{(\#)}(\theta_{ps},\varphi_{ps})\sum_{\#'}B^{(\#\#')}(x_p,z_s,y_m)A^{(\#')}(\theta_{pm})V_j^{(\#)}(\theta_{pm},\varphi_{pm})\Big)\Big\}
\end{aligned}
\tag{4.3.57}
$$

and as a result

$$\begin{aligned}\left(D_{ikl}^{(\#)}(\theta_{ps},\varphi_{ps})\right) &\notin \operatorname{ran}\mathscr{A}_{ij}^{\infty(\#)}(\boldsymbol{z}_s)\\ \left(D_{ik}^{(\#)}(\theta_{ps},\varphi_{ps})\right) &\notin \operatorname{ran}\mathscr{A}_{ij}^{\infty(\#)}(\boldsymbol{z}_s)\end{aligned} \tag{4.3.58}$$

Therefore, we have

$$\boldsymbol{z}_s = \boldsymbol{y}_{m^*} \in \{\boldsymbol{y}_m\}_{\boldsymbol{y}_m\in E} \Leftarrow \mathscr{M}^{(\#)}(\boldsymbol{z}_s) \subset \operatorname{ran}\mathscr{A}_{ij}^{\infty(\#)}(\boldsymbol{z}_s) \tag{4.3.59}$$

The result then follows from Eqs. (4.3.54) and (4.3.59). □

Now, let us formulate the indicator function used to identify the location of the point-like scatterers. Consider the basis $\left\{\left(\Psi_n^{(\#)}(\boldsymbol{z}_s)\right)\right\}_n$ for ker $\left(\mathscr{A}_{ij}^{(\#)}(\boldsymbol{z}_s)\right)^H$. Then, in accordance with Theorem 4.2, we find that

$$\left\{\left(\Psi_n^{(\#)}(\boldsymbol{z}_s)\right)\right\}_n \perp \left(D_{ijk}^{(\#)}(\theta_{ps},\varphi_{ps})\right),\ (k=0,1,2,3) \tag{4.3.60}$$

when $\boldsymbol{z}_s \in E$. For Eq. (4.3.60), we assume that

$$\left(D_{ij0}^{(\#)}(\theta_{ps},\varphi_{ps})\right) = \left(D_{ij}^{(\#)}(\theta_{ps},\varphi_{ps})\right), \tag{4.3.61}$$

where $\left(D_{ij0}^{(\#)}(\theta_{ps},\varphi_{ps})\right)$ denotes the directivity tensor for the monopole Green's function.

Based on the above convention, we can have the following four kinds of indicator function with respect to the use of the monopole and dipole Green's functions:

$$\phi_k(\boldsymbol{z}_s) = \prod_{\#\in K}\left[\sum_n\sum_{j=1}^{3}\left|\left(\Psi_n^{(\#)}(\boldsymbol{z}_s)\right)^H\left(D_{ijk}^{(\#)}(\theta_{ps},\varphi_{ps})\right)\right|^2\right]^{-1},\ (k=0,1,2,3) \tag{4.3.62}$$

The indicator function in Eq. (4.3.62) has the following properties:

$$\lim_{\boldsymbol{z}_s\to\boldsymbol{y}_m\in E}\phi_k(\boldsymbol{z}_s) = \infty \tag{4.3.63}$$

As can be seen in Theorem 4.2, the present method is related to the MUSIC algorithm presented Appendix -E. The present indicator function, however, is based on the properties of the far-field operator $\mathscr{A}_{ij}^{\infty(\#)}(\boldsymbol{z}_s)$ defined in Eq. (4.3.41), which is defined for each probe point $\boldsymbol{z}_s$. As a result, the basis of the kernel of $\left(\mathscr{A}_{ij}^{\infty(\#)}\right)^H$ has to be also calculated with respect to each probe point. Although this procedure is complicated, it is expected to improve the accuracy of the reconstruction of the locations of many point-like scatterers based on a small number of source and sensor grid points at the free surface.

Note 4.7 Kernel and range of an operator in matrix form

Let $\boldsymbol{x}, \boldsymbol{y} \in \mathbb{C}^N$ and A be a matrix of $N \times N$. We define $\ker A$ and $\operatorname{ran} A$ as follows:

$$\begin{aligned} \ker A &= \{\boldsymbol{x} \in \mathbb{C}^N \,|\, A\boldsymbol{x} = \boldsymbol{0}\} \\ \operatorname{ran} A &= \{\boldsymbol{y} \in \mathbb{C}^N \,|\, \boldsymbol{y} = A\boldsymbol{x},\ \forall \boldsymbol{x} \in \mathbb{C}^N\} \end{aligned} \tag{N4.7.1}$$

In addition, Hermitian adjoint of A, A^H, satisfies

$$(\boldsymbol{x}, A\boldsymbol{y}) = (A^H\boldsymbol{x}, \boldsymbol{y}) \tag{N4.7.2}$$

where $(\cdot\,,\cdot)$ denotes the scalar product.

Now, let $\boldsymbol{y} \in \ker A^H$. Then, we have the following:

$$\begin{aligned} 0 &= (A^H\boldsymbol{y}, \boldsymbol{x}) \\ &= (\boldsymbol{y}, A\boldsymbol{x}),\ \ \forall \boldsymbol{x} \in \mathbb{C}^N \end{aligned} \tag{N4.7.3}$$

which implys

$$\ker A^H \perp \operatorname{ran} A \tag{N4.7.4}$$

4.3.3 NUMERICAL EXAMPLES

4.3.3.1 Analysis model

The analyzed model, used to validate the previously presented formulation, is illustrated in Figs. 4.3.4(a) and 4.3.4(b). These figures depict grid points located at the free surface and point-like scatterers within the elastic half-space. The grid points at the free surface serve as both source and receiver sensors, with a grid interval of 2.0 km and a total of 121 points. These source and sensor grid points are distributed across a 20 km $\times$ 20 km area, which corresponds to the horizontal coverage of the point-like scatterers.

In contrast, the analyzed model incorporates a significantly larger number of point-like scatterers, totaling 1618, compared to the count of surface grid points. These point-like scatterers are horizontally distributed over a 10 km $\times$ 10 km area, placed at 0.25 km intervals.

For the background structure of the wavefield, the P and S wave velocities are 2 km/s and 1 km/s, respectively, along with a mass density of 2 g/cm^3. The set of point-like scatterers collectively assumes the shape of the object, which adheres to the salt model described in Abubakar et al. (2011). The amplitudes of the fluctuations of the Lam'e parameters and mass density, as defined in Eq. (4.3.1), are as follows:

$$\begin{aligned} \tilde{\lambda}_m &= 0.175\ [\mathrm{GPa}\cdot\mathrm{km}^3] \\ \tilde{\mu}_m &= 0.164\ [\mathrm{GPa}\cdot\mathrm{km}^3] \\ \tilde{\rho}_m &= 7.81 \times 10^9\ [\mathrm{kg}] \end{aligned} \tag{4.3.64}$$

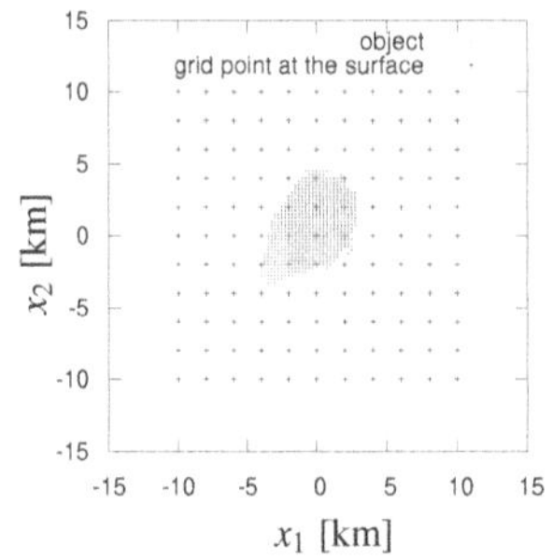

(a) Planar view

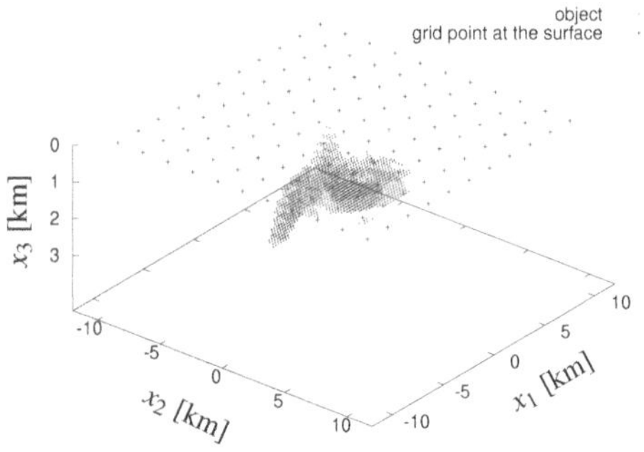

(b) Isometric view

Figure 4.3.4 Analysis model showing source and receiver sensor grids at free surface and object represented by set of point-like scatterers in elastic half-space.

In the following, Green's functions needed for constructing the kernel of the near-field operator shown in Eq. (4.3.21) are calculated using the direct wavenumber integrals shown in Eq. (3.2.91) in the previous chapter. For the direct wavenumber integral, the trapezoidal formula is employed after the effects of the singularity of the Rayleigh pole are removed, as is described in Appendix D.

4.3.3.2 Numerical results of identification of point-like scatterers

For the identification of the location of the point-like scatterers, we have four kinds of indicator function due to the choice of the directivity tensor for the Green's function, as specified by Eq. (4.3.62). We will now assess the reconstruction accuracy with respect to the chosen directivity tensor.

Figure 4.3.5 shows the spatial distribution of the amplitude of the indicator functions around the point-like scatterers (colored white). Specifically, the cloud of probe points is represented in gray scale color according to the amplitude of the indicator function. The analysis frequency is 0.5 Hz. The figure shows the differences between four kind of directivity tensors.

Figure 4.3.5 demonstrates that the high-amplitude regions of the indicator functions consistently coincide with the point-like scatterer locations, regardless of the chosen directivity tensor. However, upon closer examination of the spatial distribution of ϕ_3, which employs the directivity tensor of the dipole source around the x_3 axis, we observe significantly smaller amplitudes compared to other cases. Nevertheless, the alignment between the high-amplitude regions and the scatterer locations affirms the general validity of our method. Consequently, these indicator functions effectively serve the purpose of pinpointing the point-like scatterer locations

We should still have lots of tasks to examine the properties of the present approach. For example, the sensitivity of grid intervals, that of frequencies for the anal-

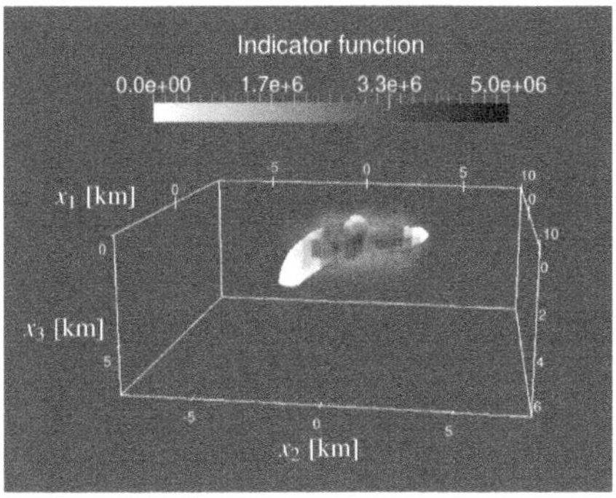

(a) spatial distribution of ϕ_0

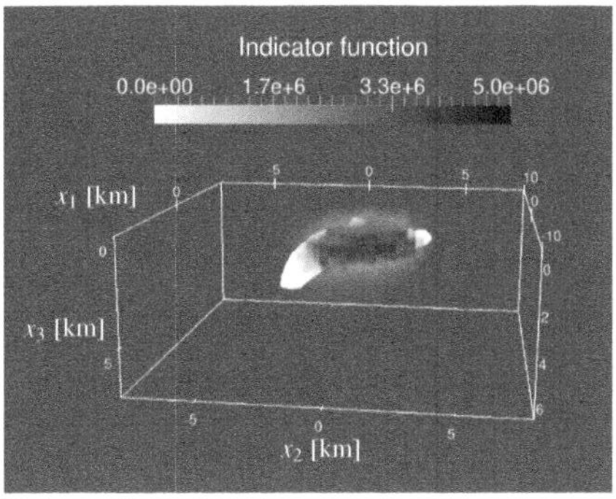

(b) spatial distribution of ϕ_1

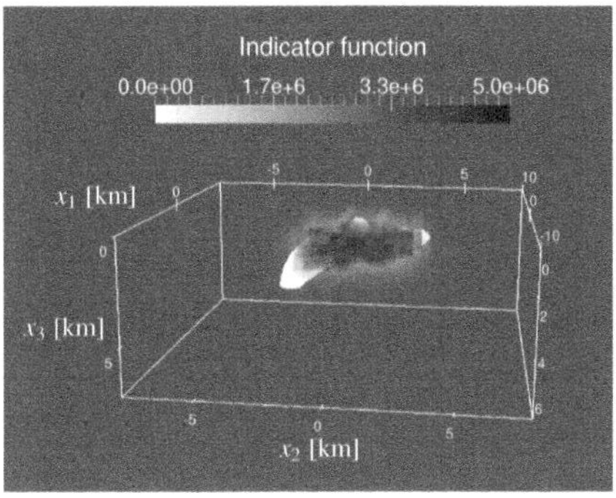

(c) spatial distribution of ϕ_2

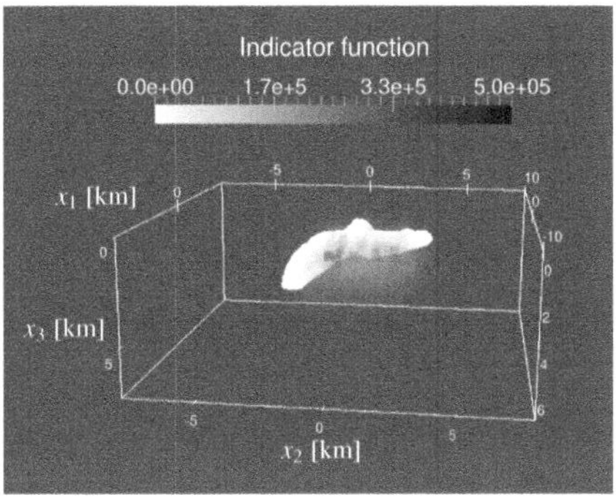

(d) spatial distribution of ϕ_3

Figure 4.3.5 Sensitivity of indicator functions to choice of directivity tensor.

ysis and so are important factors for the usefulness of the inverse analysis. For the further investigations, we can refer the articles (Touhei, 2018, 2019).

A Tensor algebra for continuum mechanics

A.1 GENERAL REMARKS

In the field of continuum mechanics, tensors are important in the analysis of the mechanical behavior of a continuous medium. In this appendix, tensor algebra is explained, which helps to clarify strain and stress tensors.

First, we prepare two different coordinate systems, $\mathscr{O}\{\boldsymbol{e}_1, \boldsymbol{e}_2, \boldsymbol{e}_3\}$ and $\mathscr{O}'\{\boldsymbol{e}'_1, \boldsymbol{e}'_2, \boldsymbol{e}'_3\}$, respectively, where $\boldsymbol{e}_j$ and $\boldsymbol{e}'_j$ are the orthonormal bases that span the entire $\mathbb{R}^3$ space, as shown in Fig. A.1. We will discuss the transformation rule for components of vectors and tensors between these two different coordinate systems. As shown in Fig. A.1, the origins of the two coordinate systems do not have to be identical. The properties of the orthonormal basis are presented by

$$\boldsymbol{e}_i \cdot \boldsymbol{e}_j = \delta_{ij}, \quad \boldsymbol{e}'_i \cdot \boldsymbol{e}'_j = \delta_{ij} \tag{A.1.1}$$

where $\boldsymbol{e}_i \cdot \boldsymbol{e}_j$ denotes the scalar product of vectors $\boldsymbol{e}_i$ and $\boldsymbol{e}_j$. In addition, δ_{ij} is the Kronecker delta defined by:

$$\delta_{ij} = \begin{cases} 1 & (i = j) \\ 0 & (i \neq j) \end{cases} \tag{A.1.2}$$

The discussion starts with the transformation rule for the vector components.

A.2 TRANSFORMATION RULES FOR VECTOR COMPONENTS

Let $\boldsymbol{v}$ be a (free) vector in $\mathbb{R}^3$. For the analysis of a vector, a coordinate system is necessary to describe the vector, and the representation of $\boldsymbol{v}$ in terms of the two different coordinate systems $\mathscr{O}$ and $\mathscr{O}'$ becomes

$$\boldsymbol{v} = \sum_{i=1}^{3} v_i \boldsymbol{e}_i = \sum_{i=1}^{3} v'_i \boldsymbol{e}'_i \tag{A.2.1}$$

where (v_1, v_2, v_3) and (v'_1, v'_2, v'_3) are the sets of the components of a vector $\boldsymbol{v}$ in the two different coordinate systems. Equation (A.2.1) indicates that a vector changes its components depending on the coordinate system, whereas the vector itself is a geometrical object that is independent of the coordinate system. For the case in which

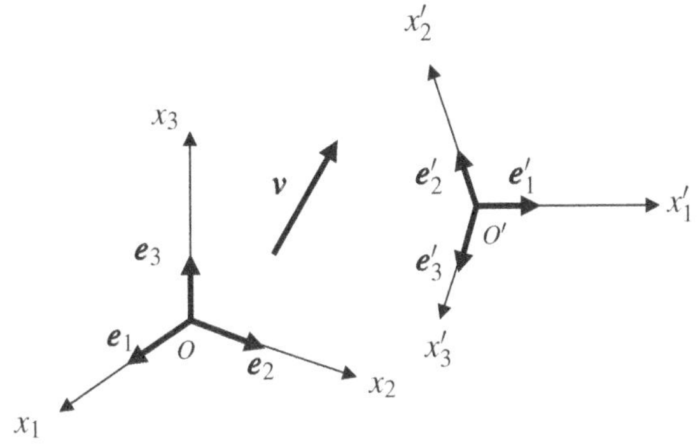

Figure A.1 Two different coordinate systems used to investigate transformation rule for vectors and tensors. A vector $\boldsymbol{v}$ as a geometrical object is represented by $\boldsymbol{v} = \sum_{i=1}^{3} v_i \boldsymbol{e}_i = \sum_{i=1}^{3} v_i' \boldsymbol{e}_i$ according to the two different coordinate systems.

it is simply necessary to display the components of a vector, these components can be expressed in the following form:

$$\boldsymbol{v} \underset{\mathscr{O}}{\rightarrow} (v_1, v_2, v_3), \ \boldsymbol{v} \underset{\mathscr{O}'}{\rightarrow} (v_1', v_2', v_3') \tag{A.2.2}$$

by clarifying the coordinate system used.

At this point, let us establish the relationship between the orthonormal bases $\mathscr{O}'\{\boldsymbol{e}_1', \boldsymbol{e}_2', \boldsymbol{e}_3'\}$ and $\mathscr{O}\{\boldsymbol{e}_1, \boldsymbol{e}_2, \boldsymbol{e}_3\}$. Since these base vectors are free vectors, the relationship between these vectors can be described in the following form:

$$\boldsymbol{e}_k' = \sum_{i=1}^{3} a_{ki} \boldsymbol{e}_i, \ (k = 1, 2, 3) \tag{A.2.3}$$

where a_{ki} can be expressed by

$$a_{ki} = \boldsymbol{e}_k' \cdot \boldsymbol{e}_i \tag{A.2.4}$$

because

$$\begin{aligned} \boldsymbol{e}_j \cdot \boldsymbol{e}_k' &= \sum_{i=1}^{3} a_{ki} \boldsymbol{e}_j \cdot \boldsymbol{e}_i \\ &= \sum_{i=1}^{3} a_{ki} \delta_{ij} = a_{kj} \end{aligned} \tag{A.2.5}$$

In a similar manner, we also have

$$\boldsymbol{e}_i = \sum_{k=1}^{3} a_{ki} \boldsymbol{e}_k', \ (i = 1, 2, 3) \tag{A.2.6}$$

Note that a_{ki} is also defined by Eq. (A.2.4).

Substitution of Eq. (A.2.3) into Eq. (A.2.1) yields:

$$\sum_{i=1}^{3} v_i \boldsymbol{e}_i = \sum_{k=1}^{3} v'_k \sum_{i=1}^{3} a_{ki} \boldsymbol{e}_i \tag{A.2.7}$$

As a result, we obtain the transformation rule for a vector component:

$$v_i = \sum_{k=1}^{3} a_{ki} v'_k, \quad (i = 1, 2, 3) \tag{A.2.8}$$

At this point, we would like to simplify the expression for the transformation rule by introducing the *summation convention*. By means of the summation convention, Eq. (A.2.8) can be written simply in the following form:

$$v_i = a_{ki} v'_k \tag{A.2.9}$$

Namely the, summation rule is adopted for the repeated index, which is k for this case. We call this index a *dummy index*. In contrast to a dummy index, the index i for Eq. (A.2.9) is called a *free index*. Equation (A.2.9) must hold for all possible values of the free index ($i = 1, 2, 3$ in this case). The following equation is a result of application of the summation convention to Eq. (A.2.7):

$$v_i \boldsymbol{e}_i = v'_k a_{ki} \boldsymbol{e}_i \tag{A.2.10}$$

where k and i are dummy indices. Once, we understand the use of the summation convention, we can simplify complicated equations such as Eq. (A.2.7). In the following, we always use the summation convention unless otherwise stated.

Equations (A.2.3) and (A.2.6) show that

$$a_{ki} a_{kj} = \delta_{ij}, \quad a_{ki} a_{li} = \delta_{kl} \tag{A.2.11}$$

and the array of (a_{ki})

$$A = \begin{pmatrix} a_{11} & a_{12} & a_{13} \\ a_{21} & a_{22} & a_{23} \\ a_{31} & a_{32} & a_{33} \end{pmatrix} \tag{A.2.12}$$

forms an orthogonal matrix:

$$AA^T = A^T A = I \tag{A.2.13}$$

where I is the identity matrix, and A^T is a transpose of A. As a result, we have

$$v'_k = a_{kj} v_j \tag{A.2.14}$$

for the inverse of Eq. (A.2.9).

Now, let $|A|$ be the determinant of the matrix A. Then, we know that

$$|A| = |A^T|, \quad |I| = 1 \tag{A.2.15}$$

Therefore, according to Eq. (A.2.13), we have

$$|A| = \pm 1 \tag{A.2.16}$$

as an orthogonal matrix of A.

A.3 DEFINITION AND BASIC PROPERTIES OF TENSORS

Based on the transformation rule for vector components, we can extend our discussion to tensors. The following are the definitions of a tensor and its components:

Definition A.1 *(Tensor) A rank-p tensor is a function of p vectors into real numbers, and is linear in each of its p arguments. Namely, a rank-p tensor* $\boldsymbol{T}$ *is characterized as:*

$$\begin{aligned} & \boldsymbol{T}(\boldsymbol{v}_1, \boldsymbol{v}_2, \ldots, \alpha \boldsymbol{v}_k^{(1)} + \beta \boldsymbol{v}_k^{(2)}, \ldots, \boldsymbol{v}_p) \\ = \quad & \alpha \boldsymbol{T}(\boldsymbol{v}_1, \boldsymbol{v}_2, \ldots, \boldsymbol{v}_k^{(1)}, \ldots, \boldsymbol{v}_p) + \beta \boldsymbol{T}(\boldsymbol{v}_1, \boldsymbol{v}_2, \ldots, \boldsymbol{v}_k^{(2)}, \ldots, \boldsymbol{v}_p) \in \mathbb{R} \\ & (k = 1, \ldots, p), (\alpha, \beta \in \mathbb{R}) \end{aligned} \tag{A.3.1}$$

[Remark] It is of course that we can define a rank-p tensor as a linear map from p vectors into complex numbers. For simplicity, however, we restrict ourselves to the definition of a tensor as a linear map into real numbers in this Appendix.

Definition A.2 *(Components of a tensor) The components of a rank-p tensor* $\boldsymbol{T}$ *are defined as*

$$T_{i_1, i_2, \ldots, i_p} = \boldsymbol{T}(\boldsymbol{e}_{i_1}, \boldsymbol{e}_{i_2}, \ldots, \boldsymbol{e}_{i_p}) \tag{A.3.2}$$

where

$$\begin{aligned} \boldsymbol{e}_{i_k} &\in \{\boldsymbol{e}_1,\ \boldsymbol{e}_2,\ \boldsymbol{e}_3\}, \\ i_k &\in \{1, 2, 3\},\ \ (k = 1, 2, \ldots, p) \end{aligned} \tag{A.3.3}$$

The number of components for a rank-p tensor is 3^p.

By means of the components of the tensor, a map of p vectors into real number can be computed. Let $\boldsymbol{v}_k$, $(k = 1, 2, \ldots, p)$ be expressed by

$$\boldsymbol{v}_k = v_{i_k}^{(k)} \boldsymbol{e}_{i_k},\ \ (k = 1, \ldots, p) \tag{A.3.4}$$

Then, we have

$$\boldsymbol{T}(\boldsymbol{v}_1, \boldsymbol{v}_2, \ldots, \boldsymbol{v}_p) = v_{i_1}^{(1)} v_{i_2}^{(2)} \cdots v_{i_p}^{(p)} T_{i_1, i_2, \ldots, i_p} \tag{A.3.5}$$

A transformation rule for the components of a tensor can now be established. Let

$$\begin{aligned} T'_{j_1, j_2, \ldots, j_p} \quad = \quad & \boldsymbol{T}(\boldsymbol{e}'_{j_1}, \boldsymbol{e}'_{j_2}, \ldots, \boldsymbol{e}'_{j_p}), \\ & (\boldsymbol{e}'_{j_k} \in \{\boldsymbol{e}'_1,\ \boldsymbol{e}'_2,\ \boldsymbol{e}'_3\},\ \ k = 1, 2, \ldots, p) \end{aligned} \tag{A.3.6}$$

remembering that

$$\boldsymbol{e}'_{j_k} = a_{j_k i_k} \boldsymbol{e}_{i_k},\ \ (k = 1, 2, \ldots, p) \tag{A.3.7}$$

Then, we have the following transformation rule for components of tensors:

$$\begin{aligned} T'_{j_1,j_2,\ldots,j_p} \\ &= \boldsymbol{T}(a_{j_1i_1}\boldsymbol{e}_{i_1}, a_{j_2i_2}\boldsymbol{e}_{i_2}, \ldots, a_{j_pi_p}\boldsymbol{e}_{i_p}), \\ &= a_{j_1i_1}a_{j_2i_2}\cdots a_{j_pi_p}T_{i_1,i_2,\ldots,i_p} \end{aligned} \tag{A.3.8}$$

The sum of tensors of the same rank can be defined. Let $\boldsymbol{T}$ and $\boldsymbol{S}$ be rank-p tensors. Then, the sum of these tensors $(\boldsymbol{T}+\boldsymbol{S})$ is defined as

$$\begin{aligned} (\boldsymbol{T}+\boldsymbol{S})(\boldsymbol{v}_1, \ldots, \boldsymbol{v}_p) \\ &= \boldsymbol{T}(\boldsymbol{v}_1, \ldots, \boldsymbol{v}_p) + \boldsymbol{S}(\boldsymbol{v}_1, \ldots, \boldsymbol{v}_p) \end{aligned} \tag{A.3.9}$$

Multiplication of a scalar by a tensor can also be defined. Let α be a scalar, and let multiplication of α and a rank-p tensor $\boldsymbol{T}$ (i.e., $(\alpha\boldsymbol{T})$) be defined as

$$(\alpha\boldsymbol{T})(\boldsymbol{v}_1, \ldots, \boldsymbol{v}_p) = \alpha\boldsymbol{T}(\boldsymbol{v}_1, \ldots, \boldsymbol{v}_p) \tag{A.3.10}$$

A.4 METRIC TENSOR

Here we present examples of tensors in order to develop a concrete understanding of the properties of tensors presented in the previous section. A typical example of a tensor might be a metric tensor, which is a rank-2 tensor defined as:

$$\boldsymbol{g}(\boldsymbol{u}, \boldsymbol{v}) = \boldsymbol{u} \cdot \boldsymbol{v} \tag{A.4.1}$$

It is evident that $\boldsymbol{g}$ satisfies the tensor properties. We also refer to $\boldsymbol{g}$ as the metric tensor in an Euclidean space, since it provides the length of a vector as:

$$\sqrt{\boldsymbol{g}(\boldsymbol{u}, \boldsymbol{u})} = |\boldsymbol{u}| \tag{A.4.2}$$

In terms of vector components $\boldsymbol{u}$ and $\boldsymbol{v}$, Eq. (A.4.1) can also be expressed as

$$\boldsymbol{g}(\boldsymbol{u}, \boldsymbol{v}) = \delta_{ij}u_iv_j = u_iv_i \tag{A.4.3}$$

Let us investigate the transformation rule for the components of the tensor. We have

$$\begin{aligned} g'_{kl} &= a_{ki}a_{lj}g_{ij} = a_{ki}a_{lj}\delta_{ij} \\ &= a_{ki}a_{li} = \delta_{kl} \end{aligned} \tag{A.4.4}$$

We see that the components of the metric tensor are invariant under coordinate transformation. A tensor having the above properties is called an *isotropic tensor*, which is used when considering an isotropic medium.

A map of two vectors in a coordinate system $\mathscr{O}'\{\boldsymbol{e}'_1, \boldsymbol{e}'_2, \boldsymbol{e}'_3\}$,

$$\boldsymbol{u} \underset{\mathscr{O}'}{\rightarrow} (u'_1, u'_2, u'_3), \quad \boldsymbol{v} \underset{\mathscr{O}'}{\rightarrow} (v'_1, v'_2, v'_3) \tag{A.4.5}$$

into real numbers in terms of components of the metric tensor and vectors is given by

$$g(\boldsymbol{u}, \boldsymbol{v}) = \delta_{ij} u'_i v'_j = u'_i v'_i \tag{A.4.6}$$

An important point here is that $u_i v_i = u'_i v'_i$, because

$$\begin{aligned} u'_k v'_k &= a_{ki} u_i a_{kj} v_j \\ &= \delta_{ij} u_i v_j = u_i v_i \end{aligned} \tag{A.4.7}$$

Note that a map of a tensor from vectors into real numbers is invariant with respect to the coordinate system.

A.5 TRIPLE SCALAR PRODUCT

Next, we investigate an example of a linear map of three vectors into real numbers in the following form:

$$\boldsymbol{T}(\boldsymbol{u}, \boldsymbol{v}, \boldsymbol{w}) = \boldsymbol{u} \times \boldsymbol{v} \cdot \boldsymbol{w} \tag{A.5.1}$$

where $\boldsymbol{u} \times \boldsymbol{v}$ denotes a vector product of these two vectors. The vector product $\boldsymbol{u} \times \boldsymbol{v}$ is characterized by the following four points:

(i) $(\boldsymbol{u} \times \boldsymbol{v}) \perp \boldsymbol{u}$, (ii) $(\boldsymbol{u} \times \boldsymbol{v}) \perp \boldsymbol{v}$, (iii) $|\boldsymbol{u} \times \boldsymbol{v}|$ is equal to the area of a parallelogram spanned by $\boldsymbol{u}$ and $\boldsymbol{v}$, and (iv) the direction of $(\boldsymbol{u} \times \boldsymbol{v})$ is determined by the right-hand rule, as shown in Fig. A.2.

Due to the characterization of the vector product, for the case in which

$$\begin{aligned} \boldsymbol{u} &\underset{\mathcal{O}}{\rightarrow} (u_1, u_2, u_3) \\ \boldsymbol{v} &\underset{\mathcal{O}}{\rightarrow} (v_1, v_2, v_3) \end{aligned} \tag{A.5.2}$$

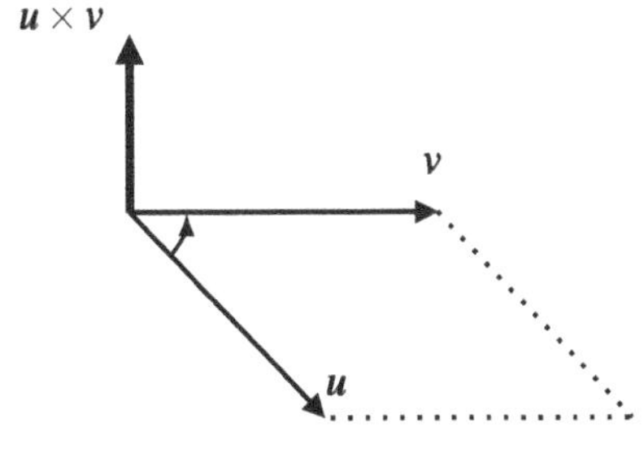

Figure A.2 Vector product of $\boldsymbol{u} \times \boldsymbol{v}$. The direction of the vector product is defined by the right-hand rule.

the result of the vector product becomes

$$\begin{aligned} \boldsymbol{u} \times \boldsymbol{v} & \\ \underset{\mathscr{O}}{\rightarrow} & \quad (u_2 v_3 - u_3 v_2, u_3 v_1 - u_1 v_3, u_1 v_2 - u_2 v_1) \end{aligned} \tag{A.5.3}$$

The operation involving three vectors shown in Eq. (A.5.1) is known as the triple scalar product and is also expressed as

$$\begin{aligned} \boldsymbol{T}(\boldsymbol{u}, \boldsymbol{v}, \boldsymbol{w}) &= \begin{vmatrix} u_1 & v_1 & w_1 \\ u_2 & v_2 & w_2 \\ u_3 & v_3 & w_3 \end{vmatrix} \\ &= e_{ijk} u_i v_j w_k \end{aligned} \tag{A.5.4}$$

where e_{ijk} is the Eddington symbol defined by

$$e_{ijk} = \begin{cases} +1 & (i, j, k) \in \{(1,2,3), (2,3,1), (3,1,2)\} \\ -1 & (i, j, k) \in \{(3,2,1), (2,1,3), (1,3,2)\} \\ 0 & \text{otherwise} \end{cases} \tag{A.5.5}$$

The triple scalar product is the signed volume of a parallelepiped spanned by the three vectors.

The component of the triple scalar product is

$$\boldsymbol{T}(\boldsymbol{e}_i, \boldsymbol{e}_j, \boldsymbol{e}_k) = e_{ijk} = T_{ijk} \tag{A.5.6}$$

The application of the transformation rule for components of tensor to T_{ijk} becomes

$$\begin{aligned} T_{i'j'k'} &= a_{i'l} a_{j'm} a_{k'n} e_{lmn} \\ &= \begin{vmatrix} a_{i'1} & a_{j'1} & a_{k'1} \\ a_{i'2} & a_{j'2} & a_{k'2} \\ a_{i'3} & a_{j'3} & a_{k'3} \end{vmatrix} = |A| e_{i'j'k'} \end{aligned} \tag{A.5.7}$$

where $|A|$ is the determinant of the matrix of the array of $(a_{k'i})$, which takes values of ± 1 for the orthogonal matrix. For the case of

$$|\boldsymbol{e}_1, \boldsymbol{e}_2, \boldsymbol{e}_3| = |\boldsymbol{e}_{1'}, \boldsymbol{e}_{2'}, \boldsymbol{e}_{3'}| \tag{A.5.8}$$

namely, both bases are in the same direction, $|A| = +1$. Conversely, when the coordinate system direction is reversed, $|A| = -1$. The result of the coordinate system transformation for the triple scalar product depends on the direction of the coordinate system. We refer to the linear map of vectors to real numbers that exhibits a dependence on the coordinate system direction as a *pseudo tensor*.

A.6 ISOTROPIC RANK-4 TENSOR

In this section, we will examine the mathematical representation of a rank-4 isotropic tensor. This tensor is utilized to express Hooke's law for an isotropic medium, as discussed in the main text of Chapter 1. Prior to seeking the details of an isotropic rank-4 tensor, we will first establish the mathematical form of a rank-2 isotropic tensor. As previously mentioned, a metric tensor remains invariant under coordinate transformations, which is a characteristic exhibited by an isotropic tensor. The following lemma for an isotropic rank-2 tensor plays an important role in the proof for characterizing an isotropic rank-4 tensor.

Lemma A.1 *A rank-2 isotropic tensor C_{ij} is expressed in the form of*

$$C_{ij} = \alpha\delta_{ij} \tag{A.6.1}$$

where α is a scalar.

[Proof]
For the proof, we consider two distinct orthonormal bases $\mathcal{O}\{\boldsymbol{e}_1, \boldsymbol{e}_2, \boldsymbol{e}_3\}$ and $\mathcal{O}'\{\boldsymbol{e}'_1, \boldsymbol{e}'_2, \boldsymbol{e}'_3\}$.

By introducing a parameter $t \in \mathbb{R}$, we make the assumption that $\mathcal{O}'$ is a function of t, represented as $\mathcal{O} = \mathcal{O}'(0)$. We proceed to parameterize the orthogonal matrix A defined in Eq. (A.2.12) as $A = A(t)$. This parameterization reveals the following relationships:

$$a_{ik}(t)a_{ij}(t) = \delta_{kj} \tag{A.6.2}$$

$$a_{ij}(0) = \delta_{ij} \tag{A.6.3}$$

Now, we define ω_{ij} such that

$$\omega_{ij} = \left(\frac{da_{ij}(t)}{dt}\right)_{t=0} \tag{A.6.4}$$

Using Eqs. (A.6.2) and (A.6.3), we obtain

$$\omega_{jk} + \omega_{kj} = 0 \tag{A.6.5}$$

We refer to the properties of ω_{jk} shown in Eq. (A.6.5) as *alternative*, where we can express ω_{jk} as

$$\omega_{jk} = e_{ijk}\omega_i \tag{A.6.6}$$

Note that ω_i can be uniquely defined from ω_{jk} by

$$\omega_i = (1/2)e_{ijk}\omega_{jk} \tag{A.6.7}$$

To derive Eq. (A.6.7) from Eq. (A.6.6), it is sufficient to use

$$e_{ijk}e_{ilm} = \delta_{jl}\delta_{km} - \delta_{jm}\delta_{kl} \tag{A.6.8}$$

It is evident that

$$C_{ij} = a_{ip}a_{jq}C_{pq} \tag{A.6.9}$$

because C_{ij} is isotropic. Differentiating both sides of Eq. (A.6.9), we obtain

$$\omega_{ip}C_{pj} + \omega_{jq}C_{iq} = 0 \tag{A.6.10}$$

Equation (A.6.10) can be reformulated as

$$\omega_a(e_{aip}C_{pj} + e_{ajq}C_{iq}) = 0 \tag{A.6.11}$$

using Eq. (A.6.6). Hence, we also arrive at

$$e_{akl}(e_{aip}C_{pj} + e_{ajq}C_{iq}) = 0 \tag{A.6.12}$$

which leads to

$$\delta_{ki}C_{lj} - \delta_{il}C_{kj} + \delta_{kj}C_{il} - \delta_{lj}C_{ik} = 0 \tag{A.6.13}$$

In the case where $j = k$, Eq. (A.6.13) simplifies to

$$C_{li} + 2C_{il} = \delta_{il}C_{jj} \tag{A.6.14}$$

Interchanging the subscripts i and l for Eq. (A.6.14), we also obtain

$$C_{il} + 2C_{li} = \delta_{il}C_{jj} \tag{A.6.15}$$

By combining Eqs. (A.6.14) and (A.6.15) to eliminate C_{il}, we arrive at

$$C_{li} = \alpha\delta_{il} \tag{A.6.16}$$

where α is give as

$$\alpha = \frac{1}{3}C_{jj}\left(= \frac{1}{3}\sum_{j=1}^{3} C_{jj}\right) \tag{A.6.17}$$

This establishes the conclusion we have sought. □

The characterization of an isotropic rank-4 tensor can be extended from the results of an isotropic rank-2 tensor. Consider a rank-4 isotropic tensor C_{ijkl} with the property $C_{ijkl} = C_{jikl}$. We can state the following theorem:

Theorem A.1 *A rank-4 isotropic tensor C_{ijkl}, for which $C_{ijkl} = C_{jikl}$, is expressed in the form of*

$$C_{ijkl} = \alpha\delta_{ij}\delta_{kl} + \beta(\delta_{ik}\delta_{jl} + \delta_{il}\delta_{jk}) \tag{A.6.18}$$

where α and β are scalars. Consequently, the tensor also exhibits the following symmetries:

$$\begin{aligned} C_{ijkl} &= C_{ijlk} \\ C_{ijkl} &= C_{klij} \end{aligned} \tag{A.6.19}$$

[Proof]
Similarly, for the proof of Lemma A.1, we employ two distinct orthonormal bases: $\mathcal{O}\{\boldsymbol{e}_1, \boldsymbol{e}_2, \boldsymbol{e}_3\}$ and $\mathcal{O}'\{\boldsymbol{e}'_1, \boldsymbol{e}'_2, \boldsymbol{e}'_3\}$, along with the parametrized orthogonal matrix $A = A(t)$ and ω_{ij} defined by Eq. (A.6.4).

It is evident that

$$C_{ijkl} = a_{ip}a_{jq}a_{kr}a_{ls}C_{pqrs} \tag{A.6.20}$$

Differentiating both sides of Eq. (A.6.20), we obtain

$$\omega_{ip}C_{pjkl} + \omega_{jq}C_{iqkl} + \omega_{kr}C_{ijrl} + \omega_{ls}C_{ijks} = 0 \tag{A.6.21}$$

By employing Eq. (A.6.6), Eq. (A.6.21) takes the form:

$$\omega_h(e_{hia}C_{ajkl} + e_{hja}C_{iakl} + e_{hka}C_{ijal} + e_{hla}C_{ijka}) = 0 \tag{A.6.22}$$

Consequently, we arrive at

$$e_{hjm}e_{hia}C_{ajkl} + e_{hjm}e_{hja}C_{iakl} + e_{hjm}e_{hka}C_{ijal} + e_{hjm}e_{hla}C_{ijka} = 0 \tag{A.6.23}$$

After applying Eq. (A.6.8) to Eq. (A.6.23), we replace the indices j with a and m with j. This yields the following equation:

$$\begin{aligned} C_{jikl} + 2C_{ijkl} + C_{ikjl} + C_{ilkj} \\ = \delta_{ij}C_{aakl} + \delta_{jk}C_{iaal} + \delta_{jl}C_{iaka} \end{aligned} \tag{A.6.24}$$

Regarding Eq. (A.6.24), it is important to recognize that C_{aakl}, C_{iaal} and C_{iaka} are isotropic rank-2 tensors. This is almost evident, as shown by

$$C_{aakl} = \sum_{a=1}^{3} C_{aakl} \tag{A.6.25}$$

Consequently, in accordance with Lemma A.1, we are able to establish the following relationships:

$$\begin{aligned} C_{aakl} &= A\delta_{kl} \\ C_{iaal} &= B\delta_{il} \\ C_{iaka} &= C\delta_{ik} \end{aligned} \tag{A.6.26}$$

where A, B and C are scalars. Subsequently, Eq. (A.6.24) transforms into the expression:

$$\begin{aligned} 2C_{ijkl} + (C_{ijkl} + C_{ikjl} + C_{ilkj}) \\ = A\delta_{ij}\delta_{kl} + B\delta_{il}\delta_{jk} + C\delta_{ik}\delta_{jl} \end{aligned} \tag{A.6.27}$$

For Eq. (A.6.27), we have used the assumption of $C_{ijkl} = C_{jikl}$. Proceeding to cyclically substitute the subscript indices j, k, and l for Eq. (A.6.27), we derive the following expressions:

$$\begin{aligned} 2C_{iklj} + (C_{iklj} + C_{ilkj} + C_{ijlk}) \\ = A\delta_{ik}\delta_{lj} + B\delta_{ij}\delta_{kl} + C\delta_{il}\delta_{kj} \end{aligned} \tag{A.6.28}$$

$$\begin{aligned} 2C_{iljk} + (C_{iljk} + C_{ijlk} + C_{ikjl}) \\ = A\delta_{il}\delta_{jk} + B\delta_{ik}\delta_{jl} + C\delta_{ij}\delta_{lk} \end{aligned} \tag{A.6.29}$$

Addition of Eqs. (A.6.27), (A.6.28) and (A.6.29) yields the resulting equation:

$$\begin{aligned}&3(C_{ijkl}+C_{iklj}+C_{iljk})+2(C_{ijlk}+C_{ikjl}+C_{ilkj})\\&=\quad(A+B+C)(\delta_{ij}\delta_{kl}+\delta_{ik}\delta_{jl}+\delta_{il}\delta_{jk})\end{aligned}\tag{A.6.30}$$

Interchanging the subscripted indices k and l in Eq. (A.6.30) yields

$$\begin{aligned}&3(C_{ijlk}+C_{ilkj}+C_{ikjl})+2(C_{ijkl}+C_{iljk}+C_{iklj})\\&=\quad(A+B+C)(\delta_{ij}\delta_{kl}+\delta_{ik}\delta_{jl}+\delta_{il}\delta_{jk})\end{aligned}\tag{A.6.31}$$

Subtraction of three times Eq. (A.6.30) from two times Eq. (A.6.31) and vice versa yields

$$\begin{aligned}C_{ijkl}+C_{iklj}+C_{iljk}&=C_{ijlk}+C_{ikjl}+C_{ilkj}\\&=\frac{1}{5}(A+B+C)(\delta_{ij}\delta_{kl}+\delta_{ik}\delta_{jl}+\delta_{il}\delta_{jk})\end{aligned}\tag{A.6.32}$$

To complete the proof, we begin by rewriting Eq. (A.6.27) as follows:

$$\begin{aligned}&3C_{ijkl}+C_{ikjl}+C_{ilkj}\\&=\quad A\delta_{ij}\delta_{kl}+B\delta_{il}\delta_{jk}+C\delta_{ik}\delta_{jl}\end{aligned}\tag{A.6.33}$$

We also use the second side of Eq. (A.6.32) as:

$$\begin{aligned}&C_{ijlk}+C_{ikjl}+C_{ilkj}\\&=\quad\frac{1}{5}(A+B+C)(\delta_{ij}\delta_{kl}+\delta_{ik}\delta_{jl}+\delta_{il}\delta_{jk})\end{aligned}\tag{A.6.34}$$

Subtracting Eq. (A.6.34) from Eq. (A.6.33), we obtain:

$$\begin{aligned}&3C_{ijkl}-C_{ijlk}\\&=\quad\frac{1}{5}\Big[(4A-B-C)\delta_{ij}\delta_{kl}+(-A+4B-C)\delta_{il}\delta_{jk}\\&\qquad+(-A-B+4C)\delta_{ik}\delta_{jl}\Big]\end{aligned}\tag{A.6.35}$$

By interchanging the subscripted indices k and l, we derive from Eq. (A.6.35):

$$\begin{aligned}&3C_{ijlk}-C_{ijkl}\\&=\quad\frac{1}{5}\Big[(4A-B-C)\delta_{ij}\delta_{kl}+(-A-B+4C)\delta_{il}\delta_{jk}\\&\qquad+(-A+4B+C)\delta_{ik}\delta_{jl}\Big]\end{aligned}\tag{A.6.36}$$

Combining Eqs. (A.6.35) and (A.6.36), we can eliminate C_{ijlk}, yielding the derived expression for a rank-4 isotropic tensor:

$$C_{ijkl}=\alpha\delta_{ij}\delta_{kl}+\beta\delta_{il}\delta_{jk}+\gamma\delta_{ik}\delta_{jl}\tag{A.6.37}$$

where

$$\begin{aligned}\alpha &= \frac{1}{10}(4A - B - C)\\ \beta &= \frac{1}{40}(-4A + 11B + C)\\ \gamma &= \frac{1}{40}(-4A + B + 11C)\end{aligned} \tag{A.6.38}$$

Furthermore, under our assumption that $C_{ijkl} = C_{jikl}$, we have $\beta = \gamma$. This concludes the proof. □

B Fourier transform, Fourier-Hankel transform, and Dirac delta function

B.1 FOURIER INTEGRAL TRANSFORM IN $\mathbb{R}^N$

Let f be a function that maps $\mathbb{R}^n$ to $\mathbb{C}$. We also assume that $\boldsymbol{x}$ and $\boldsymbol{\xi} \in \mathbb{R}^n$. The Fourier transform of f is expressed as

$$\hat{f}(\boldsymbol{\xi}) = \frac{1}{\sqrt{2\pi}^n} \int_{\mathbb{R}^n} f(\boldsymbol{x}) \exp(-i\boldsymbol{x} \cdot \boldsymbol{\xi}) d\boldsymbol{x} \tag{B.1.1}$$

where $\hat{f}$ is the Fourier transform of f and $\boldsymbol{x} \cdot \boldsymbol{\xi}$ is defined by

$$\boldsymbol{x} \cdot \boldsymbol{\xi} = x_1\xi_1 + x_2\xi_2 + \cdots + x_n\xi_n \tag{B.1.2}$$

We express the inverse Fourier transform of f as

$$\check{f}(\boldsymbol{\xi}) = \frac{1}{\sqrt{2\pi}^n} \int_{\mathbb{R}^2} f(\boldsymbol{x}) \exp(i\boldsymbol{x} \cdot \boldsymbol{\xi}) d\boldsymbol{x} \tag{B.1.3}$$

where $\check{f}$ denotes the inverse Fourier transform of f. Occasionally, we express the Fourier transform and its inverse as

$$\hat{f} = \mathscr{F} f, \; \check{f} = \overline{\mathscr{F}} f \tag{B.1.4}$$

An important property of the Fourier transform and its inverse is the following equation:

$$f = \mathscr{F}\overline{\mathscr{F}} = \overline{\mathscr{F}}\mathscr{F} f \tag{B.1.5}$$

is valid for a set of rapidly decreasing functions (Reed and Simon 1975). That is, we have

$$\overline{\mathscr{F}} = \mathscr{F}^{-1} \tag{B.1.6}$$

for the set of decreasing functions. In addition, the following equation

$$\mathscr{F}\left((\partial_1^{\alpha_1} \partial_2^{\alpha_2} \cdots \partial_n^{\alpha_n}) f\right) = (i\xi_1)^{\alpha_1} (i\xi_2)^{\alpha_2} \cdots (i\xi_n)^{\alpha_n} \hat{f} \tag{B.1.7}$$

is also valid for the rapidly decreasing functions. An example of using Eq. (B.1.7) can be expressed as

$$\mathscr{F}(\nabla^2 f)(\boldsymbol{\xi}) = \mathscr{F}\left((\partial_1^2 + \partial_2^2 + \cdots + \partial_n^2) f\right)(\boldsymbol{\xi}) = -\xi^2 \hat{f}(\boldsymbol{\xi}) \tag{B.1.8}$$

DOI: 10.1201/9781003251729-B

where $\xi^2 = \xi_1^2 + \xi_2^2 + \cdots + \xi_n^2$, which is used on the left-hand side of Eq. (2.4.13) in the main text. The properties of the Fourier transform shown in Eqs. (B.1.6) and (B.1.7) together with the Dirac delta function to obtain Green's function in the discussion in the main text. Note that the Dirac delta function and Green's function are known not to be in the class of rapidly decreasing functions, but in the class of distribution. Despite this, we can also assume that Eqs. (B.1.6) and (B.1.7) are available for the Dirac delta and Green's functions.

The forms of the Fourier transform and its inverse transform are expressed in Eqs. (2.4.10) and (2.4.12) in Chapter 2 are slightly different from Eqs. (B.1.1) and (B.1.3), respectively. However, we can derive Eqs. (2.4.10) and (2.4.12) from Eqs. (B.1.1) and (B.1.3). For example, as shown in Eq. (2.4.12), we define the Fourier transform as

$$\begin{aligned} \hat{f}_M(\boldsymbol{\xi}) &= \int_{\mathbb{R}^3} f(\boldsymbol{x}) \exp(-i\boldsymbol{x}\cdot\boldsymbol{\xi})d\boldsymbol{x} \\ &\left(= \sqrt{2\pi}^3 \mathscr{F} f = \sqrt{2\pi}^3 \hat{f}(\boldsymbol{\xi})\right) \end{aligned} \tag{B.1.9}$$

Subsequently, based on the inverse Fourier transform of Eq. (B.1.3), we obtain:

$$\begin{aligned} f(\boldsymbol{x}) &= \frac{1}{\sqrt{2\pi}^3} \int_{\mathbb{R}^3} \hat{f}(\boldsymbol{\xi}) \exp(i\boldsymbol{x}\cdot\boldsymbol{\xi})d\boldsymbol{\xi} \; (= \overline{\mathscr{F}} \hat{f}) \\ &= \frac{1}{8\pi^3} \int_{\mathbb{R}^3} \hat{f}_M(\boldsymbol{\xi}) \exp(i\boldsymbol{x}\cdot\boldsymbol{\xi})d\boldsymbol{\xi} \end{aligned} \tag{B.1.10}$$

which is the expression of the inverse Fourier transform given in Eq. (2.4.12).

B.2 HANKEL TRANSFORM AND FOURIER-HANKEL TRANSFORM

When Green's function is expressed in terms of cylindrical waves, we must introduce the following Fourier-Bessel integral transform (Morse and Feshbach, §6.3, 1953):

$$f(r) = \int_0^\infty J_m(\xi r)\xi d\xi \int_0^\infty r' f(r') J_m(\xi r')dr' \tag{B.2.1}$$

which yields the following Hankel and inverse Hankel transforms (Duffy 1994):

$$\hat{f}(\xi) = \int_0^\infty r J_m(\xi r) f(r) dr \tag{B.2.2}$$

$$f(r) = \int_0^\infty \xi J_m(\xi r) \hat{f}(\xi) d\xi \tag{B.2.3}$$

For a function defined in a cylindrical coordinate system $f(r,\theta)$, we define the Fourier-Hankel and its inverse transforms as

$$\begin{aligned} g^m(\xi) &= \int_0^{2\pi} d\theta \int_0^\infty r f(r,\theta) Y_\xi^{m*}(r,\theta) dr \\ f(r,\theta) &= \frac{1}{2\pi} \int_0^\infty \xi g^m(\xi) Y_\xi^m(r,\theta) d\xi \end{aligned} \tag{B.2.4}$$

where

$$Y_{\xi}^{m*}(r,\theta) = \exp(-im\theta)J_m(\xi r), \quad Y_{\xi}^{m}(r,\theta) = \exp(im\theta)J_m(\xi r) \tag{B.2.5}$$

B.3 DIRAC DELTA FUNCTION

The Dirac delta function $\delta(x)$, $x \in \mathbb{R}$, is defined as

$$\int_{-\infty}^{\infty} \varphi(x)\delta(x)dx = \varphi(0) \tag{B.3.1}$$

where φ is a continuous function. Equation (B.3.1) yields

$$\int_{-\infty}^{\infty} x\delta(x)\varphi(x)dx = 0 \tag{B.3.2}$$

which gives

$$x\delta(x) = 0 \tag{B.3.3}$$

Note that Eq. (B.3.3) was used in Eq. (2.4.15).

In the main text, when describing the relationship between Green's functions in the frequency and time domains using Fourier integral transforms, as shown in Eq. (2.4.10), the Dirac delta function exhibits the following properties in terms of these transforms:

$$\begin{aligned} \delta(t) &= \frac{1}{2\pi}\int_{-\infty}^{\infty} \exp(i\omega t)d\omega \\ 1 &= \int_{-\infty}^{\infty} \delta(t)\exp(-i\omega t)dt \end{aligned} \tag{B.3.4}$$

An important property of the Dirac delta function is the Cauchy principal integral, which is used for Eqs. (2.4.16) and (2.4.17) in the main text. We notice that

$$\begin{aligned} \int \frac{\varphi(x)}{x \pm i\varepsilon}dx &= \int_{R_\pm} \frac{\varphi(x)}{x}dx \\ &= \text{P.V.}\int \frac{\varphi(x)}{x}dx \mp \frac{1}{2}\int_C \frac{\varphi(x)}{x}dx \\ &= \text{P.V.}\int \frac{\varphi(x)}{x}dx \mp i\pi\varphi(0), \quad (\text{double-sign corresponds}) \end{aligned} \tag{B.3.5}$$

where P.V. denotes the Cauchy principal value, defined as

$$\text{P.V.}\int \frac{\varphi(x)}{x}dx = \lim_{\varepsilon \searrow 0}\int_{|x|\geq\varepsilon} \frac{\varphi(x)}{x}dx \tag{B.3.6}$$

In addition, the paths of the integral, namely R_+, R_-, and C, are shown in Fig. B.1. Equation (B.3.5) can be written as

$$\int \frac{1}{x \pm i\varepsilon}\varphi(x)dx = \left(\text{P.V.}\int \frac{1}{x} \mp \int i\pi\delta(x)\right)\varphi(x)dx \tag{B.3.7}$$

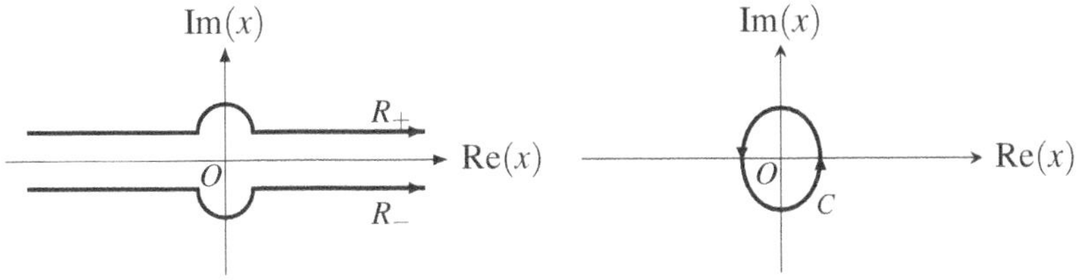

(a) Paths R_+ and R_- (b) Path C

Figure B.1 Paths of integral, namely R_+, R_-, and C, for Eq. (B.3.5).

and we have

$$\frac{1}{x \pm i\varepsilon} = \text{P.V.}\frac{1}{x} \mp i\pi\delta(x) \tag{B.3.8}$$

Note that Eq. (B.3.8) has an important role in the derivation of Green's function, as shown in Eqs. (2.4.13), (2.4.16) and (2.4.17).

In the context of Eq. (B.3.5), φ is expected to be a well-behaved function. Many textbooks, including Ikawa (1997), provide a proof for Eq. (B.3.5) under the assumption that φ belongs to a class of functions that are infinitely differentiable and possess compact support.

The relationship between the Dirac delta function and the Heaviside unit step function defined by

$$H(x) = \begin{cases} 1 & (x > 0) \\ 0 & (x < 0) \end{cases} \tag{B.3.9}$$

is also important. Let $\varphi(\cdot)$ be a rapidly decreasing function. Then, we have

$$\begin{aligned} &\int_0^{\infty} \varphi(x)\frac{dH(x)}{dx}dx \\ &= \left[\varphi(x)H(x)\right]_{-\infty}^{\infty} - \int_{-\infty}^{\infty} \frac{d\varphi(x)}{dx}H(x)dx \\ &= -\int_0^{\infty} \frac{d\varphi(x)}{dx}dx = \varphi(0) \end{aligned} \tag{B.3.10}$$

by integral by parts. Equations (B.3.10) and (B.3.1) show that

$$\delta(x) = \frac{dH(x)}{dx} \tag{B.3.11}$$

In general, the Dirac delta function $\delta(\boldsymbol{x}), \boldsymbol{x} \in \mathbb{R}^n$, is also defined by

$$\int_{\mathbb{R}^n} \varphi(\boldsymbol{x})\delta(\boldsymbol{x})d\boldsymbol{x} = \varphi(0) \tag{B.3.12}$$

Equation (2.2.29) in the main text uses the properties of Eq. (B.3.12). Namely, we have

$$\int_V \delta(\boldsymbol{x}-\boldsymbol{y})\boldsymbol{u}(\boldsymbol{y})d\boldsymbol{y} = \boldsymbol{u}(\boldsymbol{x}), \quad (\boldsymbol{x} \in V) \tag{B.3.13}$$

In the Cartesian coordinate system, $\delta(\boldsymbol{x}), \boldsymbol{x} \in \mathbb{R}^n$ can be factorized as

$$\delta(\boldsymbol{x}) = \delta(x_1)\delta(x_2)\cdots\delta(x_n) \tag{B.3.14}$$

where

$$\boldsymbol{x} = (x_1, x_2, \ldots, x_n) \tag{B.3.15}$$

For the case $n = 3$, the Dirac delta function in terms of the cylindrical coordinate system is expressed as

$$\delta(\boldsymbol{x} - \boldsymbol{x}') = \frac{1}{r}\delta(r - r')\delta(\theta - \theta')\delta(x_3 - x_3') \tag{B.3.16}$$

where

$$\begin{aligned} \boldsymbol{x} &= (r, \theta, x_3) \\ \boldsymbol{x}' &= (r', \theta', x_3) \end{aligned} \tag{B.3.17}$$

If $r' = 0$ for Eq. (B.3.16), Eq. (B.3.16) becomes

$$\delta(\boldsymbol{x} - \boldsymbol{x}') = \frac{1}{2\pi r}\delta(r)\delta(x_3 - x_3') \tag{B.3.18}$$

Equations (B.3.16) and (B.3.18) are important for deriving the expression of Green's function for an elastic half-space in terms of the superposition of cylindrical waves.

C Green's function in the wavenumber domain

C.1 OVERVIEW OF TASK

In Chapter 3, we define Green's function in the wavenumber domain using Eqs. (3.2.49) and (3.2.50), which are expressed as

$$\left(-\mathscr{A}_{m'l'}(\partial_3,\xi_r)+\rho\omega^2\delta_{m'l'}\right)\hat{g}_{l'k'}(x_3,y_3,\xi_r) = -\delta_{k'm'}\delta(x_3-y_3) \quad \text{(C.1.1)}$$

$$\lim_{x_3\to+0}\mathscr{P}_{m'l'}(\partial_3,\xi_r)\hat{g}_{l'k'}(x_3,y_3,\xi_r) = 0 \quad \text{(C.1.2)}$$

We encountered problems similar to those presented in Eqs. (3.5.19) and (3.5.20) as follows:

$$\left(\mathscr{A}_{i'j'}(\partial_3,\xi_r)-\eta^2\mu\delta_{i'j'}\right)g_{j'k'}(x_3,y_3,\xi_r,\eta) = \delta_{i'k'}\delta(x_3-y_3) \quad \text{(C.1.3)}$$

$$\lim_{x_3\to+0}\mathscr{P}_{i'j'}(\partial_3,\xi_r)g_{j'k'}(x_3,y_3,\xi_r,\eta) = 0, \quad \text{(C.1.4)}$$

We used the solution for Eqs. (C.1.3) and (C.1.4) as the resolvent kernel, which plays an important role in the derivation of the spectral representation of Green's function.

This appendix presents the derivation process for the solutions to these equations. We also derive eigenfunctions for the point and continuous spectra for operator $\mathscr{A}_{i'j'}$ based on Eqs. (C.1.3) and (C.1.4). For the discussion in this appendix, we review the following explicit forms of operators, $\mathscr{A}$ and $\mathscr{P}$:

$$\mathscr{A}_{i'j'}(\partial_3,\xi_r) = \begin{bmatrix} -(\lambda+2\mu)\partial_3^2+\mu\xi_r^2 & (\lambda+\mu)\xi_r\partial_3 & 0 \\ -(\lambda+\mu)\xi_r\partial_3 & -\mu\partial_3^2+(\lambda+2\mu)\xi_r^2 & 0 \\ 0 & 0 & -\mu\partial_3^2+\mu\xi_r^2 \end{bmatrix}$$

$$\mathscr{P}_{i'j'}(\partial_3,\xi_r) = \begin{bmatrix} (\lambda+2\mu)\partial_3 & -\lambda\xi_r & 0 \\ \mu\xi_r & \mu\partial_3 & 0 \\ 0 & 0 & \mu\partial_3 \end{bmatrix} \quad \text{(C.1.5)}$$

C.2 CONSTRUCTION OF GREEN'S FUNCTION IN THE WAVENUMBER DOMAIN

C.2.1 HOMOGENEOUS SOLUTION FOR ELASTIC WAVE EQUATION IN THE WAVENUMBER DOMAIN

The definition of Green's function given in Eq. (C.1.1) includes

$$(-\mathscr{A}_{i'j'}(\partial_3,\xi_r)+\rho\omega^2\delta_{i'j'})\hat{g}_{j'k'}(x_3,y_3,\xi_r)=0, \quad (x_3\neq y_3) \quad \text{(C.2.1)}$$

DOI: 10.1201/9781003251729-C

Therefore, we can construct Green's function by means of a solution that satisfies

$$\left(-\mathscr{A}_{i'j'}(\partial_3,\xi_r)+\delta_{i'j'}\rho\omega^2\right)\hat{u}_{j'}(\hat{\boldsymbol{x}})=0, \tag{C.2.2}$$

for the same reason described in Note 3.1.

According to Eq. (3.2.31), the solution of Eq. (C.2.2) can be expressed as

$$\begin{aligned}
\hat{u}_{1'}(\hat{\boldsymbol{x}}) &= \partial_3\hat{\phi}(\hat{\boldsymbol{x}})+\xi_r^2\hat{\psi}(\hat{\boldsymbol{x}})\\
\hat{u}_{2'}(\hat{\boldsymbol{x}}) &= \xi_r\hat{\phi}(\hat{\boldsymbol{x}})+\xi_r\partial_3\hat{\psi}(\hat{\boldsymbol{x}})\\
\hat{u}_{3'}(\hat{\boldsymbol{x}}) &= \xi_r\hat{\chi}(\hat{\boldsymbol{x}})
\end{aligned} \tag{C.2.3}$$

where $\hat{\phi}$, $\hat{\psi}$ and $\hat{\chi}$ satisfy

$$\begin{aligned}
\left(\partial_3^2-\xi_r^2+k_L^2\right)\hat{\phi}(\hat{\boldsymbol{x}}) &= 0\\
\left(\partial_3^2-\xi_r^2+k_T^2\right)\hat{\psi}(\hat{\boldsymbol{x}}) &= 0\\
\left(\partial_3^2-\xi_r^2+k_T^2\right)\hat{\chi}(\hat{\boldsymbol{x}}) &= 0
\end{aligned} \tag{C.2.4}$$

and $\hat{\boldsymbol{x}}$ is defined by Eq. (3.2.6).

The solutions for Eq. (C.2.4) for the scalar potentials are expressed as

$$\begin{aligned}
\hat{\phi}(\hat{\boldsymbol{x}}) &= e^{-\gamma x_3}\Delta_1+e^{\gamma x_3}\Delta_2\\
\hat{\psi}(\hat{\boldsymbol{x}}) &= e^{-\nu x_3}\Delta_3+e^{\nu x_3}\Delta_4\\
\hat{\chi}(\hat{\boldsymbol{x}}) &= e^{-\nu x_3}\Delta_5+e^{\nu x_3}\Delta_6
\end{aligned} \tag{C.2.5}$$

where

$$\begin{aligned}
\gamma &= \sqrt{\xi_r^2-k_L^2}\\
\nu &= \sqrt{\xi_r^2-k_T^2}
\end{aligned} \tag{C.2.6}$$

and Δ_j, $(j=1,\ldots,6)$ are potential coefficients. The branches of γ and ν are chosen such that $\mathrm{Re}(\gamma)>0$ and $\mathrm{Re}(\nu)>0$. The substitution of Eq. (C.2.5) into Eq. (C.2.3) yields:

$$\begin{aligned}
\hat{u}_{1'}(\hat{\boldsymbol{x}}) &= -\gamma e^{-\gamma x_3}\Delta_1+\gamma e^{\gamma x_3}\Delta_2+\xi_r^2e^{-\nu x_3}\Delta_3+\xi_r^2e^{\nu x_3}\Delta_4\\
\hat{u}_{2'}(\hat{\boldsymbol{x}}) &= \xi_re^{-\gamma x_3}\Delta_1+\xi_re^{\gamma x_3}\Delta_2-\xi_r\nu e^{-\nu x_3}\Delta_3+\xi_r\nu e^{\nu x_3}\Delta_4\\
\hat{u}_{3'}(\hat{\boldsymbol{x}}) &= \xi_re^{-\nu x_3}\Delta_5+\xi_re^{\nu x_3}\Delta_6
\end{aligned} \tag{C.2.7}$$

which are the solutions to Eq. (C.2.2), obtained from the undetermined coefficients without considering the boundary conditions.

C.2.2 GREEN'S FUNCTION IN THE WAVENUMBER DOMAIN EXPRESSED AS UNDETERMINED COEFFICIENTS

Figure C.1 shows the location of the source point that divides an elastic half-space into two regions $x_3<y_3$ and $x_3>y_3$, where y_3 is the location of the source point.

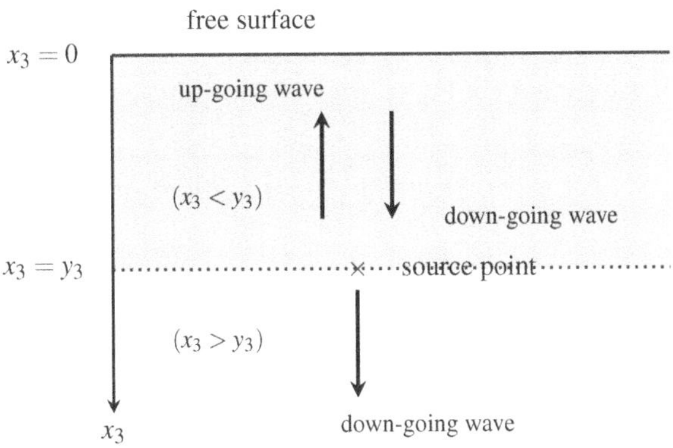

Figure C.1 Location of source point for Green's function in the wavenumber domain. In the region $x_3 < y_3$, up- and down-going waves have to be observed, whereas in the region $x_3 > y_3$, only down-going waves have to be observed.

In the region $x_3 < y_3$, up- and down-going waves must be observed, whereas in the region $x_3 > y_3$, only down-going waves that satisfy the radiation conditions, must be observed. Based on Eq. (C.2.7), for the region $x_3 < y_3$, Green's function is expressed as follows:

$$\begin{aligned} \hat{g}_{1'k'}(x_3, y_3, \xi_r) &= -\gamma e^{-\gamma x_3}\Delta_{1k'} + \gamma e^{\gamma x_3}\Delta_{2k'} + \xi_r^2 e^{-\nu x_3}\Delta_{3k'} + \xi_r^2 e^{\nu x_3}\Delta_{4k'} \\ \hat{g}_{2'k'}(x_3, y_3, \xi_r) &= \xi_r e^{-\gamma x_3}\Delta_{1k'} + \xi_r e^{\gamma x_3}\Delta_{2k'} - \xi_r \nu e^{-\nu x_3}\Delta_{3k'} + \xi_r \nu e^{\nu x_3}\Delta_{4k'} \\ \hat{g}_{3'k'}(x_3, y_3, \xi_r) &= \xi_r e^{-\nu x_3}\Delta_{5k'} + \xi_r e^{\nu x_3}\Delta_{6k'} \end{aligned} \quad \text{(C.2.8)}$$

For the case $x_3 > y_3$, Green's function is expressed as

$$\begin{aligned} \hat{g}_{1'k'}(x_3, y_3, \xi_r) &= -\gamma e^{-\gamma x_3}\Delta_{7k'} + \xi_r^2 e^{-\nu x_3}\Delta_{8k'} \\ \hat{g}_{2'k'}(x_3, y_3, \xi_r) &= \xi_r e^{-\gamma x_3}\Delta_{7k'} - \xi_r \nu e^{-\nu x_3}\Delta_{8k'} \\ \hat{g}_{3'k'}(x_3, y_3, \xi_r) &= \xi_r e^{-\nu x_3}\Delta_{9k'} \end{aligned} \quad \text{(C.2.9)}$$

which is found to satisfy the radiation condition

$$\lim_{x_3 \to \infty} \hat{g}_{i'k'}(x_3, y_3, \xi_r) = 0 \quad \text{(C.2.10)}$$

because the branches are selected such that $\mathrm{Re}(\gamma) > 0$ and $\mathrm{Re}(\nu) > 0$.

C.2.3 CONDITIONS FOR DETERMINING COEFFICIENTS OF GREEN'S FUNCTION IN THE WAVENUMBER DOMAIN

We now have nine coefficients $\Delta_{jk'}$ $(j = 1, \ldots, 9)$ (for a fixed k') for Green's function. To determine these coefficients, we have three free surface conditions shown in Eq. (C.1.2) and three continuity conditions at $x_3 = y_3$ expressed as

$$\left[\hat{g}_{i'k'}(x_3, y_3, \xi_r) \right]_{x_3 = y_3 - \varepsilon}^{x_3 = y_3 + \varepsilon} \longrightarrow 0, \quad (\varepsilon \to 0) \quad \text{(C.2.11)}$$

based on the results presented in Note 3.1. The remaining three conditions are the jump conditions for traction at $x_3 = y_3$, as follows:

Let $u_{i'}(\hat{\boldsymbol{x}})$ be the solution of Eq. (C.2.2). We investigate the following integral:

$$\begin{aligned} J &= \int_{y_3-\varepsilon}^{y_3+\varepsilon} \Big[\hat{u}_{i'}(\hat{\boldsymbol{x}})\big(-\mathscr{A}_{i'j'}(\partial_3, x_3) + \rho\omega^2\delta_{i'j'}\big)\hat{g}_{j'k'}(x_3, y_3, \xi_r) \\ &\quad -\hat{g}_{i'k'}(x_3, y_3, \xi_r)\big(-\mathscr{A}_{i'j'}(\partial_3, \xi_r) + \rho\omega^2\delta_{i'j'}\big)\hat{u}_{j'}(\hat{\boldsymbol{x}})\Big] dx_3 \end{aligned} \tag{C.2.12}$$

Note that the integral can also be expressed as

$$J = \int_{y_3-\varepsilon}^{y_3+\varepsilon} -\hat{u}_{i'}(\hat{\boldsymbol{x}})\delta_{i'k'}\delta(x_3 - y_3)dx_3 \tag{C.2.13}$$

based on Eq. (C.1.1). The application of integration by parts to Eq. (C.2.12) yields

$$\begin{aligned} J &= \Big[\hat{u}_{j'}(\hat{\boldsymbol{x}})\mathscr{P}_{i'j'}(\partial_3, \xi_r)\hat{g}_{j'k'}(x_3, y_3, \xi_r) - \hat{g}_{i'k'}(x_3, y_3, \xi_r)\mathscr{P}_{i'j'}(\partial_3, \xi_r)\hat{u}_{j'}(\hat{\boldsymbol{x}})\Big]_{y_3-\varepsilon}^{y_3+\varepsilon} \\ &= \hat{u}_{i'}(\xi_1, \xi_2, y_3)\Big[\mathscr{P}_{i'j'}\,\hat{g}_{j'k'}(x_3, y_3, \xi_r)\Big]_{x_3=y_3-\varepsilon}^{x_3=y_3+\varepsilon} + O(\varepsilon) \end{aligned} \tag{C.2.14}$$

from Eq. (C.2.11). Equation (C.2.13) yields

$$J = -\hat{u}_{i'}(\xi_1, \xi_2, y_3)\delta_{i'k'} \tag{C.2.15}$$

From Eqs. (C.2.14) and (C.2.15), we obtain the jump conditions for traction at $x_3 = y_3$ as follows:

$$\Big[\mathscr{P}_{i'j'}\,\hat{g}_{j'k'}(x_3, y_3, \xi_r)\Big]_{x_3=y_3-\varepsilon}^{x_3=y_3+\varepsilon} \to -\delta_{i'k'}, \quad (\varepsilon \to 0) \tag{C.2.16}$$

which yields three conditions for the nine undetermined conditions.

We now have nine conditions for the nine undetermined coefficients of Green's function. We can determine all coefficients using the above conditions, except for a point ξ_r that satisfies

$$F(\xi_r) = (2\xi_r^2 - k_T^2)^2 - 4\xi_r^2\gamma\nu = 0 \tag{C.2.17}$$

where $F(\cdot)$ denotes the Rayleigh function defined in Eq. (3.1.54) in § 3.1.

After complicated calculations to determine all coefficients, Green's function is expressed as follows:

$$\begin{aligned} \hat{g}_{i'j'}(x_3, y_3, \xi_r) &= \hat{p}_{i'k'}(x_3, \xi_r)\hat{q}_{k'j'}(y_3, \xi_r) \quad, (0 \le x_3 < y_3) \\ \hat{g}_{i'j'}(x_3, y_3, \xi_r) &= \hat{p}_{j'k'}(y_3, \xi_r)\hat{q}_{k'i'}(x_3, \xi_r) \quad, (y_3 < x_3) \end{aligned} \tag{C.2.18}$$

where $\hat{p}_{i'k}$ is defined by

$$
\begin{aligned}
\hat{p}_{1'1'}(x_3,\xi_r) &= \frac{1}{k_T^2}\left[-(2\xi_r^2-k_T^2)\cosh(\gamma x_3)+2\xi_r^2\cosh(\nu x_3)\right] \\
\hat{p}_{1'2'}(x_3,\xi_r) &= \frac{1}{k_T^2}\left[2\gamma\xi_r\sinh(\gamma x_3)-\frac{\xi_r(2\xi_r^2-k_T^2)}{\nu}\sinh(\nu x_3)\right] \\
\hat{p}_{2'1'}(x_3,\xi_r) &= \frac{1}{k_T^2}\left[-\frac{\xi_r}{\gamma}(2\xi_r^2-k_T^2)\sinh(\gamma x_3)+2\xi_r\nu\sinh(\nu x_3)\right] \\
\hat{p}_{2'2'}(x_3,\xi_r) &= \frac{1}{k_T^2}\left[2\xi_r^2\cosh(\gamma x_3)-(2\xi_r^2-k_T^2)\cosh(\nu x_3)\right] \\
\hat{p}_{3'3'}(x_3,\xi_r) &= \cosh(\nu x_3)
\end{aligned}
\tag{C.2.19}
$$

and $\hat{q}_{kj'}$ is defined by

$$
\begin{aligned}
\hat{q}_{1'1'}(y_3,\xi_r) &= \frac{1}{\mu F(\xi_r)}\left[\gamma(2\xi_r^2-k_T^2)e^{-\gamma y_3}-2\xi_r^2\gamma e^{-\nu y_3}\right] \\
\hat{q}_{1'2'}(y_3,\xi_r) &= \frac{\xi_r}{\mu F(\xi_r)}\left[-(2\xi_r^2-k_T^2)e^{-\gamma y_3}+2\gamma\nu e^{-\nu y_3}\right] \\
\hat{q}_{2'1'}(y_3,\xi_r) &= \frac{\xi_r}{\mu F(\xi_r)}\left[2\gamma\nu e^{-\gamma y_3}-(2\xi_r^2-k_T^2)e^{-\nu y_3}\right] \\
\hat{q}_{2'2'}(y_3,\xi_r) &= \frac{1}{\mu F(\xi_r)}\left[-2\nu\xi_r^2e^{-\gamma y_3}+\nu(2\xi_r^2-k_T^2)e^{-\nu y_3}\right] \\
\hat{q}_{3'3'}(y_3,\xi_r) &= \frac{1}{\mu\nu}e^{-\nu y_3}
\end{aligned}
\tag{C.2.20}
$$

Green's function in the wavenumber domain is found to have a very complex form; however, for the case $x_3 = 0$, the representation of Green's function is simpler and is expressed as

$$
\hat{g}_{i'j'}(x_3,y_3,\xi_r) = \hat{q}_{i'j'}(y_3,\xi_r), \quad (\text{at } x_3 = 0) \tag{C.2.21}
$$

This form is used for the application of the approximation method for Green's function for an elastic half-space.

C.2.4 HOMOGENEOUS SOLUTION AT THE ROOT OF THE RAYLEIGH FUNCTION

Equation (C.2.20) explains why we can not construct Green's function in the wavenumber domain for the P-SV components at the root of the Rayleigh function. As discussed in §3.1, we obtain a homogeneous solution that can be regarded as an eigenfunction at the root of the Rayleigh function. The homogeneous solution for the elastic wave equation in the wavenumber domain satisfies

$$
\begin{aligned}
(-\mathscr{A}_{i'j'}(\partial_3,\kappa_R)+\rho\omega^2\delta_{i'j'})\hat{u}_{j'}(\hat{\boldsymbol{x}}) &= 0, \\
\lim_{x_3\to 0}\mathscr{P}_{i'j'}\hat{u}_{j'}(\hat{\boldsymbol{x}}) &= 0
\end{aligned}
\tag{C.2.22}
$$

where κ_R is the root of the Rayleigh function.

According to Eq. (C.2.7), the homogeneous solution at the root of the Rayleigh function can be expressed as

$$\begin{aligned}\hat{u}_{1'}(x_3) &= -\gamma e^{-\gamma x_3}\Delta_1 + \kappa_R^2 e^{-\nu x_3}\Delta_3\\ \hat{u}_{2'}(x_3) &= \kappa_R e^{-\gamma x_3}\Delta_1 - \kappa_R \nu e^{-\nu x_3}\Delta_3\\ \hat{u}_{3'}(x_3) &= -\kappa_R \nu e^{-\nu x_3}\Delta_5\end{aligned} \tag{C.2.23}$$

where

$$\begin{aligned}\gamma &= \sqrt{\kappa_R^2 - k_L^2}\\ \nu &= \sqrt{\kappa_R^2 - k_T^2}\end{aligned} \tag{C.2.24}$$

The free boundary conditions for Eq. (C.2.23) are expressed as

$$\lim_{x_3 \to +0} \mathscr{P}_{i'j'}(\partial_3, \xi_r)\hat{u}_{j'}(\hat{\boldsymbol{x}}) = 0 \tag{C.2.25}$$

which yields:

$$\begin{bmatrix}(2\kappa_R^2 - k_T^2) & -2\nu\kappa_R^2 & 0\\ -2\gamma\kappa_R & \kappa_R(2\kappa_R^2 - k_T^2) & 0\\ 0 & 0 & -\xi_r\nu\end{bmatrix}\begin{bmatrix}\Delta_1\\ \Delta_3\\ \Delta_5\end{bmatrix} = \begin{bmatrix}0\\0\\0\end{bmatrix} \tag{C.2.26}$$

The ratio Δ_1/Δ_3 can be determined using Eq. (C.2.26), which yields the expression of the eigenfunction as the Rayleigh wave mode for Eq. (C.2.23). We also observe that the SH wave component vanished as a homogeneous solution.

C.3 RESOLVENT KERNEL, SPECTRUM, AND EIGENFUNCTIONS

C.3.1 RESOLVENT KERNEL

To obtain the solution for Eqs. (C.1.3) and (C.1.4), we discuss Green's function in the wavenumber domain presented in the previous section. For the solution, we replace γ and ν given in Eq. (C.2.6) by

$$\begin{aligned}\gamma &= \sqrt{\xi_r^2 - (c_T/c_L)^2\eta^2}\\ \nu &= \sqrt{\xi_r^2 - \eta^2}\end{aligned} \tag{C.3.1}$$

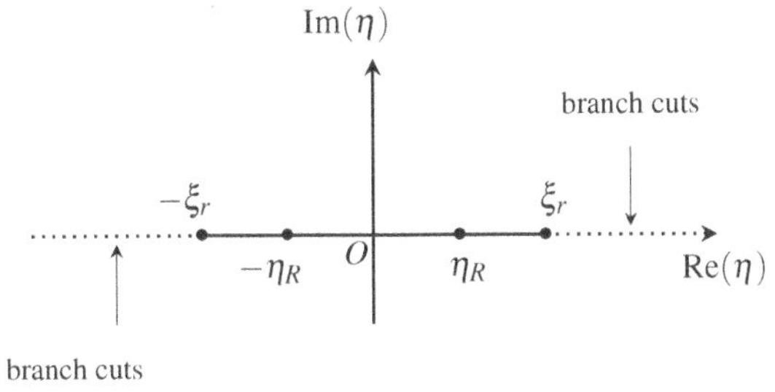

Figure C.2 Location of branch cuts and η_R in complex η plane used to define resolvent kernel.

and the Rayleigh function $F(\xi)$ given in Eq. (C.2.17) by

$$F_R(\xi_r, \eta) = (2\xi_r^2 - \eta^2)^2 - 4\xi_r^2 \gamma \nu \tag{C.3.2}$$

From Eq. (C.3.1), the branch points for $g_{i'k'}(x_3, y_3, \xi_r, \eta)$ as the solution for Eqs. (C.1.3) and (C.1.4) in the complex η plane are $\eta = \pm\xi_r$ and $\eta = \pm(c_L/c_T)\xi_r$. Recall that κ_R, which satisfies $F(\kappa_R) = 0$, has the property $|\kappa_R| > k_T$. Therefore, we see that η_R, which satisfies $F(\xi_r, \eta_R) = 0$, has the property $|\eta_R| < \xi_r$. We set the branch cuts for the complex η plane on the real axis of $|\eta| \geq \xi_r$ to realize $\mathrm{Re}\,(\gamma) > 0$ and $\mathrm{Re}\,(\nu) > 0$ in the permissible sheets.

Now, we define sets of η such that

$$\begin{aligned} B_p &= \{\eta_R \in \mathbb{R} \,|\, F_R(\xi_r, \eta_R) = 0\} \\ B_c &= \{\eta \in \mathbb{R} \,|\, |\eta| \geq \xi_r\} \end{aligned} \tag{C.3.3}$$

and construct $g_{i'j'}(x_3, y_3, \xi_r, \eta)$ for $\eta \in \mathbb{C} \setminus (B_p \cup B_c)$. Based on Eq. (C.2.18), we can define $g_{i'j'}(x_3, y_3, \xi_r, \eta)$ for $\eta \in \mathbb{C} \setminus (B_p \cup B_c)$ that satisfies the radiation condition:

$$g_{i'j'}(x_3, y_3, \xi_r, \eta) \to 0, \quad (x_3 \to \infty) \tag{C.3.4}$$

As mentioned in the main text, $g_{i'j'}(x_3, y_3, \xi_r, \eta)$ defined for $\eta \in \mathbb{C} \setminus (B_p \cup B_c)$ is called the *resolvent kernel* and the set of η defined by $B_p \cup B_c$ is called the *spectrum of* $\mathscr{A}_{i'j'}$. As we will see later, we can construct the eigenfunctions for operator $\mathscr{A}_{i'j'}$ for $\eta \in B_p \cup B_c$.

Equation (C.2.18) shows that the resolvent kernel is expressed as

$$\begin{aligned} g_{i'j'}(x_3, y_3, \xi_r, \eta) &= p_{i'k'}(x_3, \xi_r, \eta) q_{k'j'}(y_3, \xi_r, \eta), \quad (0 \leq x_3 < y_3) \\ g_{i'j'}(x_3, y_3, \xi_r, \eta) &= p_{j'k'}(y_3, \xi_r, \eta) q_{k'i'}(y_3, \xi_r, \eta), \quad (y_3 < x_3) \end{aligned} \tag{C.3.5}$$

where $p_{i'j'}$ and $q_{i'j'}$ are expressed as

$$
\begin{aligned}
p_{1'1'}(x_3,\xi_r,\eta) &= \frac{1}{\eta^2}\left[-(2\xi_r^2-\eta^2)\cosh(\gamma x_3)+2\xi_r^2\cosh(\nu x_3)\right] \\
p_{1'2'}(x_3,\xi_r,\eta) &= \frac{1}{\eta^2}\left[2\gamma\xi_r\sinh(\gamma x_3)-\frac{\xi_r(2\xi_r^2-\eta^2)}{\nu}\sinh(\nu x_3)\right] \\
p_{2'1'}(x_3,\xi_r,\eta) &= \frac{1}{\eta^2}\left[-\frac{\xi_r}{\gamma}(2\xi_r^2-\eta^2)\sinh(\gamma x_3)+2\xi_r\nu\sinh(\nu x_3)\right] \\
p_{2'2'}(x_3,\xi_r,\eta) &= \frac{1}{\eta^2}\left[2\xi_r^2\cosh(\gamma x_3)-(2\xi_r^2-\eta^2)\cosh(\nu x_3)\right] \\
p_{3'3'}(x_3,\xi_r,\eta) &= \cosh(\nu x_3) \qquad \text{(C.3.6)} \\
q_{1'1'}(y_3,\xi_r,\eta) &= \frac{1}{\mu F_R(\xi_r,\eta)}\left[\gamma(2\xi_r^2-\eta^2)e^{-\gamma y_3}-2\xi_r^2\gamma e^{-\nu y_3}\right] \\
q_{1'2'}(y_3,\xi_r,\eta) &= \frac{\xi_r}{\mu F_R(\xi_r,\eta)}\left[-(2\xi_r^2-\eta^2)e^{-\gamma y_3}+2\gamma\nu e^{-\nu y_3}\right] \\
q_{2'1'}(y_3,\xi_r,\eta) &= \frac{\xi_r}{\mu F_R(\xi_r,\eta)}\left[2\gamma\nu e^{-\gamma y_3}-(2\xi_r^2-\eta^2)e^{-\nu y_3}\right] \\
q_{2'2'}(y_3,\xi_r,\eta) &= \frac{1}{\mu F_R(\xi_r,\eta)}\left[-2\nu\xi_r^2 e^{-\gamma y_3}+\nu(2\xi_r^2-\eta^2)e^{-\nu y_3}\right] \\
q_{3'3'}(y_3,\xi_r,\eta) &= \frac{1}{\mu\nu}e^{-\nu y_3} \qquad \text{(C.3.7)}
\end{aligned}
$$

Note that we cannot define $p_{i'j'}$ for the case $\eta = 0$, as shown in Eq. (C.3.6). In this case, the definition of $p_{i'j'}$ must be based on the limit $\eta \to 0$.

The important properties of $g_{i'j'}(x_3,y_3,\xi_r,\eta)$ for $\eta \in \mathbb{C}\setminus(B_p \cup B_c)$ are as follows:

$$
\begin{aligned}
\lim_{y_3\to\infty} y_3^m |\mathscr{P}_{i'j'} g_{j'k'}(x_3,y_3,\xi_r,\eta)| &= 0 \\
\sup_{y_3\in\mathbb{R}_+} \left|\mathscr{P}_{i'j'} g_{j'k'}(x_3,y_3,\xi_r,\eta)\right| &< \infty \\
\sup_{x_3\in\mathbb{R}_+} \int_{\mathbb{R}_+} |g_{i'j'}(x_3,y_3,\xi_r,\eta)|dy_3 &< \infty \\
\sup_{y_3\in\mathbb{R}_+} \int_{\mathbb{R}_+} |g_{i'j'}(x_3,y_3,\xi_r,\eta)|dx_3 &< \infty \qquad \text{(C.3.8)}
\end{aligned}
$$

which are used in the discussion of Theorem 3.1.

The resolvent kernel also plays an important role in deriving Lemmas 3.3 and 3.4. For the derivations, we use the following expression for the resolvent kernel:

$$
g_{i'j'}(x_3,y_3,\xi_r,\eta) = w_{i'j'}(x_3,y_3,\xi_r,\eta) + v_{i'm''}(x_3,\xi_r,\eta)\Delta_{j'm''}(\eta) \qquad \text{(C.3.9)}
$$

where $w_{i'j'}$ and $v_{i'm''}$ are respectively defined by

$$\begin{aligned}
(-\mathscr{A}_{i'j'}(\partial_3,\xi_r)+\eta^2\mu\delta_{i'j'})w_{j'k'}(x_3,y_3,\xi_r,\eta) &= 0, \ (0\le x_3<y_3) \\
w_{i'j'}(x_3,y_3,\xi_r,\eta) &= 0, \ (y_3\le x_3) \\
\mathscr{P}_{i'j'}w_{j'k'}(x_3,y_3,\xi_r,\eta) &= \delta_{i'k'}, \ (x_3+\varepsilon=y_3,\varepsilon\to 0) \\
(-\mathscr{A}_{i'j'}(\partial_3,\xi_r)+\eta^2\mu\delta_{i'j'})v_{j'm''}(x_3,\xi_r,\eta) &= 0, \ (0\le x_3<\infty) \qquad \text{(C.3.10)}
\end{aligned}$$

We can construct $w_{i'j'}$ and $v_{i'm''}$ by using Eqs. (C.2.8) and (C.2.9). The subscript m'' is used for the multiplicity of the solution. In addition, $\Delta_{j'm''}$ is used to establish the free boundary conditions for the resolvent kernel, using the following equation:

$$\begin{aligned}
&\mathscr{P}_{i'j'}(\partial_3,\xi_r)w_{j'k'}(x_3,y_3,\xi_r,\eta) \\
&+ \ \mathscr{P}_{i'j'}(\partial_3,\xi_r)v_{i'm''}(x_3,\xi_r,\eta)\Delta_{j'm''}(\eta)=0, \ (\text{at } x_3=0) \qquad \text{(C.3.11)}
\end{aligned}$$

C.3.2 EIGENFUNCTION FOR POINT SPECTRUM

We define the eigenfunction and eigenvalue for operator $\mathscr{A}_{i'j'}$ using the following equation:

$$\begin{aligned}
\mathscr{A}_{i'j'}(\partial_3,\xi_r)\psi_{j'm''}(x_3,\xi_r,\eta) &= \mu\xi_3^2\psi_{i'm''}(x_3,\xi_r,\eta) \\
\lim_{x_3\to+0}\mathscr{P}_{i'j'}(\partial_3,\xi_r)\psi_{j'm''}(x_3,\xi_r,\eta) &= 0 \qquad \text{(C.3.12)}
\end{aligned}$$

where ψ is the eigenfunction, subscript m'' for the eigenfunction denotes the multiplicity of the eigenfunction, and η is the eigenvalue. The domain of operator $\mathscr{A}_{i'j'}$ is defined by Eq. (3.5.16), which is expressed as

$$D(\mathscr{A}_{i'j'})=\left\{u_{i'}\in L_2(\mathbb{R}_+)\mid \mathscr{P}_{i'j'}u_{j'}=0 \text{ at } x_3=0\right\} \qquad \text{(C.3.13)}$$

Next, we show that η_R, which is the root of $F_R(\xi_r,\eta_R)=0$, is the eigenvalue of operator $\mathscr{A}_{i'j'}$. As mentioned previously, we know that $|\eta_R|<\xi_r$. In this case, based on Eq. (C.2.22), a homogeneous solution that satisfies the following equation

$$(-\mathscr{A}_{i'j'}(\partial_3,\xi_r)+\eta_R^2\mu\delta_{i'j'})u_{j'}(x_3)=0 \qquad \text{(C.3.14)}$$

is expressed as

$$\begin{aligned}
u_{1'}(x_3) &= -\gamma e^{-\gamma x_3}\Delta_1+\xi_r^2e^{-\nu x_3}\Delta_3 \\
u_{2'}(x_3) &= \xi_r e^{-\gamma x_3}\Delta_1-\xi_r\nu e^{-\nu x_3}\Delta_3 \\
u_{3'}(x_3) &= \xi_r e^{-\nu x_3}\Delta_5 \qquad \text{(C.3.15)}
\end{aligned}$$

Based on the discussion presented for Eq. (C.2.22), the free boundary conditions

$$\mathscr{P}_{i'j'}(\partial_3,\xi_r)u_{j'}(x_3)=0, \ (\text{at } x_3=0) \qquad \text{(C.3.16)}$$

yields $u_{3'}=0$ and a nontrivial solution for $u_{1'}$ and $u_{2'}$, which is the eigenfunction. Therefore, $\xi_3=\eta_R$ is an eigenvalue of operator $\mathscr{A}_{i'j'}$. For the nontrivial solution, we can define the ratio of the coefficients of Δ_1/Δ_2 of Eq. (C.3.15), which is uniquely determined by the normalization condition:

$$\int_0^\infty \left(|u_{1'}(x_3)|^2 + |u_{2'}(x_3)|^2\right) dx_3 = 1 \tag{C.3.17}$$

The array for the eigenfunction $\psi_{i'j''}$ used in the main text of this book is expressed as

$$\left[\psi_{i'j''}(x_3,\xi_r,\eta_R)\right] = \begin{bmatrix} u_{1'}(x_3) & u_{2'}(x_3) & 0 \\ 0 & 0 & 0 \\ 0 & 0 & u_{3'}(x_3) \end{bmatrix}^T \tag{C.3.18}$$

This is referred to as the eigenfunction for the point spectrum. $u_{3'}$ in Eq. (C.3.18) is zero. Explicit forms of the eigenfunctions are also necessary to describe the algorithm for a fast generalized Fourier transform. In the main text, in Note 4.6, we use C_1 for Δ_1 and C_2 for Δ_3, where Δ_1 and Δ_3 are used to describe the eigenfunction for the point spectrum shown in Eq. (C.3.15).

C.3.3 EIGENFUNCTION FOR CONTINUOUS SPECTRUM

Based on Eq. (C.2.7), we can construct a bounded solution that satisfies

$$\begin{aligned} \mathscr{A}_{i'j'}\phi_{j'}(x_3,\xi_r,s) &= \mu\xi_3^2\phi_{i'}(x_3,\xi_r,s) \\ \lim_{x_3\to 0}\mathscr{P}_{i'j'}\phi_{j'}(x_3,\xi_r,s) &= 0 \end{aligned} \tag{C.3.19}$$

for $\xi_r < s$ despite the fact that $\phi_{i'} \notin D(\mathscr{A}_{i'j'}) \subset L_2(\mathbb{R}_+)$. We refer to the solutions $\phi_{i'}$ *improper eigenfunctions*. We investigate improper eigenfunctions for the regions $\xi_r < s < \xi_r(c_L/c_T)$ and $\xi_r(c_L/c_T) < s$ separately.

C.3.3.1 Improper eigenfunctions for the region $\xi_r < s < \xi_r(c_L/c_T)$

For the region $\xi_r < s < \xi_r(c_L/c_T)$, the following form of the bounded homogeneous solution of Eq. (C.3.19) is possible:

$$\begin{aligned} \phi_{1'}(x_3,\xi_r,s) &= -\gamma e^{-\gamma x_3}\Delta_1 + \xi_r^2\cos(\bar{\nu}x_3)\Delta_3 + \xi_r^2\sin(\bar{\nu}x_3)\Delta_4 \\ \phi_{2'}(x_3,\xi_r,s) &= \xi_r e^{-\gamma x_3}\Delta_1 - \xi_r\bar{\nu}\sin(\bar{\nu}x_3)\Delta_3 + \xi_r\bar{\nu}\cos(\bar{\nu}x_3)\Delta_4 \\ \phi_{3'}(x_3,\xi_r,s) &= \xi_r\cos(\bar{\nu}x_3)\Delta_5 + \xi_r\sin(\bar{\nu}x_3)\Delta_6 \end{aligned} \tag{C.3.20}$$

where

$$\bar{\nu} = \sqrt{s^2 - \xi_r^2} \tag{C.3.21}$$

The application of the free boundary conditions to Eq. (C.3.20) solves the coefficients as:

$$\begin{aligned} \Delta_3 &= \alpha(\xi_r,s)\Delta_1 \\ \Delta_4 &= \beta(\xi_r,s)\Delta_1 \\ \Delta_6 &= 0 \end{aligned} \tag{C.3.22}$$

where

$$\alpha(\xi_r, s) = \frac{2\gamma}{\xi_r^2 - \bar{\nu}^2}, \quad \beta(\xi_r, s) = \frac{\bar{\nu}^2 - \xi_r^2}{2\xi_r^2 \bar{\nu}} \tag{C.3.23}$$

Therefore, the construction of the homogeneous solution of Eq. (C.3.19) that satisfies Eq. (C.3.19) is possible because of an improper eigenfunction.

The normalization conditions for the eigenfunction

$$\begin{aligned}
&\int_0^\infty \Big(\phi_{1'}(x_3, \xi_r, s)\phi_{1'}(x_3, \xi_r', s') \\
&+ \quad \phi_{2'}(x_3, \xi_r, s)\phi_{2'}(x_3, \xi_r', s')\Big)dx_3 = \delta(s - s')\delta_{\xi_r, \xi_r'} \\
&\int_0^\infty \phi_{3'}(x_3, \xi_r, s)\phi_{3'}(x_3, \xi_r', s')dx_3 \\
&= \quad \delta(s - s')\delta_{\xi_r, \xi_r'}
\end{aligned} \tag{C.3.24}$$

lead to

$$\Delta_3^2 + \Delta_4^2 = \frac{2}{\pi}\frac{1}{\bar{\nu}\xi_r^2 s} \tag{C.3.25}$$

and

$$\Delta_5 = \frac{1}{\xi_r}\sqrt{\frac{2s}{\pi\bar{\nu}}} \tag{C.3.26}$$

where

$$\delta_{\xi_r, \xi_r'} = \begin{cases} 1 & (\text{when } \xi_r = \xi_r') \\ 0 & (\text{otherwise}) \end{cases} \tag{C.3.27}$$

The array for the eigenfunction $\psi_{i'j''}$ used in the main text is expressed as:

$$[\psi_{i'j''}(x_3, \xi_r, s)] = \begin{bmatrix} \phi_{1'}(x_3, \xi_r, s) & \phi_{2'}(x_3, \xi_r, s) & 0 \\ 0 & 0 & 0 \\ 0 & 0 & \phi_{3'}(x_3, \xi_r, s) \end{bmatrix}^T \tag{C.3.28}$$

In the main text, in Note 4.6, we use C_3 for Δ_1, C_4 for Δ_3, C_5 for Δ_4, and C_6 for Δ_5, where Δ_1, Δ_3, Δ_4 and Δ_5 are used in Eq. (C.3.20).

C.3.3.2 Improper eigenfunction for the region $\xi_r(c_L/c_T) < s$

For the region $\xi_r(c_L/c_T) < s$, the homogeneous solution for Eq. (C.3.19) becomes

$$\begin{aligned}
\phi_{1'}(x_3, \xi_r, s) &= \gamma\cos(\gamma x_3)\Delta_1 - \gamma\sin(\gamma x_3)\Delta_2 + \xi_r^2\cos(\bar{\nu}x_3)\Delta_3 + \xi_r^2\sin(\bar{\nu}x_3)\Delta_4 \\
\phi_{2'}(x_3, \xi_r, s) &= \xi_r\sin(\gamma x_3)\Delta_1 + \xi_r\cos(\gamma x_3)\Delta_2 - \xi_r\bar{\nu}\sin(\bar{\nu}x_3)\Delta_3 + \xi_r\bar{\nu}\cos(\bar{\nu}x_3)\Delta_4 \\
\phi_{3'}(x_3, \xi_r, s) &= \xi_r\cos(\bar{\nu}x_3)\Delta_5 + \xi_r\sin(\bar{\nu}x_3)\Delta_6
\end{aligned} \tag{C.3.29}$$

where

$$\bar{\gamma} = \sqrt{s^2(c_T/c_L)^2 - \xi_r^2} \tag{C.3.30}$$

The free boundary conditions lead to the following relationship between coefficients:

$$\Delta_3 = \bar{\alpha}(\xi_r, s)\Delta_1, \quad \Delta_4 = \beta(\xi_r, s)\Delta_2 \tag{C.3.31}$$

where

$$\bar{\alpha}(\xi_r, s) = \frac{2\bar{\gamma}}{\bar{\nu}^2 - \xi_r^2} \tag{C.3.32}$$

Equations (C.3.29) and (C.3.31) imply that there are two linearly independent eigenfunctions for the P-SV wave components. They are

$$\begin{aligned}
\phi_{1'}^{(1)}(x_3, \xi_r, s) &= \gamma\cos(\gamma x_3)\Delta_1 + \bar{\alpha}(\xi_r, s)\xi_r^2\cos(\bar{\nu}x_3)\Delta_1 \\
\phi_{2'}^{(1)}(x_3, \xi_r, s) &= \xi_r\sin(\gamma x_3)\Delta_1 - \bar{\alpha}(\xi_r, s)\xi_r\bar{\nu}\sin(\bar{\nu}x_3)\Delta_1 \\
\phi_{1'}^{(2)}(x_3, \xi_r, s) &= -\gamma\sin(\gamma x_3)\Delta_2 + \beta(\xi_r, s)\xi_r^2\sin(\bar{\nu}x_3)\Delta_2 \\
\phi_{2'}^{(2)}(x_3, \xi_r, s) &= \xi_r\cos(\gamma x_3)\Delta_2 + \beta(\xi_r, s)\xi_r\bar{\nu}\cos(\bar{\nu}x_3)\Delta_2
\end{aligned} \tag{C.3.33}$$

where the superscripts (1) and (2) are used to distinguish the two types of linearly independent eigenfunctions. The amplitudes Δ_1 and Δ_2 can be determined using the following normalization conditions:

$$\sum_{j'=1}^{2}\int_0^{\infty}\phi_{j'}^{(\tau)}(x_3, \xi_r, s)\phi_{j'}^{(\tau)}(x_3, \xi_r', s')dx_3 = \delta(s - s')\delta_{\xi_r, \xi_r'}, \quad (\tau = 1, 2) \tag{C.3.34}$$

The results are

$$\begin{aligned}
\Delta_1^2 &= \frac{2}{\pi}\frac{1}{s(\bar{\gamma} + \bar{\alpha}^2\bar{\nu}\xi_r^2)} \\
\Delta_2^2 &= \frac{2}{\pi}\frac{1}{s(\bar{\gamma} + \beta^2\bar{\nu}\xi_r^2)}
\end{aligned} \tag{C.3.35}$$

The array for the eigenfunction $\psi_{i'j''}$ for the region $\xi_r(c_L/c_T) < s$ is expressed as follows:

$$[\psi_{i'j''}(x_3, \xi_r, s)] = \begin{bmatrix} \phi_{1'}^{(1)}(x_3, \xi_r, s) & \phi_{2'}^{(1)}(x_3, \xi_r, s) & 0 \\ \phi_{1'}^{(2)}(x_3, \xi_r, s) & \phi_{2'}^{(2)}(x_3, \xi_r, s) & 0 \\ 0 & 0 & \phi_{3'}(x_3, \xi_r, s) \end{bmatrix}^T \tag{C.3.36}$$

The eigenfunctions defined by Eqs. (C.3.28) and (C.3.36) are the *eigenfunction of the continuous spectrum.*

According to Eqs. (C.3.9) and (C.3.10), we can establish the following equations:

$$\left(-\mathscr{A}_{i'j'}(\partial_3, \xi_r) + s^2\mu\delta_{i'j'}\right)$$
$$\left(v_{j'k''}(x_3, \xi_r, s + i\varepsilon) - v_{j'k''}(x_3, \xi_r, s - i\varepsilon)\right) = 0, \quad (\varepsilon \to 0) \tag{C.3.37}$$

$$\mathscr{P}_{i'j'}(\partial_3, \xi_r) \left(v_{j'k''}(x_3, \xi_r, s+i\varepsilon) - v_{j'k''}(x_3, \xi_r, s-i\varepsilon) \right) = 0, \quad (\text{at } x_3 = 0,\ \varepsilon \to 0) \tag{C.3.38}$$

Therefore, the eigenfunction for the continuous spectrum can be constructed by

$$\psi_{i'k''}(x_3, \xi_r, s) = v_{i'k''}(x_3, \xi_r, s+i\varepsilon) - v_{i'k''}(x_3, \xi_r, s-i\varepsilon), \quad (\varepsilon \to 0) \tag{C.3.39}$$

for $s > \xi_r$. For this reason, we refer to $v_{i'k''}(x_3, \xi_r, s \pm i\varepsilon), (s > \xi_r)$ as *the definition function for the eigenfunctions associated with the continuous spectrum.*

C.3.4 EXPLICIT FORM OF THE DEFINITION FUNCTION

The explicit form of $v_{i'k''}$ defined by Eq. (C.3.10) is important for deriving Eq. (3.5.70). Based on Eqs. (C.3.20) and (C.3.29), the form of $v_{i'k''}$ can be set as

$$v_{i'k''}(x_3, \xi_r, \eta) = a_{i'j'}(x_3, \xi_r, \eta) d_{j'k''}(\xi_r, \eta) \tag{C.3.40}$$

where $a_{i'j'}$ denotes

$$[a_{i'j'}(x_3, \xi_r, \eta)] = \begin{bmatrix} -\gamma \exp(-\gamma x_3) & \xi_r^2 \exp(-\nu x_3) & 0 \\ \xi_r \exp(-\gamma x_3) & -\xi_r \nu \exp(-\nu x_3) & 0 \\ 0 & 0 & \xi_r \exp(-\nu x_3) \end{bmatrix} \tag{C.3.41}$$

and $d_{j'k''}$ is the matrix of the coefficients that realize the free boundary conditions and the normalization conditions for the improper eigenfunction for the case where $\varepsilon \to 0$. For the region $\xi_r < s < (c_L/c_T)\xi_r$, $d_{j'k''}$ can be expressed as

$$\begin{aligned} [d_{j'k''}(\xi_r, s+i\varepsilon)] &= \begin{bmatrix} \Delta & 0 & 0 \\ \theta_1 \Delta & 0 & 0 \\ 0 & 0 & D_3 \end{bmatrix} \\ [d_{j'k''}(\xi_r, s-i\varepsilon)] &= \begin{bmatrix} 0 & 0 & 0 \\ \theta_2 \Delta & 0 & 0 \\ 0 & 0 & -D_3 \end{bmatrix} \end{aligned} \tag{C.3.42}$$

where

$$\begin{aligned} \theta_1 &= \frac{4\xi_r^2 \bar{\nu}\gamma + i(\xi_r^2 - \bar{\nu}^2)^2}{4\xi_r^2 \bar{\nu}(\xi_r^2 - \bar{\nu}^2)} \\ \theta_2 &= \frac{-4\xi_r^2 \bar{\nu}\gamma + i(\xi_r^2 - \bar{\nu}^2)^2}{4\xi_r^2 \bar{\nu}(\xi_r^2 - \bar{\nu}^2)} \end{aligned} \tag{C.3.43}$$

$$D_3 = \frac{1}{2\xi_r} \sqrt{\frac{2s}{\pi \bar{\nu}}} \tag{C.3.44}$$

$$\Delta^2 = \frac{8\bar{\nu}\xi_r^2 (2\xi_r^2 - s^2)}{\pi s [(2\xi_r^2 - s^2)^4 + 16\xi_r^4 \bar{\nu}^2 \gamma^2]} \tag{C.3.45}$$

For the region $(c_L/c_T)\xi_r < s$, $d_{j'k''}$ is expressed as follows:

$$\left[d_{j'k''}(\xi_r, s+i\varepsilon)\right] = \begin{bmatrix} D_1 & D_2 & 0 \\ i\bar{\alpha}D_1 & -i\beta D_2 & 0 \\ 0 & 0 & D_3 \end{bmatrix}$$

$$\left[d_{j'k''}(\xi_r, s-i\varepsilon)\right] = \begin{bmatrix} D_1 & -D_2 & 0 \\ -i\bar{\alpha}D_1 & -i\beta D_2 & 0 \\ 0 & 0 & -D_3 \end{bmatrix} \qquad \text{(C.3.46)}$$

where

$$D_1 = \frac{\Delta_1}{2i}, \quad D_2 = \frac{\Delta_2}{2} \qquad \text{(C.3.47)}$$

Note that Δ_1 and Δ_2 are given by Eq. (C.3.35).

D Comparison of Green's function obtained using various computational methods

D.1 COMPUTATIONAL METHODS FOR GREEN'S FUNCTION FOR COMPARISONS AND ANALYZED MODEL

D.1.1 FORMULATIONS FOR GREEN'S FUNCTION FOR COMPARISONS

In the main text, we present several formulations of Green's function for an elastic half-space. In this appendix, we compare the numerical results obtained using different formulations of Green's function. Table D.1 summarizes the formulations for Green's function and the corresponding equations presented in the main text and the Appendix.

Table D.1
Formulations for Green's function to be examined.

	Formulation	Equations
1)	Fourier-Hankel transform	(3.2.91), (D.1)
2)	Steepest descent path method	(3.4.36) , (3.4.55) (3.4.56), (D.5)
3)	Spectral theory	(3.5.98), (D.6)

Figure D.1.1 shows the analyzed model for computing Green's function. A buried point source is located on the x_3 axis. The P and S wave velocities are $c_L = 2$ km/s and $c_T = 1$ km/s, respectively, and the mass density is $\rho = 2$ g/cm^3. In addition, the analysis frequency is 1 Hz and the amplitude of the point force is 1×10^7 kN, with excitation in the vertical direction at a depth of 1 km.

DOI: 10.1201/9781003251729-D

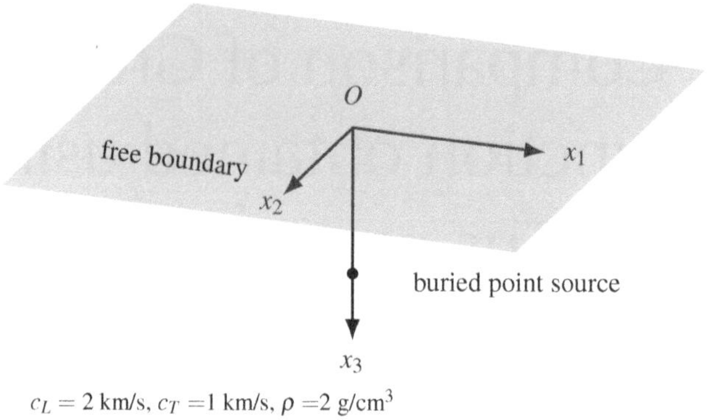

Figure D.1.1 Buried point source model for examining accuracy of Green's function for elastic half-space.

D.1.2 FOURIER-HANKEL TRANSFORM FOR GREEN'S FUNCTION

Based on Eq. (3.2.91), the Fourier-Hankel transform for Green's function is expressed as follows:

$$G_{ij}(\boldsymbol{x},\boldsymbol{y}) = \frac{1}{2\pi} C_{i\alpha}(\theta) \sum_{m=-1}^{+1} \int_0^\infty \xi H^m_{\alpha l'}(\xi, r\theta) \hat{g}_{l'p'}(x_3, y_3) d\xi \, \hat{F}^m_{p'j} \tag{D.1}$$

For the computation of Eq. (D.1), the trapezoidal formula is used throughout after removing the effects of the singularity of the Rayleigh pole. This removal is carried out using

$$\begin{aligned}
&\int_0^\infty \xi H^m_{\alpha l'}(\xi, r\theta) \hat{g}_{l'p'}(x_3, y_3) d\xi \\
&= \int_0^\infty \left(\xi H^m_{\alpha l'}(\xi, r\theta) \hat{g}_{l'p'}(x_3, y_3) - \chi(\delta_1, \kappa_R) \frac{A_R}{\xi - \kappa_R} \right) d\xi \\
&\quad + \text{P.V.} \int_{\kappa_R - \delta_1}^{\kappa_R + \delta_1} \frac{A_R}{\xi - \kappa_R} d\xi + \pi i A_R
\end{aligned} \tag{D.2}$$

where κ_R is the Rayleigh pole, A_R is defined by

$$A_R = \lim_{\xi \to \kappa_R} (\xi - \kappa_R) \xi \, H^m_{\alpha l'}(\xi, r\theta) \hat{g}_{l'p'}(x_3, y_3) \tag{D.3}$$

χ is the function defined by

$$\chi(\delta_1, \xi_R) = \begin{cases} 1 & \text{when } \ \xi_R - \delta_1 \le \xi \le \kappa_R + \delta_1 \\ 0 & \text{otherwise} \end{cases} \tag{D.4}$$

for a small $\delta_1 > 0$ and P.V. denotes the Cauchy principal value. The trapezoidal formula is applied to the first term on the right-hand side of Eq. (D.2). Note that a smaller wavenumber increment for the trapezoidal formula is necessary for an interval that contains k_T compared to that for other intervals due to the weak singularity of ν^{-1} in the integrand. The interval that contains $\xi^{(2)}$ is denoted by $[k_T - \delta_2, k_T + \delta_2]$ for small $\delta_2 > 0$. In the main text, the values for δ_1 and δ_2 and the increment of the wavenumber for the trapezoidal formula are specified for numerical integration. For the direct wavenumber integral, the trapezoidal formula is employed after the effects of the singularity of the Rayleigh pole are removed. The parameters δ_1 and δ_2 are set to 1.0 and 0.7 km^{-1}, respectively. For the discretization of the interval $[\xi^{(2)} - \delta_2, \xi^{(2)} + \delta_2]$ for the trapezoidal formula, the increment of the wavenumber $\Delta\xi$ is set to 1×10^{-4} km^{-1}. Otherwise, $\Delta\xi$ is set to 1×10^{-2} km^{-1}.

D.1.3 STEEPEST DESCENT PATH METHOD FOR GREEN'S FUNCTION

Based on Eqs. (3.4.36) , (3.4.55), and (3.4.56), Green's function approximated using the steepest descent path method and the Rayleigh wave is expressed as

$$\begin{aligned} G_{ij}(\boldsymbol{x},\boldsymbol{y}) &= \frac{\exp(ik_TR)}{4\pi R} D_{ij}^{(T_V)}(\theta,\varphi) + \frac{\exp(ik_TR)}{4\pi R} D_{ij}^{(T_H)}(\theta,\varphi) \\ &\quad + \frac{\exp(ik_LR)}{4\pi R} D_{ij}^{(L)}(\theta,\varphi) + G_{ij}^{(R)}(\boldsymbol{x},\boldsymbol{y}) + O(R^{-2}) \end{aligned} \tag{D.5}$$

For the computation of Eq. (D.5), we do not need to specify a detailed numerical implementation.

D.1.4 SPECTRAL FORM OF GREEN'S FUNCTION

Based on Eq. (3.5.98), the spectral form of Green's function is expressed as

$$\begin{aligned} G_{ij}(\boldsymbol{x},\boldsymbol{y}) &= \int_{\mathbb{R}^2} \sum_{\xi \in \sigma_p} \frac{\Lambda_{im''}(\boldsymbol{\xi},\boldsymbol{x})\Lambda_{jm''}(\boldsymbol{\xi},\boldsymbol{y})}{\mu\xi_3^2 - \rho\omega^2 + i\varepsilon} d\xi_1 d\xi_2 \\ &\quad + \int_{\mathbb{R}^2} \int_{\xi_r}^{\infty} \frac{\Lambda_{im''}(\boldsymbol{\xi},\boldsymbol{x})\Lambda_{jm''}(\boldsymbol{\xi},\boldsymbol{y})}{\mu\xi_3^2 - \rho\omega^2 + i\varepsilon} d\xi_3 d\xi_1 d\xi_2 \end{aligned} \tag{D.6}$$

For the computation of Eq. (D.6), we must define (N_1, N_2, N_3) in Eq. (4.2.23), grid intervals $(\Delta x_1, \Delta x_2, \Delta x_3)$ for Eq. (4.2.23), and $(\Delta\xi_1, \Delta\xi_2, \Delta\xi_3)$ for Eq. (4.2.22). For the numerical examples, we set $N_1 = N_2 = 256$, $N_3 = 128$, and $\Delta x_1 = \Delta x_2 = \Delta x_3 = 0.25$ km. As a result, $\Delta\xi_j$ $(j = 1,2,3)$ becomes:

$$\begin{aligned} \Delta\xi_j &= \frac{2\pi}{N_j \Delta x_j} = 0.09817 \text{ km}^{-1}, \ (j = 1,2) \\ \Delta\xi_3 &= \frac{\pi}{N_3 \Delta x_3} = 0.09817 \text{ km}^{-1} \end{aligned} \tag{D.7}$$

In addition, we set $\varepsilon = 0.6$ in Eq. (D.6).

D.2 NUMERICAL RESULTS

Figures D.2.1 (a) and (b) show comparisons of Green's function along the x_1 axis. Note that λ_T in the figures is the wavelength of the S wave, which is 1 km. In addition, the displacement denotes the vertical component of the vector.

Figure D.2.1 (a) shows a comparison of Green's function expressed by the Fourier-Hankel transform and the steepest descent path methods. As can be seen, the results of the steepest descent path method for Green's function well approximate those of the Fourier-Hankel transform, as far as $|x_1/\lambda_T| > 2$. Figure D.2.1 (b) shows a comparison of Green's function expressed by the Fourier-Hankel transform and the spectral theory. As can be seen, there is almost complete agreement between these results. Figures D.2.1 (a) and (b) validate the accuracy of Green's function expressed by the Fourier-Hankel transform and the spectral theory at the free surface. In addition, we can verify how the steepest descent path method approximates Green's function. Figures D.2.2 (a) and (b) show the displacement amplitude of Green's function in the vertical plane $x_2 = 0$ computed using the Fourier-Hankel transform and its spectral theory. These numerical results are characterized by high-displacement-amplitude regions, which are close to the point source, along the free surface, and below the point source. The high-displacement-amplitude region along the free surface is due to the Rayleigh wave propagation and those below the point source are due to P and S wave propagation. The very-high-amplitude region at a depth of $x_3/\lambda_T = 1$ is due to the singularity of Green's function at the point source. The results obtained by the two methods match well, which validates the accuracy of the generalized Fourier transform for the computation of Green's function.

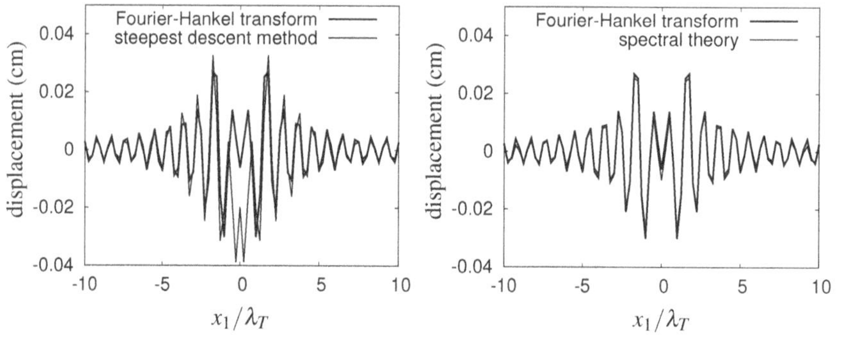

(a) Comparison of Eqs. (D.1) and (D.5) (b) Comparison of Eqs. (D.1) and (D.6)

Figure D.2.1 Comparison of Green's function on x_1 axis, where λ_T is the wavelength of S wave. Green's function are compared in terms of the vertical displacement with respect to the vertical excitation force applied at the buried point source.

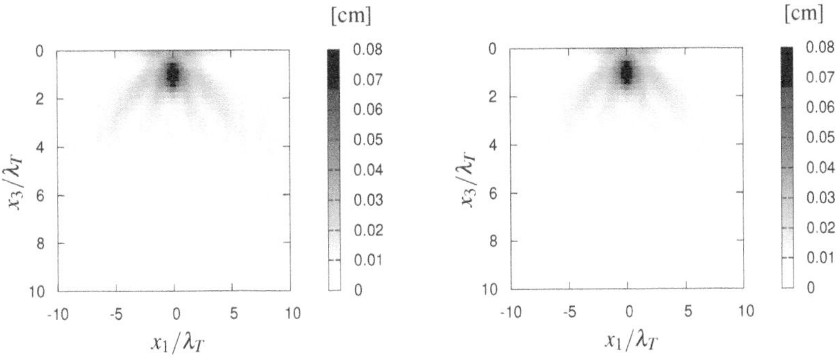

(a) Fourier-Hankel transform (Eq. (D.1)) (b) Spectral theory (Eq. (D.6))

Figure D.2.2 Displacement amplitude of Green's function at vertical plane $x_2 = 0$.

E Music algorithm for detecting location of point-like scatterers

We consider a scalar wavefield in 3D full space with a finite number of point-like scatterers[1] The equation for the wavefield is expressed as

$$\nabla^2 u(\boldsymbol{x}) + k^2 u(\boldsymbol{x}) = -\sum_{m=1}^{M} q_m \delta(\boldsymbol{x} - \boldsymbol{x}_m) u(\boldsymbol{x}) \tag{E.1}$$

For the case where a plane wave propagates toward point-like scatterers, the expression of the total wavefield becomes

$$\begin{aligned} u(\boldsymbol{x}) &= \int_{\mathbb{R}^3} \frac{\exp(ik|\boldsymbol{x}-\boldsymbol{y}|)}{4\pi|\boldsymbol{x}-\boldsymbol{y}|} \sum_{m=1}^{M} q_m \delta(\boldsymbol{y}-\boldsymbol{x}_m) u(\boldsymbol{y}) d\boldsymbol{y} + u^{(I)}(\boldsymbol{x}) \\ &= \sum_{m=1}^{M} \frac{\exp(ik|\boldsymbol{x}-\boldsymbol{x}_m|)}{4\pi|\boldsymbol{x}-\boldsymbol{x}_m|} q_m u(\boldsymbol{x}_m) + u^{(I)}(\boldsymbol{x}) \end{aligned} \tag{E.2}$$

where $u^{(I)}$ is the plane incident wavefield expressed as

$$u^{(I)}(\boldsymbol{x}) = \exp(ik\boldsymbol{p}\cdot\boldsymbol{x}) \tag{E.3}$$

Note that $|\boldsymbol{p}| = 1$, which describes the direction of the plane wave propagation. If we apply the Born approximation to Eq. (E.2), the expression of the solution becomes

$$u(\boldsymbol{x}) = \sum_{m=1}^{M} \frac{\exp(ik|\boldsymbol{x}-\boldsymbol{x}_m|)}{4\pi|\boldsymbol{x}-\boldsymbol{x}_m|} q_m u^{(I)}(\boldsymbol{x}_m) + u^{(I)}(\boldsymbol{x}) \tag{E.4}$$

Since the left-hand side of Eq. (E.4) describes the total wavefield, the expression of the scattered wavefield becomes

$$u^{(s)}(\boldsymbol{x}) = \sum_{m=1}^{M} \frac{\exp(ik|\boldsymbol{x}-\boldsymbol{x}_m|)}{4\pi|\boldsymbol{x}-\boldsymbol{x}_m|} q_m u^{(I)}(\boldsymbol{x}_m) \tag{E.5}$$

We now let

$$r = |\boldsymbol{x}|, \quad \boldsymbol{s} = \boldsymbol{x}/r \tag{E.6}$$

---

[1]The content of this appendix is according to "Kirsch, A.: The factorization method for inverse problems, Newton Institute, `http://www.newton.ac.uk/files/seminar/20110728090009451-152765.pdf`, (2011)."

DOI: 10.1201/9781003251729-E

and investigate the behavior of Eq. (E.5) for the case $r \to \infty$. We see that

$$\begin{aligned}\frac{\exp\left(ik|\boldsymbol{x}-\boldsymbol{x}_m|\right)}{|\boldsymbol{x}-\boldsymbol{x}_m|} \\ &= \frac{\exp(ikr)}{r}\frac{r}{|\boldsymbol{x}-\boldsymbol{x}_m|}\exp\left(ik(|\boldsymbol{x}-\boldsymbol{x}_m|-r\right)\end{aligned} \quad \text{(E.7)}$$

For Eq. (E.7), we notice that

$$\begin{aligned}|\boldsymbol{x}-\boldsymbol{x}_m|-r \\ &= r\left(|\boldsymbol{s}-\boldsymbol{x}_m/r|-1\right) \\ &= r\left(\left[|\boldsymbol{s}|^2-2\boldsymbol{s}\cdot\boldsymbol{x}_m/r+|\boldsymbol{x}_m/r|^2\right]^{1/2}-1\right) \\ &= -\boldsymbol{s}\cdot\boldsymbol{x}_m+O(r^{-1})\end{aligned} \quad \text{(E.8)}$$

and

$$\begin{aligned}\frac{r}{|\boldsymbol{x}-\boldsymbol{x}_m|} &= \left[|\boldsymbol{s}|^2-2\boldsymbol{s}\cdot\boldsymbol{x}_m/r+|\boldsymbol{x}_m/r|^2\right]^{-1/2} \\ &= 1+O(r^{-1})\end{aligned} \quad \text{(E.9)}$$

As result, Eq. (E.5) can be expressed as

$$u^{(s)}(\boldsymbol{x}) = \frac{1}{4\pi|\boldsymbol{x}|}\sum_{m=1}^{M} q_m e^{-ik\boldsymbol{s}\cdot\boldsymbol{x}_m}e^{ik\boldsymbol{p}\cdot\boldsymbol{x}_m}+O(r^{-1}) \quad \text{(E.10)}$$

which yields

$$u^{\infty}(\boldsymbol{s},\boldsymbol{p}) = \sum_{m=1}^{M} q_m e^{-ik\boldsymbol{s}\cdot\boldsymbol{x}_m}e^{ik\boldsymbol{p}\cdot\boldsymbol{x}_m} \quad \text{(E.11)}$$

where u^{∞} is called the *far-field pattern.*

Recall that $|\boldsymbol{s}| = |\boldsymbol{p}| = 1$. That is, $\boldsymbol{s}$ and $\boldsymbol{p}$ are on the unit sphere. Now, we define N points on S, which are described by

$$\{\boldsymbol{\eta}_1,\boldsymbol{\eta}_2,\ldots,\boldsymbol{\eta}_N\}\subset S \quad \text{(E.12)}$$

where S is the unit sphere. We also define an $N\times N$ matrix for the far-field pattern $\boldsymbol{F}$ such that

$$\begin{aligned}\boldsymbol{F} &= \left[\sum_{m=1}^{M} q_m e^{-ik\boldsymbol{\eta}_i\cdot\boldsymbol{x}_m}e^{ik\boldsymbol{\eta}_j\cdot\boldsymbol{x}_m}\right]_{i,j=1}^{N} \\ &= \boldsymbol{HQH}^*\end{aligned} \quad \text{(E.13)}$$

where

$$\boldsymbol{Q} = \begin{bmatrix} q_1 & & & \\ & q_2 & & \\ & & \ddots & \\ & & & q_M \end{bmatrix} \quad \text{(E.14)}$$

and

$$\boldsymbol{H} = \begin{bmatrix} \boldsymbol{\phi}(\boldsymbol{x}_1) & \boldsymbol{\phi}(\boldsymbol{x}_2) & \cdots & \boldsymbol{\phi}(\boldsymbol{x}_m) \end{bmatrix} \tag{E.15}$$

and $\boldsymbol{H}^*$ is the Hermitian adjoint of $\boldsymbol{H}$. Note that $\boldsymbol{\phi}(\boldsymbol{x})$ in Eq. (E.15) is defined by

$$\boldsymbol{\phi} = \begin{pmatrix} e^{-ikr_1} & e^{-ikr_2} & \cdots & e^{-ikr_N} \end{pmatrix}^T \tag{E.16}$$

where

$$r_n = \boldsymbol{\eta}_n \cdot \boldsymbol{x} \tag{E.17}$$

and the superscript T denotes the transpose of the vector.

Based on the factorization of the matrix $\boldsymbol{F}$ shown in Eq. (E.13), we see that

$$\begin{array}{lcl} \boldsymbol{\phi}(\boldsymbol{x}) & \in & \operatorname{ran}\boldsymbol{F}, \ (\text{when}\,\boldsymbol{x} \in E) \\ \boldsymbol{\phi}(\boldsymbol{x}) & \notin & \operatorname{ran}\boldsymbol{F}, \ (\text{when}\,\boldsymbol{x} \notin E) \end{array} \tag{E.18}$$

for the case where $N \geq M$, where

$$E = \{\boldsymbol{x}_1, \boldsymbol{x}_2, \ldots, \boldsymbol{x}_M\} \tag{E.19}$$

and $\operatorname{ran}\boldsymbol{F}$ is a subspace of $\mathbb{C}^N$ defined by

$$\operatorname{ran}\boldsymbol{F} = \{\boldsymbol{x} \in \mathbb{C}^N \,|\, \boldsymbol{x} = \boldsymbol{F}\boldsymbol{y}, \ \forall \boldsymbol{y} \in \mathbb{C}^N\} \tag{E.20}$$

We know that

$$\operatorname{ran}\boldsymbol{F} \perp \ker\boldsymbol{F}^* \tag{E.21}$$

where $\ker\boldsymbol{F}^*$ is a subspace of $\mathbb{C}^N$ defined by

$$\ker\boldsymbol{F}^* = \{\boldsymbol{x} \in \mathbb{C}^N \,|\, \boldsymbol{F}^*\boldsymbol{x} = 0\} \tag{E.22}$$

The proof of Eq. (E.21) is not very complicated. Let $\boldsymbol{y}_k \in \ker\boldsymbol{F}^*$ and $\boldsymbol{y}_r \in \operatorname{ran}\boldsymbol{F}$. Then, we have $\boldsymbol{x}_a$ such that

$$\boldsymbol{y}_r = \boldsymbol{F}\boldsymbol{x}_a \tag{E.23}$$

Therefore,

$$\begin{array}{rcl} \boldsymbol{y}_r \cdot \boldsymbol{y}_k & = & \boldsymbol{F}\boldsymbol{x}_a \cdot \boldsymbol{y}_k \\ & = & \boldsymbol{x}_a \cdot \boldsymbol{F}^*\boldsymbol{y}_k = 0 \end{array} \tag{E.24}$$

As a result, we have $\boldsymbol{y}_k \perp \boldsymbol{y}_r$.

Now, let us define the orthogonal basis $\{\boldsymbol{\psi}_j\}_{j=1}^{N-M}$ such that

$$\operatorname{span}\{\boldsymbol{\psi}_j\}_{j=1}^{N-M} = \ker\boldsymbol{F}^* \tag{E.25}$$

and define the following scalar function

$$W(\boldsymbol{x}) = \left[\sum_{j=1}^{N-M} \left|\boldsymbol{\phi}(\boldsymbol{x}) \cdot \boldsymbol{\psi}_j\right|^2\right]^{-1} \tag{E.26}$$

Then, based on Eqs. (E.15) and (E.25), we find the following properties of $W(\boldsymbol{x})$:

$$\begin{aligned} W(\boldsymbol{x}) &= \infty, \text{ when } \boldsymbol{x} \in E \\ W(\boldsymbol{x}) &< \infty, \text{ when } \boldsymbol{x} \notin E \end{aligned} \tag{E.27}$$

That is, we can identify the locations of point-like scatterers from the divergence of the function $W(\boldsymbol{x})$, which is defined by the properties of Eq. (E.21). In general, the construction of $\boldsymbol{F}$ from observations of the scattered wavefield. The algorithm for identifying the location of point-like scatterers based on Eq. (E.21) is referred to as the *MUSIC* algorithm.

Answers

CHAPTER 1

1. a. According to Fig. 1.5.1 in the main text, the relationship between $\mathcal{O}(\boldsymbol{e}_1, \boldsymbol{e}_2)$ and $\mathcal{O}'(\boldsymbol{e}'_1, \boldsymbol{e}'_2)$ is expressed by the following equations:

$$\boldsymbol{e}'_1 = \cos\theta \boldsymbol{e}_1 + \sin\theta \boldsymbol{e}2$$
$$\boldsymbol{e}2' = -\sin\theta \boldsymbol{e}_1 + \cos\theta \boldsymbol{e}_2$$

Therefore, the elements a_{ij} are:
$a_{11} = \cos\theta$, $a_{12} = \sin\theta$, $a_{21} = -\sin\theta$, $a_{22} = \cos\theta$

b. The transformation rule for the components of a vector can be derived from the relationship of the base vectors. Specifically, we have:

$$\begin{aligned}
\boldsymbol{v} &= v'_i \boldsymbol{e}'_i \\
&= v'_1(\cos\theta \boldsymbol{e}_1 + \sin\theta \boldsymbol{e}_2) + v'_2(-\sin\theta \boldsymbol{e}_1 + \cos\theta \boldsymbol{e}_2) \\
&= (v'_1 \cos\theta - v'_2 \sin\theta)\boldsymbol{e}_1 + (v'_1 \sin\theta + v'_2 \cos\theta)\boldsymbol{e}_2 = v_i \boldsymbol{e}_i
\end{aligned}$$

This yields the relationship:

$$v_i = a_{ki} v'_k \iff v'_k = a_{kj} v_j$$

c. The linearity of the tensor gives us:

$$\begin{aligned}
T'_{ij} &= \boldsymbol{T}(\boldsymbol{e}'_i, \boldsymbol{e}'_j) \\
&= \boldsymbol{T}(a_{il}\boldsymbol{e}_l, a_{jm}\boldsymbol{e}_m) \\
&= a_{il} a_{jm} T_{lm}
\end{aligned}$$

This is equivalent to

$$\begin{bmatrix} T'_{11} & T'_{12} \\ T'_{21} & T'_{22} \end{bmatrix} = \begin{bmatrix} \cos\theta & \sin\theta \\ -\sin\theta & \cos\theta \end{bmatrix} \begin{bmatrix} T_{11} & T_{12} \\ T_{21} & T'_{22} \end{bmatrix} \begin{bmatrix} \cos\theta & -\sin\theta \\ \sin\theta & \cos\theta \end{bmatrix}$$

d.

$$\begin{aligned}
T'_{11} &= T_{11}\cos^2\theta + 2T_{12}\sin\theta\cos\theta + T_{22}\sin^2\theta \\
T'_{12} &= -T_{11}\cos\theta\sin\theta + T_{12}(\cos^2\theta - \sin^2\theta) + T_{22}\cos\theta\sin\theta \\
T'_{22} &= T_{11}\sin^2\theta - 2T_{12}\sin\theta\cos\theta + T_{22}\cos^2\theta
\end{aligned}$$

2. a. We define an n-axis with $\boldsymbol{n}$. A position vector on the n-axis is expressed by the parameter n as:

$$\boldsymbol{x} \underset{\mathcal{O}}{\rightarrow} (x_1(n), x_2(n), x_3(n))$$

We then define an average rate of change of φ along the axis as:

$$\begin{aligned}
\frac{\Delta\varphi}{\Delta n} &= \frac{\varphi(x_1(n+\Delta n), x_2(n+\Delta n), x_3(n+\Delta n)) - \varphi(x_1(n), x_2(n), x_3(n))}{\Delta n} \\
&= \frac{\varphi(x_1(n+\Delta n), x_2(n+\Delta n), x_3(n+\Delta n)) - \varphi(x_1(n), x_2(n+\Delta n), x_3(n+\Delta n))}{\Delta x_1}\frac{\Delta x_1}{\Delta n} \\
&+ \frac{\varphi(x_1(n), x_2(n+\Delta n), x_3(n+\Delta n)) - \varphi(x_1(n), x_2(n), x_3(n+\Delta n))}{\Delta x_2}\frac{\Delta x_2}{\Delta n} \\
&+ \frac{\varphi(x_1(n), x_2(n), x_3(n+\Delta n)) - \varphi(x_1(n), x_2(n), x_3(n))}{\Delta x_3}\frac{\Delta x_3}{\Delta n}
\end{aligned} \tag{S.1.1}$$

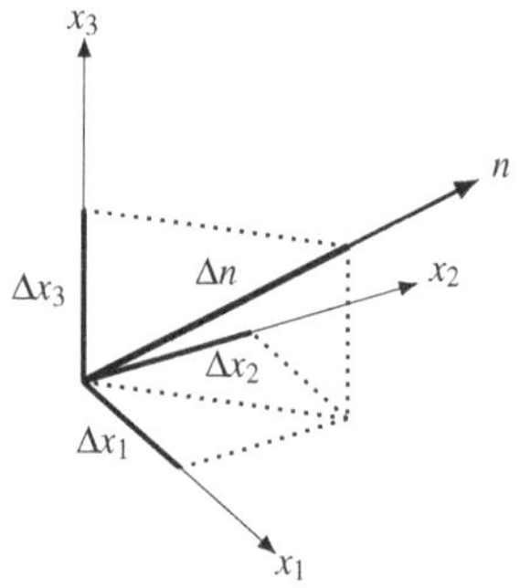

Figure S.1 The relationship between Δn and Δx_j, $(j = 1, 2, 3)$, where Δn is the increment of n. We see that $\Delta x_j / \Delta n$, $(j = 1, 2, 3)$ defines the directional cosine of the n-axis.

where Δx_1, Δx_2 and Δx_3 are defined by

$$\Delta x_j = x_j(n + \Delta n) - x_j(n), \ (j = 1, 2, 3)$$

From this, $\Delta x_j / \Delta n$ is found to be the direction cosine with respect to x_j and the n axes, as shown in Fig. S.1. Therefore, $\boldsymbol{n}$ is expressed as:

$$\boldsymbol{n} \underset{\mathcal{O}}{\rightarrow} \left(\frac{\Delta x_1}{\Delta n}, \frac{\Delta x_2}{\Delta n}, \frac{\Delta x_3}{\Delta n} \right)$$

$$\underset{\Delta n \to 0}{\rightarrow} \left(\frac{\partial x_1}{\partial n}, \frac{\partial x_2}{\partial n}, \frac{\partial x_3}{\partial n} \right)$$

As a result, from Eq. (S.1.1), the directional derivative of a scalar function along the n-axis can be expressed as:

$$\frac{\partial \varphi(\boldsymbol{x})}{\partial n} = \lim_{\Delta n \to 0} \frac{\Delta \varphi(\boldsymbol{x})}{\Delta n} = \nabla \varphi(\boldsymbol{x}) \cdot \boldsymbol{n}$$

b. We can apply the Gauss divergence theorem, which states:

$$\int_V \operatorname{div} \boldsymbol{f} dV = \int_S \boldsymbol{n} \cdot \boldsymbol{f} dS$$

This theorem is derived from Eq. (1.3.51). If we express the vector field $\boldsymbol{f}$ as:

$$\boldsymbol{f} = \nabla \varphi$$

then the Gauss divergence theorem simplifies to:

$$\int_V \nabla^2 \varphi(\boldsymbol{x}) dV = \int_S \frac{\partial \varphi(\boldsymbol{x})}{\partial n} dS$$

c. We apply Green's theorem to the Laplace operator, as shown in Eq. (1.3.54). Consequently, Eq. (1.5.7) becomes nearly self-evident.

3. Firstly, we observe that the material derivative of the Jacobian defined by Eq. (1.3.16) can be expressed as follows:

$$\frac{DJ}{Dt} = \begin{vmatrix} \frac{D}{Dt}\left(\frac{\partial x_1}{\partial X_1}\right) & \frac{D}{Dt}\left(\frac{\partial x_1}{\partial X_2}\right) & \frac{D}{Dt}\left(\frac{\partial x_1}{\partial X_3}\right) \\ \frac{\partial x_2}{\partial X_1} & \frac{\partial x_2}{\partial X_2} & \frac{\partial x_2}{\partial X_3} \\ \frac{\partial x_3}{\partial X_1} & \frac{\partial x_3}{\partial X_2} & \frac{\partial x_3}{\partial X_3} \end{vmatrix} + \begin{vmatrix} \frac{\partial x_1}{\partial X_1} & \frac{\partial x_1}{\partial X_2} & \frac{\partial x_1}{\partial X_3} \\ \frac{D}{Dt}\left(\frac{\partial x_2}{\partial X_1}\right) & \frac{D}{Dt}\left(\frac{\partial x_2}{\partial X_2}\right) & \frac{D}{Dt}\left(\frac{\partial x_2}{\partial X_3}\right) \\ \frac{\partial x_3}{\partial X_1} & \frac{\partial x_3}{\partial X_2} & \frac{\partial x_3}{\partial X_3} \end{vmatrix}$$
$$+ \begin{vmatrix} \frac{\partial x_1}{\partial X_1} & \frac{\partial x_1}{\partial X_2} & \frac{\partial x_1}{\partial X_3} \\ \frac{\partial x_2}{\partial X_1} & \frac{\partial x_2}{\partial X_2} & \frac{\partial x_2}{\partial X_3} \\ \frac{D}{Dt}\left(\frac{\partial x_3}{\partial X_1}\right) & \frac{D}{Dt}\left(\frac{\partial x_3}{\partial X_2}\right) & \frac{D}{Dt}\left(\frac{\partial x_3}{\partial X_3}\right) \end{vmatrix} \quad \text{(S.1.2)}$$

This is due to the following properties of the material derivative:

$$\frac{D}{Dt}(fgh) = gh\frac{D}{Dt}f + fh\frac{D}{Dt}g + fg\frac{D}{Dt}h$$

Using Eq. (1.3.20), we can also derive:

$$\begin{vmatrix} \frac{D}{Dt}\left(\frac{\partial x_1}{\partial X_1}\right) & \frac{D}{Dt}\left(\frac{\partial x_1}{\partial X_2}\right) & \frac{D}{Dt}\left(\frac{\partial x_1}{\partial X_3}\right) \\ \frac{\partial x_2}{\partial X_1} & \frac{\partial x_2}{\partial X_2} & \frac{\partial x_2}{\partial X_3} \\ \frac{\partial x_3}{\partial X_1} & \frac{\partial x_3}{\partial X_2} & \frac{\partial x_3}{\partial X_3} \end{vmatrix}$$

$$= \frac{\partial v_1}{\partial x_1}\begin{vmatrix} \frac{\partial x_1}{\partial X_1} & \frac{\partial x_1}{\partial X_2} & \frac{\partial x_1}{\partial X_3} \\ \frac{\partial x_2}{\partial X_1} & \frac{\partial x_2}{\partial X_2} & \frac{\partial x_2}{\partial X_3} \\ \frac{\partial x_3}{\partial X_1} & \frac{\partial x_3}{\partial X_2} & \frac{\partial x_3}{\partial X_3} \end{vmatrix} + \frac{\partial v_1}{\partial x_2}\begin{vmatrix} \frac{\partial x_2}{\partial X_1} & \frac{\partial x_2}{\partial X_2} & \frac{\partial x_2}{\partial X_3} \\ \frac{\partial x_2}{\partial X_1} & \frac{\partial x_2}{\partial X_2} & \frac{\partial x_2}{\partial X_3} \\ \frac{\partial x_3}{\partial X_1} & \frac{\partial x_3}{\partial X_2} & \frac{\partial x_3}{\partial X_3} \end{vmatrix}$$

$$+\frac{\partial v_1}{\partial x_3}\begin{vmatrix} \frac{\partial x_3}{\partial X_1} & \frac{\partial x_3}{\partial X_2} & \frac{\partial x_3}{\partial X_3} \\ \frac{\partial x_2}{\partial X_1} & \frac{\partial x_2}{\partial X_2} & \frac{\partial x_2}{\partial X_3} \\ \frac{\partial x_3}{\partial X_1} & \frac{\partial x_3}{\partial X_2} & \frac{\partial x_3}{\partial X_3} \end{vmatrix} = \frac{\partial v_1}{\partial x_1}J$$

Therefore, Eq. (S.1.2) simplifies to:

$$\frac{DJ}{Dt} = \left(\frac{\partial v_1}{\partial x_1} + \frac{\partial v_2}{\partial x_2} + \frac{\partial v_3}{\partial x_3}\right)J = \operatorname{div}\boldsymbol{v}\,J$$

4. Let the coordinates of the vertices of the tetrahedron shown in Fig. S.2 be $A(a,0,0)$, $B(0,b,0)$, and $C(0,0,c)$, where a, b, and c are positive real numbers. The equation of a plane that includes $\triangle$ABC is given by:

$$\frac{x_1}{a} + \frac{x_2}{b} + \frac{x_3}{c} = 1 \tag{S.1.3}$$

Note that a vector $\left(\frac{1}{a}, \frac{1}{b}, \frac{1}{c}\right)$ is orthogonal to $\triangle$ABC. Therefore, the equation for the straight line l, shown in Fig. S.2, which originates from the origin and is orthogonal to $\triangle$ABC, is given by:

$$x_1 = \frac{1}{a}t, \quad x_2 = \frac{1}{b}t, \quad x_3 = \frac{1}{c}t \tag{S.1.4}$$

where t is a real parameter. Substituting Eq. (S.1.4) into Eq. (S.1.3), we get:

$$t = \frac{a^2b^2c^2}{b^2c^2 + c^2a^2 + a^2b^2} \tag{S.1.5}$$

This t value corresponds to the intersection point of $\triangle$ABC and the straight line l, denoted as P in Fig. S.2. By using this parameter, the distance between the origin of the coordinate system and $\triangle$ ABC can be calculated as:

$$\overline{OP} = \frac{abc}{\sqrt{b^2c^2 + c^2a^2 + a^2b^2}}$$

Based on the relationship between the volume of the tetrahedron and the area $\triangle$ABC expressed by:

$$\frac{1}{6}abc = \frac{1}{3}\,\overline{OP} \times \triangle\mathrm{ABC}$$

we can find the area of $\triangle$ABC as:

$$\triangle\mathrm{ABC} = (1/2)\sqrt{b^2c^2 + c^2a^2 + a^2b^2}$$

Since the unit normal vector of a plane which includes $\triangle$ABC is expressed as:

$$\underset{\mathcal{O}}{\boldsymbol{n}\rightarrow} \frac{1}{\sqrt{\frac{1}{a^2} + \frac{1}{b^2} + \frac{1}{c^2}}}\left(\frac{1}{a}, \frac{1}{b}, \frac{1}{c}\right)$$

we can see that:

$$\underset{\mathcal{O}}{\boldsymbol{n}\rightarrow} \frac{1}{\triangle\mathrm{ABC}}\left(\triangle\mathrm{OBC}, \triangle\mathrm{OCA}, \triangle\mathrm{OAB}\right)$$

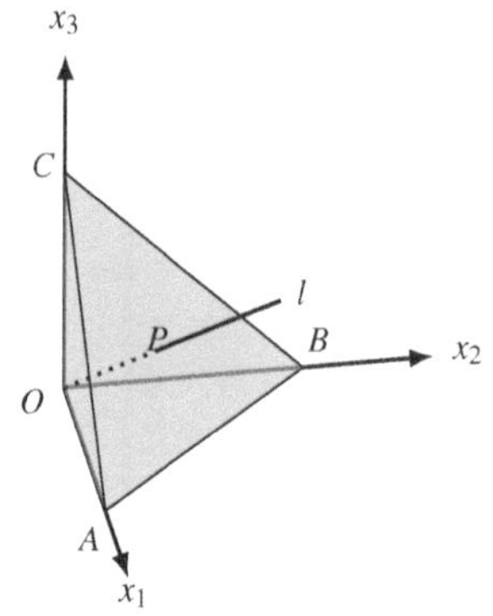

Figure S.2 The tetrahedron element $O-$ABC. Note that the point P is in $\triangle$ABC and $OP\perp\triangle$ABC

CHAPTER 2

1. a. Let us assume that a solution of the Helmholtz equation can be expressed as

$$u(\boldsymbol{x}) = U_1(x_1)U_2(x_2)U_3(x_3)$$

Substituting this into the Helmholtz equation:

$$\Big(\frac{\partial^2}{\partial x_1^2}+\frac{\partial^2}{\partial x_2^2}+\frac{\partial^2}{\partial x_3^2}+k^2\Big)u(\boldsymbol{x})=0$$

yields the following expression:

$$U_2(x_2)U_3(x_3)\frac{d^2U_1(x_1)}{dx_1^2}+U_1(x_1)U_3(x_3)\frac{d^2U_2(x_2)}{dx_2^2} \quad \text{(S.2.6)}$$

$$+U_1(x_1)U_2(x_2)\frac{d^2U_3(x_3)}{dx_3^2}+k^2U_1(x_1)U_2(x_2)U_3(x_3)=0 \quad \text{(S.2.7)}$$

We also assume that

$$U(x_1)U(x_2)U(x_3)\neq 0$$

and divide both side of Eq. (S.2.7) by $U_1(x_1)U_2(x_2)U_3(x_3)$, which yields:

$$\frac{1}{U_1(x_1)}\frac{d^2U_1(x_1)}{dx_1^2}+\frac{1}{U_2(x_2)}\frac{d^2U_2(x_2)}{dx_2^2}+\frac{1}{U_3(x_3)}\frac{d^2U_3(x_3)}{dx_3^2}+k^2=0 \quad \text{(S.2.8)}$$

Equation (S.2.8) shows that each term of the left-hand side of Eq. (S.2.8) must be constant. Therefore, we have:

$$\begin{aligned}\frac{1}{U_1(x_1)}\frac{d^2U_1(x_1)}{dx_1^2} &= -\xi_1^2\\ \frac{1}{U_2(x_2)}\frac{d^2U_2(x_2)}{dx_2^2} &= -\xi_2^2\\ \frac{1}{U_3(x_3)}\frac{d^2U_3(x_3)}{dx_3^2} &= -\xi_3^2\end{aligned}$$

where ξ_1, ξ_2 and ξ_3 are constants satisfying:

$$\xi_1^2+\xi_2^2+\xi_3^2=k^2$$

The ordinary differential equations with respect to U_j, $(j=1,2,3)$ yield:

$$U_j(x_j) = A_j \exp(i\xi_j x_j) + B_j \exp(-i\xi_j x_j), \quad (j=1,2,3)$$

and we arrive at a solution of the Helmholtz equation.

b. The relationship between a Cartesian coordinate system (x_1, x_2, x_3) and a cylindrical coordinate system (r, φ, z) is given as:

$$r = \sqrt{x_1^2 + x_2^2}, \ \varphi = \tan^{-1}(x_2/x_1), \ z = x_3$$

which yields:

$$\frac{\partial r}{\partial x_1} = \cos\varphi, \ \frac{\partial r}{\partial x_2} = \sin\varphi, \ \frac{\partial \varphi}{\partial x_1} = -\frac{\sin\varphi}{r}, \ \frac{\partial \varphi}{\partial x_2} = \frac{\cos\varphi}{r}$$

Therefore, we have the following expressions:

$$\begin{aligned}
\frac{\partial u}{\partial x_1} &= \frac{\partial u}{\partial r}\frac{\partial r}{\partial x_1} + \frac{\partial u}{\partial \varphi}\frac{\partial \varphi}{\partial x_1} = \cos\theta \frac{\partial u}{\partial r} - \frac{\sin\varphi}{r}\frac{\partial u}{\partial \varphi} \\
\frac{\partial^2 u}{\partial x_1^2} &= \frac{\partial r}{\partial x_1}\frac{\partial}{\partial r}\left[\frac{\partial u}{\partial r}\frac{\partial r}{\partial x_1} + \frac{\partial u}{\partial \varphi}\frac{\partial \varphi}{\partial x_1}\right] + \frac{\partial \varphi}{\partial x_1}\frac{\partial}{\partial \varphi}\left[\frac{\partial u}{\partial r}\frac{\partial r}{\partial x_1} + \frac{\partial u}{\partial \varphi}\frac{\partial \varphi}{\partial x_1}\right] \\
&= \cos^2\varphi \frac{\partial^2 u}{\partial r^2} + \frac{2\sin\varphi\cos\varphi}{r^2}\frac{\partial u}{\partial \varphi} - \frac{2\sin\varphi\cos\varphi}{r}\frac{\partial^2 u}{\partial \varphi \partial r} \\
&\quad + \frac{\sin^2\varphi}{r}\frac{\partial u}{\partial r} + \frac{\sin^2\varphi}{r^2}\frac{\partial^2 u}{\partial \varphi^2} \qquad \text{(S.2.9)} \\
\frac{\partial u}{\partial x_2} &= \frac{\partial u}{\partial r}\frac{\partial r}{\partial x_2} + \frac{\partial u}{\partial \varphi}\frac{\partial \varphi}{\partial x_2} = \sin\varphi \frac{\partial u}{\partial r} + \frac{\cos\varphi}{r}\frac{\partial u}{\partial \varphi} \\
\frac{\partial^2 u}{\partial x_1^2} &= \frac{\partial r}{\partial x_2}\frac{\partial}{\partial r}\left[\frac{\partial u}{\partial r}\frac{\partial r}{\partial x_2} + \frac{\partial u}{\partial \varphi}\frac{\partial \varphi}{\partial x_2}\right] + \left[\frac{\partial u}{\partial r}\frac{\partial r}{\partial x_2} + \frac{\partial u}{\partial \varphi}\frac{\partial \varphi}{\partial x_2}\right] \\
&= \sin^2\varphi \frac{\partial^2 u}{\partial r^2} - \frac{2\sin\varphi\cos\varphi}{r^2}\frac{\partial u}{\partial \varphi} + \frac{2\sin\varphi\cos\varphi}{r}\frac{\partial^2 u}{\partial \varphi \partial r} \\
&\quad + \frac{\cos^2\varphi}{r}\frac{\partial u}{\partial r} + \frac{\cos^2\varphi}{r^2}\frac{\partial^2 u}{\partial \varphi^2} \qquad \text{(S.2.10)}
\end{aligned}$$

According to Eqs. (S.2.9) and (S.2.10), the representation of the Laplace operator in terms of the cylindrical coordinate system becomes:

$$\begin{aligned}
&\left(\frac{\partial^2}{\partial x_1^2} + \frac{\partial^2}{\partial x_2^2} + \frac{\partial^2}{\partial x_3^3}\right)u \\
&= \left(\frac{\partial^2}{\partial r^2} + \frac{1}{r}\frac{\partial}{\partial r} + \frac{1}{r^2}\frac{\partial^2}{\partial \varphi^2} + \frac{\partial^2}{\partial z^2}\right)u \qquad \text{(S.2.11)}
\end{aligned}$$

c. We assume a solution of the Helmholtz equation expressed by

$$u(\boldsymbol{x}) = U_r(r)U_\varphi(\varphi)U_z(z) \qquad \text{(S.2.12)}$$

Substituting Eq. (S.2.12) into the Helmholtz equation yields:

$$\begin{aligned}
&U_\theta U_z \frac{d^2 U_r}{dr^2} + U_\theta U_z \frac{1}{r}\frac{dU_r}{dr} \\
&\quad + U_r U_z \frac{1}{r^2}\frac{d^2 U_\varphi}{dr^2} + U_r U_\varphi \frac{d^2 U_z}{dz^2} + k^2 U_r U_\varphi U_z = 0 \qquad \text{(S.2.13)}
\end{aligned}$$

Now, we assume that

$$U_r U_\varphi U_z \neq 0$$

and divide the both side of Eq. (S.2.13) by $U_r U_\varphi U_z$, which yields:

$$\frac{1}{U_r}\frac{d^2U_r}{dr^2}+\frac{1}{r}\frac{1}{U_r}\frac{dU_r}{dr}$$
$$+\frac{1}{r^2}\frac{1}{U_\varphi}\frac{d^2U_\varphi}{d\varphi^2}+\frac{1}{U_z}\frac{d^2U_z}{dz^2}+k^2=0 \quad \text{(S.2.14)}$$

Note that Eq. (S.2.14) can be solved by assuming that U_φ, U_z and U_r satisfy the following equations:

$$\frac{1}{U_\varphi}\frac{d^2U_\varphi}{d\theta^2} = -m^2 \quad \text{(S.2.15)}$$
$$\frac{1}{U_z}\frac{d^2U_z}{dz^2} = -\xi^2 \quad \text{(S.2.16)}$$

$$\frac{d^2U_r}{dr^2}+\frac{1}{r}\frac{dU_r}{dr}-\frac{m^2}{r^2}U_r+(k^2-\xi^2)U_r = 0 \quad \text{(S.2.17)}$$

where m and ξ are the constants. The equation for U_r shown in Eq. (S.2.17) can be attributed to the Bessel equation which has the solution:

$$U_r(r)=A_rJ_m(\kappa r)+B_rY_m(\kappa r) \quad \text{(S.2.18)}$$

where $\kappa^2=k^2-\xi^2$,and J_m and Y_m are the first and second kinds of the Bessel functions for the m-th order, respectively. It is evident that U_θ and U_z are expressed by

$$U_z(z) = A_z\exp(i\xi z)+B_z\exp(-i\xi z) \quad \text{(S.2.19)}$$
$$U_\varphi(\varphi) = A_\varphi\exp(im\varphi)+B_\varphi\exp(-im\varphi) \quad \text{(S.2.20)}$$

By means of Eqs. (S.2.18) to (S.2.20), we have a solution of the Helmholtz equation in the cylindrical coordinate system.

2. According to the chain rule, we have

$$\frac{\partial u}{\partial x} = \frac{\partial u}{\partial \xi}\frac{\partial \xi}{\partial x}+\frac{\partial u}{\partial \eta}\frac{\partial \eta}{\partial x}=\frac{\partial u}{\partial \xi}+\frac{\partial u}{\partial \eta}$$
$$\frac{\partial^2 u}{\partial x^2} = \frac{\partial^2 u}{\partial \xi^2}+2\frac{\partial^2 u}{\partial \xi\partial \eta}+\frac{\partial^2 u}{\partial \eta^2}$$
$$\frac{\partial u}{\partial t} = \frac{\partial u}{\partial \xi}\frac{\partial \xi}{\partial t}+\frac{\partial u}{\partial \eta}\frac{\partial \eta}{\partial x}=-c\left(\frac{\partial u}{\partial \xi}+\frac{\partial u}{\partial \eta}\right)$$
$$\frac{\partial^2 u}{\partial t^2} = c^2\left(\frac{\partial^2 u}{\partial \xi^2}-2\frac{\partial^2 u}{\partial \xi\partial \eta}+\frac{\partial^2 u}{\partial \eta^2}\right)$$

Therefore, 1D wave equation:

$$\frac{\partial^2 u}{\partial x^2}=\frac{1}{c^2}\frac{\partial^2 u}{\partial t^2}$$

is modified into:

$$\frac{\partial^2 u}{\partial \eta\partial \xi}=0 \quad \text{(S.2.21)}$$

We see that a solution of Eq. (S.2.21) is expressed as:

$$u(x,t)=F(\xi)+G(\eta)=F(x-ct)+G(x+ct)$$

where F and G are arbitrary functions with differentiable twice.

3. Let us consider the following expressions:

$$R=|\mathbf{x}|=\sqrt{x_1^2+x_2^2+x_3^2}$$

Then, we can find the following derivatives:

$$\frac{\partial}{\partial x_1}\frac{e^{\alpha R}}{R} = -\frac{e^{\alpha R}}{R^3}x_1 + \frac{\alpha e^{\alpha R}}{R^2}x_1$$

since

$$\frac{\partial R}{\partial x_1} = \frac{x_1}{R}$$

We can also find the second derivative:

$$\frac{\partial^2}{\partial x_1^2}\frac{e^{\alpha R}}{R} = -\frac{e^{\alpha R}}{R^3} + x_1^2 e^{\alpha R}\left(\frac{3}{R^5} - \frac{3\alpha}{R^4} + \frac{\alpha^2}{R^3}\right) + \frac{\alpha}{R^2}e^{\alpha R}$$

Therefore, we obtain:

$$\left(\frac{\partial^2}{\partial x_1^2} + \frac{\partial^2}{\partial x_2^2} + \frac{\partial^2}{\partial x_2^2}\right)\frac{e^{\alpha R}}{R} = \alpha^2\frac{e^{\alpha R}}{R}$$

Next, we want to show the following:

$$\left(\frac{\partial^2}{\partial x_1^2} + \frac{\partial^2}{\partial x_2^2} + \frac{\partial^2}{\partial x_2^2}\right)\frac{f\left(t-\frac{R}{c}\right)}{R} = \frac{1}{c^2}\frac{\partial^2}{\partial t^2}\frac{f\left(t-\frac{R}{c}\right)}{R}$$

To show this, let's find the partial derivatives step by step:

$$\begin{aligned}
\frac{\partial}{\partial x_1}\frac{f(t-\frac{R}{c})}{R} &= -\frac{x_1}{R^3}f\left(t-\frac{R}{c}\right) - \frac{x_1}{cR^2}f'\left(t-\frac{R}{c}\right) \\
\frac{\partial^2}{\partial x_1^2}\frac{f(t-\frac{R}{c})}{R} &= -\frac{1}{R^3}f\left(t-\frac{R}{c}\right) + \frac{3x_1^2}{R^5}f\left(t-\frac{R}{c}\right) - \frac{1}{cR^2}f'\left(t-\frac{R}{c}\right) \\
&\quad + \frac{3x_1^2}{cR^4}f'\left(t-\frac{R}{c}\right) + \frac{x_1^2}{c^2R^3}f''\left(t-\frac{R}{c}\right)
\end{aligned}$$

This yields:

$$\begin{aligned}
\left(\frac{\partial^2}{\partial x_1^2} + \frac{\partial^2}{\partial x_2^2} + \frac{\partial^2}{\partial x_2^2}\right)\frac{f(t-\frac{R}{c})}{R} &= \frac{1}{c^2}\frac{1}{R}f''\left(t-\frac{R}{c}\right) \\
&= \frac{1}{c^2}\frac{\partial^2}{\partial t^2}\frac{f(t-\frac{R}{c})}{R}
\end{aligned}$$

4. a. According to Fig. S.3, we can determine the transformation rule for the base vectors of the cylindrical coordinate system from a Cartesian coordinate system is given by:

$$\begin{aligned}
\boldsymbol{e}_r &= \boldsymbol{e}_1\cos\varphi + \boldsymbol{e}_2\sin\varphi \\
\boldsymbol{e}_\varphi &= -\boldsymbol{e}_1\sin\varphi + \boldsymbol{e}_2\cos\varphi \\
\boldsymbol{e}_z &= \boldsymbol{e}_3
\end{aligned} \tag{S.2.22}$$

In other words, we can also express it in the following way:

$$\begin{aligned}
\boldsymbol{e}_1 &= \boldsymbol{e}_r\cos\varphi - \boldsymbol{e}_\varphi\sin\varphi \\
\boldsymbol{e}_2 &= \boldsymbol{e}_r\sin\varphi + \boldsymbol{e}_\varphi\cos\varphi \\
\boldsymbol{e}_3 &= \boldsymbol{e}_z
\end{aligned} \tag{S.2.23}$$

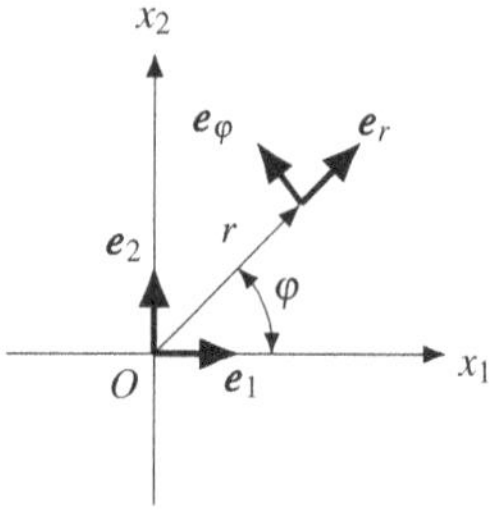

Figure S.3 Relationship of the base vectors between a Cartesian and a cylindrical coordinate system in 2D plane.

Using the transformation rule from Eqs. (S.2.22) and (S.2.23), we can express the gradient of the scalar field in terms of the cylindrical coordinate system as follows:

$$
\begin{aligned}
\nabla\phi(\boldsymbol{x}) &= \frac{\partial\phi}{\partial x_1}\boldsymbol{e}_1 + \frac{\partial\phi}{\partial x_2}\boldsymbol{e}_2 + \frac{\partial\phi}{\partial x_1}\boldsymbol{e}_3 \\
&= \left(\frac{\partial\phi}{\partial r}\frac{\partial r}{\partial x_1} + \frac{\partial\phi}{\partial\varphi}\frac{\partial\varphi}{\partial x_1}\right)(\boldsymbol{e}_r\cos\varphi - \boldsymbol{e}_\varphi\sin\varphi) \\
&\quad + \left(\frac{\partial\phi}{\partial r}\frac{\partial r}{\partial x_2} + \frac{\partial\phi}{\partial\varphi}\frac{\partial\varphi}{\partial x_2}\right)(\boldsymbol{e}_r\sin\varphi + \boldsymbol{e}_\varphi\cos\varphi) + \frac{\partial\phi}{\partial z}\boldsymbol{e}_z \\
&= \frac{\partial\phi}{\partial r}\boldsymbol{e}_r + \frac{1}{r}\frac{\partial\phi}{\partial\varphi}\boldsymbol{e}_\varphi + \frac{\partial\phi}{\partial z}\boldsymbol{e}_z
\end{aligned}
$$

Next, let us express a vector field in the following form:

$$
\begin{aligned}
\boldsymbol{u}(\boldsymbol{x}) &= u_1(\boldsymbol{x})\boldsymbol{e}_1 + u_2(\boldsymbol{x})\boldsymbol{e}_2 + u_3(\boldsymbol{x})\boldsymbol{e}_3 \\
&= u_r(\boldsymbol{x})\boldsymbol{e}_r + u_\theta(\boldsymbol{x})\boldsymbol{e}_\varphi + u_z(\boldsymbol{x})\boldsymbol{e}_z
\end{aligned}
$$

The transformation rules for the components of the vector field can be derived from Eqs. (S.2.22) and/or (S.2.23), resulting in the following forms:

$$
\begin{aligned}
u_r &= u_1\cos\varphi + u_2\sin\varphi \\
u_\varphi &= -u_1\sin\varphi + u_2\cos\varphi \\
u_z &= u_3
\end{aligned}
\tag{S.2.24}
$$

$$
\begin{aligned}
u_1 &= u_r\cos\varphi - u_\varphi\sin\varphi \\
u_2 &= u_r\sin\varphi + u_\varphi\cos\varphi \\
u_3 &= u_z
\end{aligned}
\tag{S.2.25}
$$

The divergence of the vector field in terms of the cylindrical coordinate system is expressed as:

$$
\begin{aligned}
\operatorname{div}\boldsymbol{u} &= \frac{\partial u_1}{\partial x_1} + \frac{\partial u_2}{\partial x_2} + \frac{\partial u_3}{\partial x_3} \\
&= \frac{\partial}{\partial r}\left(u_r\cos\varphi - u_\theta\sin\varphi\right)\frac{\partial r}{\partial x_1} + \frac{\partial}{\partial\varphi}\left(u_r\cos\varphi - u_\varphi\sin\varphi\right)\frac{\partial\varphi}{\partial x_1} \\
&\quad + \frac{\partial}{\partial r}\left(u_r\sin\varphi + u_\varphi\cos\varphi\right)\frac{\partial r}{\partial x_2} + \frac{\partial}{\partial\varphi}\left(u_r\sin\varphi + u_\varphi\cos\varphi\right)\frac{\partial\varphi}{\partial x_2} + \frac{\partial u_z}{\partial z} \\
&= \frac{\partial u_r}{\partial r} + \frac{u_r}{r} + \frac{1}{r}\frac{\partial u_\varphi}{\partial\varphi} + \frac{\partial u_z}{\partial z}
\end{aligned}
\tag{S.2.26}
$$

Now, let us recall that the rotation of a vector field in terms of the Cartesian coordinate system is expressed as:

$$\operatorname{rot}\boldsymbol{u} = \left(\frac{\partial u_3}{\partial x_2} - \frac{\partial u_2}{\partial x_3}\right)\boldsymbol{e}_1 + \left(\frac{\partial u_1}{\partial x_3} - \frac{\partial u_3}{\partial x_1}\right)\boldsymbol{e}_2 + \left(\frac{\partial u_2}{\partial x_1} - \frac{\partial u_1}{\partial x_2}\right)\boldsymbol{e}_2 \qquad \text{(S.2.27)}$$

If we want to derive the expression of the rotation of a vector field in terms of the cylindrical coordinate system, we need to introduce the transformation rules described by Eqs. (S.2.23) and (S.2.25). Additionally, we have to rewrite the spatial derivative in terms of the cylindrical coordinate system. These procedures are the same as for the gradient of a scalar field and the divergence of a vector field presented above. The expression of the rotation of vector field becomes:

$$\begin{aligned}\operatorname{rot}\boldsymbol{u}(\boldsymbol{x}) &= \left(\frac{1}{r}\frac{\partial u_z}{\partial \varphi} - \frac{\partial u_\varphi}{\partial z}\right)\boldsymbol{e}_r + \left(\frac{\partial u_r}{\partial z} - \frac{\partial u_z}{\partial r}\right)\boldsymbol{e}_\varphi \\ &\quad + \left(\frac{\partial u_\varphi}{\partial r} - \frac{1}{r}\frac{\partial u_r}{\partial \varphi} + \frac{u_\varphi}{r}\right)\boldsymbol{e}_z \qquad \text{(S.2.28)}\end{aligned}$$

b. In the context of the Cartesian coordinate system, we have

$$\begin{aligned}\nabla\nabla\cdot\boldsymbol{u} &= \left(\frac{\partial^2 u_1}{\partial x_1^2} + \frac{\partial^2 u_2}{\partial x_1 \partial x_2} + \frac{\partial^2 u_3}{\partial x_1 \partial x_3}\right)\boldsymbol{e}_1 \\ &\quad + \left(\frac{\partial^2 u_1}{\partial x_1 \partial x_2} + \frac{\partial^2 u_2}{\partial x_2^2} + \frac{\partial^2 u_3}{\partial x_2 \partial x_3}\right)\boldsymbol{e}_2 \\ &\quad + \left(\frac{\partial^2 u_1}{\partial x_1 \partial x_3} + \frac{\partial^2 u_2}{\partial x_2 \partial x_3} + \frac{\partial^2 u_3}{\partial x_3^2}\right)\boldsymbol{e}_3 \\ \nabla\times\nabla\times\boldsymbol{u} &= \left(\frac{\partial^2 u_2}{\partial x_1 \partial x_2} + \frac{\partial^2 u_3}{\partial x_1 \partial x_3} - \frac{\partial^2 u_1}{\partial x_2^2} - \frac{\partial^2 u_1}{\partial x_3^2}\right)\boldsymbol{e}_1 \\ &\quad + \left(\frac{\partial^2 u_1}{\partial x_2^2} + \frac{\partial^2 u_3}{\partial x_2 \partial x_3} - \frac{\partial^2 u_2}{\partial x_1^2} - \frac{\partial^2 u_2}{\partial x_3^2}\right)\boldsymbol{e}_2 \\ &\quad + \left(\frac{\partial^2 u_1}{\partial x_1 \partial x_3} + \frac{\partial^2 u_2}{\partial x_2 \partial x_3} - \frac{\partial^2 u_3}{\partial x_1^2} - \frac{\partial^2 u_3}{\partial x_2^2}\right)\boldsymbol{e}_3\end{aligned}$$

For these expressions, we can verify the following relationship:

$$\begin{aligned}&\Big(\nabla\nabla\cdot - \nabla\times\nabla\times\Big)\left(u_1\boldsymbol{e}_1 + u_2\boldsymbol{e}_2 + u_3\boldsymbol{e}_3\right) \\ &= \left(\frac{\partial^2 u_1}{\partial x_1^2} + \frac{\partial^2 u_1}{\partial x_2^2} + \frac{\partial^2 u_1}{\partial x_3^2}\right)\boldsymbol{e}_1 \\ &\quad + \left(\frac{\partial^2 u_2}{\partial x_1^2} + \frac{\partial^2 u_2}{\partial x_2^2} + \frac{\partial^2 u_2}{\partial x_3^2}\right)\boldsymbol{e}_2 \\ &\quad + \left(\frac{\partial^2 u_3}{\partial x_1^2} + \frac{\partial^2 u_3}{\partial x_2^2} + \frac{\partial^2 u_3}{\partial x_3^2}\right)\boldsymbol{e}_3 \qquad \text{(S.2.29)}\end{aligned}$$

c. By means of the cylindrical coordinate system, however, the result becomes as:

$$\begin{aligned}&\Big(\nabla\nabla\cdot - \nabla\times\nabla\times\Big)\left(u_r\boldsymbol{e}_r + u_\varphi\boldsymbol{e}_\varphi + u_z\boldsymbol{e}_z\right) \\ &= \left(\frac{\partial^2 u_r}{\partial r^2} + \frac{1}{r}\frac{\partial u_r}{\partial r} + \frac{1}{r^2}\frac{\partial^2 u_r}{\partial \varphi^2} + \frac{\partial^2 u_r}{\partial z^2} - \frac{1}{r^2}u_r - \frac{2}{r^2}\frac{\partial u_\varphi}{\partial \varphi}\right)\boldsymbol{e}_r \\ &\quad + \left(\frac{\partial^2 u_\varphi}{\partial r^2} + \frac{1}{r}\frac{\partial u_\varphi}{\partial r} + \frac{1}{r^2}\frac{\partial^2 u_\varphi}{\partial \varphi^2} + \frac{\partial^2 u_\varphi}{\partial z^2} + \frac{1}{r^2}\frac{\partial u_\varphi}{\partial \varphi} + \frac{2}{r^2}\frac{\partial u_r}{\partial \varphi} - \frac{1}{r^2}u_\varphi\right)\boldsymbol{e}_\varphi \\ &\quad + \left(\frac{\partial^2 u_z}{\partial r^2} + \frac{1}{r}\frac{\partial u_z}{\partial r} + \frac{1}{r^2}\frac{\partial^2 u_z}{\partial \varphi^2} + \frac{\partial^2 u_z}{\partial z^2}\right)\boldsymbol{e}_z \qquad \text{(S.2.30)}\end{aligned}$$

which is obtained by the use of Eqs. (S.2.26) and (S.2.28). Therefore, unfortunately, we see that

$$\begin{aligned}
&\left(\nabla\nabla\cdot-\nabla\times\nabla\times\right)\left(u_r\boldsymbol{e}_r+u_\varphi\boldsymbol{e}_\varphi+u_z\boldsymbol{e}_z\right)\\
\neq\;&\left(\frac{\partial^2u_r}{\partial r^2}+\frac{1}{r}\frac{\partial u_r}{\partial r}+\frac{1}{r^2}\frac{\partial^2u_r}{\partial\varphi^2}+\frac{\partial^2u_r}{\partial z^2}\right)\boldsymbol{e}_r\\
&+\left(\frac{\partial^2u_\varphi}{\partial r^2}+\frac{1}{r}\frac{\partial u_\varphi}{\partial r}+\frac{1}{r^2}\frac{\partial^2u_\varphi}{\partial\varphi^2}+\frac{\partial^2u_\varphi}{\partial z^2}\right)\boldsymbol{e}_\varphi\\
&+\left(\frac{\partial^2u_z}{\partial r^2}+\frac{1}{r}\frac{\partial u_z}{\partial r}+\frac{1}{r^2}\frac{\partial^2u_z}{\partial\varphi^2}+\frac{\partial^2u_z}{\partial z^2}\right)\boldsymbol{e}_z
\end{aligned}$$

d. The spatial derivatives of the base vectors for the cylindrical coordinate system are given by:

$$\begin{aligned}
\frac{\partial\boldsymbol{e}_r}{\partial r}&=0\\
\frac{\partial\boldsymbol{e}_\varphi}{\partial r}&=0\\
\frac{\partial\boldsymbol{e}_r}{\partial\varphi}&=\boldsymbol{e}_\varphi\\
\frac{\partial\boldsymbol{e}_\varphi}{\partial\varphi}&=-\boldsymbol{e}_r
\end{aligned}$$

Therefore, using Eq. (S.2.30), we arrive at the following identity:

$$\begin{aligned}
&\left(\nabla\nabla\cdot-\nabla\times\nabla\times\right)\left(u_r\boldsymbol{e}_r+u_\theta\boldsymbol{e}_\theta+u_z\boldsymbol{e}_z\right)\\
=\;&\left(\frac{\partial^2}{\partial r^2}+\frac{1}{r}\frac{\partial}{\partial r}+\frac{1}{r^2}\frac{\partial^2}{\partial\varphi^2}+\frac{\partial^2}{\partial z^2}\right)\left(u_r\boldsymbol{e}_r+u_\varphi\boldsymbol{e}_\varphi+u_z\boldsymbol{e}_z\right) \qquad \text{(S.2.31)}
\end{aligned}$$

It is of course that Eq. (2.6.11) can also be expressed by

$$\begin{aligned}
&\left(\nabla\nabla\cdot-\nabla\times\nabla\times\right)\left(u_1\boldsymbol{e}_1+u_2\boldsymbol{e}_2+u_3\boldsymbol{e}_3\right)\\
=\;&\left(\frac{\partial^2}{\partial x_1^2}+\frac{\partial^2}{\partial x_2^2}+\frac{\partial^2}{\partial x_3^2}\right)\left(u_1\boldsymbol{e}_1+u_2\boldsymbol{e}_2+u_3\boldsymbol{e}_3\right) \qquad \text{(S.2.32)}
\end{aligned}$$

In general, by taking into account the spatial derivative of the base vectors, we can establish the following identity

$$\nabla^2=\nabla\nabla\cdot-\nabla\times\nabla\times \qquad \text{(S.2.33)}$$

for the Laplace operator.

5. It is sufficient to show that

$$\int_{S_R}\left[\widehat{G}(\boldsymbol{x},\boldsymbol{y})\hat{p}(\boldsymbol{y})-\widehat{P}(\boldsymbol{x},\boldsymbol{y})\hat{u}(\boldsymbol{y})\right]dS_R(\boldsymbol{y})\to0,\quad(R(=|\boldsymbol{y}|)\to\infty) \qquad \text{(S.2.34)}$$

According to the Sommerfeld radiation condition for Green's function:

$$\widehat{P}(\boldsymbol{x},\boldsymbol{y})+ik\widehat{G}(\boldsymbol{x},\boldsymbol{y})=O(R^{-2})$$

as well as for the wavefield:

$$\hat{p}(\boldsymbol{y})+ik\hat{u}(\boldsymbol{y})=O(R^{-2})$$

the left-hand side of Eq. (S.2.34) becomes as

$$\begin{aligned}
&\int_{S_R}\left[\widehat{G}(\boldsymbol{x},\boldsymbol{y})\hat{p}(\boldsymbol{y})-\widehat{P}(\boldsymbol{x},\boldsymbol{y})\hat{u}(\boldsymbol{y})\right]dS_R(\boldsymbol{y})\\
&=\int_{S_R}\left[\widehat{G}(\boldsymbol{x},\boldsymbol{y})\hat{p}(\boldsymbol{y})-\left(-ik\widehat{G}(\boldsymbol{x},\boldsymbol{y})+O(R^{-2})\right)\hat{u}(\boldsymbol{y})\right]dS_R(\boldsymbol{y})\\
&=\int_{S_R}\left[\widehat{G}(\boldsymbol{x},\boldsymbol{y})\left(\hat{p}(\boldsymbol{y})+ik\hat{u}(\boldsymbol{y})\right)+O(R^{-2})\hat{u}(\boldsymbol{y})\right]dS_R(\boldsymbol{y})\\
&=\int_{S_R}O(R^{-3})dS_R(\boldsymbol{y})=O(R^{-1})\to0,\quad(R\to\infty)
\end{aligned}$$

For the above derivations, we have also used

$$\begin{aligned} \widehat{G}(\boldsymbol{x},\boldsymbol{y}) &= O(R^{-1}) \\ \hat{u}(,\boldsymbol{y}) &= O(R^{-1}), \quad (R \to \infty) \end{aligned}$$

6. It is sufficient to show that

$$\int_0^t d\tau \int_{S_R} \Big[G(\boldsymbol{x},\boldsymbol{y},t-\tau)p(\boldsymbol{y},\tau) - P(\boldsymbol{x},\boldsymbol{y},t-\tau)u(\boldsymbol{y},\tau) \Big] dS_R(\boldsymbol{y}) \to 0, \quad (R(=|\boldsymbol{y}|) \to \infty) \tag{S.2.35}$$

For the evaluation of the above integral, we have the radiation conditions:

$$\begin{aligned} P(\boldsymbol{x},\boldsymbol{y},t) + \frac{1}{c}\frac{\partial}{\partial t}G(\boldsymbol{x},\boldsymbol{y},t) &= O(R^{-2}) \\ p(\boldsymbol{y},t) + \frac{1}{c}\frac{\partial}{\partial t}u(\boldsymbol{y},t) &= O(R^{-2}), \quad (R \to \infty) \end{aligned}$$

The evaluation of the left-had side of Eq. (S.2.35) becomes as

$$\begin{aligned} & \int_0^t d\tau \int_{S_R} \Big[G(\boldsymbol{x},\boldsymbol{y},t-\tau)p(\boldsymbol{y},\tau) - P(\boldsymbol{x},\boldsymbol{y},t-\tau)u(\boldsymbol{y},\tau) \Big] dS_R(\boldsymbol{y}) \\ &= \int_0^t d\tau \int_{S_R} \Big[G(\boldsymbol{x},\boldsymbol{y},t-\tau)p(\boldsymbol{y},\tau) - \big(\frac{1}{c}\frac{\partial}{\partial}G(\boldsymbol{x},\boldsymbol{y},t-\tau)\big)u(\boldsymbol{y},\tau) + O(R^{-2})u(\boldsymbol{y},\tau) \Big] dS_R(\boldsymbol{y}) \\ &= \int_0^t d\tau \int_{S_R} \Big[G(\boldsymbol{x},\boldsymbol{y},t-\tau)\big(p(\boldsymbol{y},\tau) + \frac{1}{c}\dot{u}(\boldsymbol{y},\tau)\big) + O(R^{-2})u(\boldsymbol{y},\tau) \Big] dS_R(\boldsymbol{y}) \\ &= \int_0^t d\tau \int_{S_R} O(R^{-3}) dS_R(\boldsymbol{y}) \to 0, \quad (R \to \infty) \end{aligned} \tag{S.2.36}$$

For the derivation of Eq. (S.2.36), we have also used:

$$\begin{aligned} G(\boldsymbol{x},\boldsymbol{y},t) &= O(R^{-1}) \\ u(\boldsymbol{y},\tau) &= O(R^{-1}), \quad (R \to \infty) \end{aligned}$$

CHAPTER 3

1. The expression of the incident wavefield in terms of the Sommerfeld integral is

$$p^{(Inc)}(\boldsymbol{x}) = \frac{q}{4\pi}\int_0^\infty \xi \frac{e^{-\nu_1|x_3+h|}}{\nu_1} J_0(\xi r) d\xi$$

2. According to the boundary conditions shown in Eq. (3.6.2), we have

$$\begin{aligned} q + B(\xi) &= C(\xi) \\ -q + B(\xi) &= -\frac{\nu_2\rho_1}{\nu_1\rho_2}C(\xi) \end{aligned}$$

which yields

$$\begin{aligned} B(\xi) &= \frac{-\nu_2\rho_1 + \nu_1\rho_2}{\nu_2\rho_1 + \nu_1\rho_2}q \\ C(\xi) &= \frac{2\nu_1\rho_2}{\nu_2\rho_1 + \nu_1\rho_2}q \end{aligned}$$

3. By following the procedure presented by Eqs. (3.3.1) to (3.3.7), (3.4.1) to (3.4.11) and (3.4.12) to (3.4.23), we have the approximation of the reflected wavefield as

$$p^{(Ref)}(\boldsymbol{x}) \sim \frac{B(\xi_s)}{4\pi R}\exp(ik_1 R)$$

where ξ_s is the saddle point and

$$R = \sqrt{x_1^2 + x_2^2 + (x_3 - h)^2}$$

4. For the case of $\xi_s > k_2$, it is necessary to evaluate the branch line integral for the reflected wavefield. For the evaluation of the branch line integral, we start with

$$\begin{aligned} p^{(Ref)}(\boldsymbol{x}) &= \frac{q}{4\pi}\int_0^\infty \xi B(\xi)\frac{e^{-\nu|x_3+h|}}{\nu_1}J_0(\xi r)d\xi \\ &= \frac{q}{8\pi}\int_{-\infty}^\infty \xi B(\xi)\frac{e^{-\nu|x_3+h|}}{\nu_1}H_0^{(1)}(\xi r)d\xi \\ &\sim \frac{q}{8\pi}\sqrt{\frac{2}{\pi r}}e^{-i\pi/4}\int_{-\infty}^\infty G(\xi)\exp(g(\xi))d\xi \end{aligned}$$

where

$$\begin{aligned} G(\xi) &= \frac{\sqrt{\xi}}{\nu_1}\frac{\nu_1\rho_2 - \nu_2\rho_1}{\nu_2\rho_1 + \nu_1\rho_2} \\ g(\xi) &= -\sqrt{\xi^2 - k_1^2}|x_3 - h| + i\xi r \end{aligned}$$

We expand $g(\xi)$ and $G(\xi)$ around $\xi = k_2$ by

$$\begin{aligned} g(\xi) &= g(k_2) + (\xi - k_2)g'(k_2) + \cdots \\ &= ik_1R\cos(\theta - \theta_0) + (\xi - k_2)\frac{iR\sin(\theta - \theta_0)}{\cos\theta_0} + \cdots \\ G(\xi) &= \frac{\sqrt{k_2}}{\sqrt{k_2^2 - k_1^2}} - \frac{2\rho_1}{\rho_2}\frac{\sqrt{k_2}}{k_2^2 - k_1^2}\sqrt{\xi^2 - k_2^2} + \cdots \end{aligned}$$

where

$$k_2 = k_1\sin\theta_0, \quad \xi_s = k_1\sin\theta$$

Therefore, the evaluation of the branch line integral becomes as

$$\begin{aligned} &\frac{q}{8\pi}\sqrt{\frac{2}{\pi r}}e^{-i\pi/4}\int_{B_1+B_2} G(\xi)\exp(g(\xi))d\xi \\ &= -\frac{q}{8\pi}\sqrt{\frac{2k_2}{\pi r}}e^{-i\pi/4}\frac{2\rho_1}{\rho_2}\frac{1}{k_2^2 - k_1^2}\exp\left(ik_1R\cos(\theta - \theta_0)\right) \\ &\quad \int_{B_1+B_2}\sqrt{\xi^2 - k_2^2}\exp\left(i(\xi - k_2)R\frac{\sin(\theta - \theta_0)}{\cos\theta_0}\right)d\xi \\ &\sim -\frac{qi}{8\pi}\sqrt{\frac{2k_2}{\pi r}}\frac{4\rho_1}{\rho_2}\frac{1}{k_2^2 - k_1^2}\exp\left(ik_1R\cos(\theta - \theta_0)\right)\int_0^\infty \sqrt{u}\exp\left(-uR\frac{\sin(\theta - \theta_0)}{\cos\theta_0}\right)du \\ &= \frac{qi}{2\pi}\frac{\rho_1}{\rho_2}\frac{\sin\theta_0}{k_1\sqrt{\cos\theta_0\sin\theta\sin^3(\theta - \theta_0)}}\frac{e^{ik_1R\cos(\theta - \theta_0)}}{R^2} \end{aligned}$$

During the derivation, u is defined by

$$\xi = k_2 + iu$$

We have also used $r = R\sin\theta$ and

$$\int_0^\infty \sqrt{u}e^{-au}du = \frac{\sqrt{\pi}}{2}a^{-3/2}, \quad (a > 0)$$

In addition, the path of integral of B_1 and B_2 are defined by Fig. S.4.

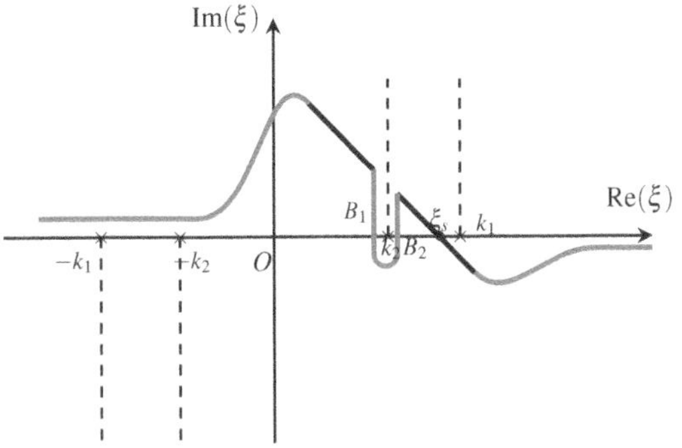

Figure S.4 Paths of integral B_1 and B_2 around the branch cut.

References

1. Abubakar, A., Pan, G., Li, M., Zhang, L. Habashy, T.M., and van den Berg, P.M, Three-dimensional seismic full-waveform inversion using the finite-difference contrast source inversion method, Geophysical prospecting, Special Issue: Modelling Methods for Geophysical Imaging: Trends and Perspective, Vol. 59, Issue 5, pp. 874-888, (2011).
2. Achenbach,J.D.: Wave propagation in elastic solids, north-holland., (1984).
3. Aki, K, and Richards, P.G.: Quantitative Seismology, second edition, University Science books, (2002).
4. Arai, A.: Hilbert Space and Quantum Mechanics, Revised and Enlarged Edition, (in Japanese), Kyoritu Shuppan, Col. Ltd. (2014).
5. Baganas, K., Guzina, B.B, Charalambopoulos, A., and Manolis,G.D.: A linear sampling method for the inverse transmission problem in near-field elastodynamics, Inverse problems, 22, 1835, (2006).
6. Barrett, R., Berry, M., Chan, T.F., Demmel, J., Donato, J.M., Dongarra, J., Eijkhout, V., Pozo, R., Romine, C. and Van der Vorst, H., 1994. Templates for the solution of Linear Systems: Building Blocks for Iterative Methods, SIAM.
7. Ben-Menahem, A. and Singh, S.J.: Seismic waves and sources, second edition, Dover, (2000).
8. Berthier, A.M.: Spectral theory and wave operators for the Scrödinger equation, Pitman advanced publishing program, (1982).
9. Bland, D.R.:Nonlinear dynamic elasticity, Blaisdell Publishing Company, (1969).
10. Brebbia,C.A., Telles, J.C.F. and Wrobel, L.C.: Boundary element techniques, Theory and applications in engineering, Springer-Verlag, (1984).
11. Cheney, M.: The linear sampling method and the MUSIC algorithm, Inverse problems, 17, 4, (2001).
12. Colton, D. and Kirsch, A.: A simple method for solving inverse scattering problems in the resonance region, Inverse problems, 12, 383-393, (1996).
13. Colton,D. and Kress, R.: Inverse acoustic and electromagnetic scattering theory, Springer, (1998).
14. De Zaeytijd,J., Bogaert,I. and Franchois, A.: An efficient hybrid MLFMA-FFT solver for the volume integral equation in case of sparse 3D inhomogeneous dielectric scatterers, Journal of Computational Physics 227, 7052-7068, (2008).
15. Duffy, D.G.: Transform methods for Solving partial differential equations, CRC press, (United States of America), (1994).
16. Eringen, A.C. and Suhubi, E.S. : Elastodynamics, volume II, linear theory, Academic Press, (1975).
17. Ewing,W.M., Jardetzky,W.S. and Press, F.: Elastic waves in layered media, McGraw-Hill, (1957).
18. Fata, S.N. and Guzina, B.B., A linear sampling method for near-field inverse problems in elastodynamics, Inverse Problems, 20, 713-736, (2004).
19. Gintides, D., Sini, M. and Thành, N.T., Detection of point-like scatterers using one type of scattered elastic waves, Journal of computational and applied mathematics, vol. 236, pp. 2137-2145, (2012).

20. Guzina, B.B., Fata, S.N. and Bonnet, M.: On the stress-wave imaging of cavities in a semi-infinite solid, International Journal of Solids and Structures 40, 1505-1523, (2003).
21. Guzina, B.B. and Madyarov, A.I.: A linear sampling approach to inverse elastic scattering in piecewise-homogeneous domains, Inverse Problems, 23, 1467, (2007).
22. Hudson, J. A. and Heritage, J. R.: The use of the Born approximation in seismic scattering problems, Geophysical Journal Royal Astronomical Society 66, 221-240, (1981).
23. Ikebe, T.: Eigenfunction expansion associated with the Schroedinger operators and their applications to scattering theory, Archive of Rational Mechanics and Analysis 5, 1-34, (1960).
24. Imamura, T.: Physics and Green's function (working title), Iwanami shoten pulishers, (1978) (in Japanese).
25. Imamura, T.: Physics and Fourier transform (working title), Iwanami shoten publishers, (1978) (in Japanese).
26. Imamura, T.: Physics and complex analysis (working title), Iwanami shoten publishers, (2004) (in Japanese).
27. Ikawa, M. : Introduction to partial differential equations, Shokabo Publishing Co.,Ltd., (1996) (in Japanese).
28. Jaswon, M.A. and Symm, G.T.: Integral equation methods in Potential theory and Elastostatics, Academic Press (1977).
29. Kirsch, A.: The factorization method for inverse problems, Newton Institute, `http://www.newton.ac.uk/files/seminar/20110728090009451-152765.pdf`, (2011).
30. Kupradze,V.D., Potential methods in the theory of elasticity, Israel program for Science translation, (1965).
31. Kupradze,V.D., Gegelia, T.G., Basheleishvili, M.O. and Burchuladze, T.V.: Three dimensional problems of the mathematical theory of elasticity and thermoelasticity, North-Holland, (1979).
32. Lamb., H. : On the propagation of tremors over the surface of an elastic solid Phil. Trans. Roy. Soc. London, A, 203, pp. 1-42, (1904).
33. Landau, L.D. and Lifshitz, E.M.:Fluid Mechanics, (Course of theoretical physics, volume 6), Pergamon Press, (1984).
34. Landau, L.D. and Lifshitz, E.M.: Quantum mechanics, non-relativistic theory, (Course of theoretical physics, volume 3), Pergamon Press, (1977).
35. Landau, L.D. and Lifshitz, E.M.: Theory of elasticity, (Course of theoretical physics, volume 7), Pergamon Press, (1986).
36. Leal, L.G.:Laminar flow and convective transport process, scaling principles and asymptotic analysis, Butterworth-Heinemann (1992).
37. Love, A.E.H.:A treatise on the mathematical theory of elasticity, Dover publication, (1944).
38. Manolis, G.D., Dineva, P.S. and Rangelov, T.V.: Wave scattering by cracks in inhomogeneous continua using BIEM, International Journal of Solids and Structures 41, 3905-3927, (2004).
39. McLachlan, N,W.: Bessel functions for engineers, Oxford University Press, (1961).
40. Morse,P. and Feshbach,H.: Methods of theoretical physics, Part 1 and 2, McGraw-Hill, Tokyo, (1953).
41. Pacheco, P.S.: Parallel programming with MPI, Morgan Kaufmann Publishers, (1997).

42. Pelekanos, G., Abubakar, A. and van den Berg, P.M.: Contrast source inversion methods in Elastodynamics, J. Acoust. Soc. Am, 114, 2825-2834, (2004).
43. Pourahmadian, F., Guzina, B.B. and Haddar, H.: Generalized linear sampling method for elastic-wave sensing of heterogeneous fractures Inverse Problems, Volume 33, Number 5, (2017).
44. Pujol, J.: Elastic wave propagation and generation in seismology, Cambridge University Press, (2003).
45. Reed, M. and Simon, B.: Functional Analysis, (Methods of modern mathematical physics I) , Academic Press, (1980)
46. Reed, M. and Simon, B.: Fourier analysis and self-adjointness, (methods of modern mathematical physics II), Academic Press (1975).
47. Romdhane, A., Grandjean, G., Brossier, R., Réjiba, F., Operto, S., and Virieux, J.: Shallow structures characterization by 2d elastic waveform inversion, Geophysics, 76(3), R81-R93, (2011).
48. Sato, Y. : Theory of elastic waves (working title), Iwanami shoten publishers (1978) (in Japanese).
49. Saito, M.: The theory of seismic wave propagation, University of Tokyo Press (2009) (in Japanese).
50. Sano, O.: Mechanics of continuum, Shokabo Publishing Co.,Ltd., (2000) (in Japanese).
51. Schutz, B.F.: A first course in general relativity, Cambridge University press, (1990).
52. Sommerfeld, A.: Partial Differential Equations in Physics, Academic Press, New York, (1949).
53. Strain, J.: A fast Laplace transform based on Laguerre functions, Mathematics of Computation 58, 275-283., (1992).
54. Sternburg, E. and Eubanks,R.A.: On stress functions for elasto-kinetics and the integration of the repeated wave equation, Quart. Appl. Math. 15, pp. 149-153, (1953),
55. Tashiro, Y.: Tensor analysis, Shokabo Publishing Co.,Ltd., (1981) (in Japanese).
56. Touhei, T. and Ohmachi, T.: A FE-BE method for dynamic analysis of dam-foundation-reservoir systems in the time domain, Earthquake engineering and structural dynamics, Vol.22, 195-209, 1993.
57. Touhei, T.: Complete eigenfunction expansion form of the Green's function for elastic layered half-space, Archive of applied mechanics, Vol. 72, 13-38, 2002.
58. Touhei, T.: Generalized Fourier transform and its application to the volume integral equation for elastic wave propagation in a half space, International Journal of Solids and Structures 46, 52-73, (2009).
59. Touhei, T.: A fast volume integral equation method for elastic wave propagation in a half space, International Journal of Solids and Structures 48, 3194-3208, (2011).
60. Touhei, T.: Inversion of point-like scatterers in an elastic half-space by the application of the far-field properties of the Green's function to the near field operator, International Journal of Solids and Structures 136-137, 112-124, (2018)
61. Touhei, T. and Maruyama, T.: Pseudo-projection approach to reconstruct locations of point-like scatterers characterized by Lamé parameters and mass densities in an elastic half-space, International Journal of Solids and Structures 169, 187-204, (2019).
62. Yang, J., Abubakar, A., van den Berg, P.M., Habashy, T.M. and Reitich, F.: A CG-FFT approach to the solution of a stress-velocity formulation of three-dimensional scattering problems, Journal of Computational physics 227, 10018-10039, (2008).

Index

azimuthal Fourier components, 102

Bessel function, 90, 102
Born approximation, 179, 181, 182, 238
boundary element, 163, 165
boundary integral equation, 162, 163
branch cut, 91, 94
branch line integral, 111, 112, 135
branch point, 91

Cauchy principal integral, 216
Cauchy principal value, 47, 216
Cauchy relation, 18
causality of Green's function, 64
complex wavenumber plane, 91
constitutive equation, 24
continuity equation, 15
continuous spectrum, 150
continuum body, 1
continuum mechanics, 1
critical angle, 81, 82, 133
cylindrical wave, 34

d'Alembert formula, 33
definition function for the eigenfunction associated with the continuous spectrum, 231
definition of a tensor and its components, 205
dipole Green's function, 198
Dirac delta function, 39, 45, 93, 216
directional derivative, 35
directivity tensor, 134
discrete Fourier transform, 176
discrete Laplace transform, 175, 176
divergence-free, 38
divergence-free vector field, 38
dummy index, 204

Eddington symbol, 36
eigenfunction for the continuous spectrum, 150, 231
eigenfunction for the point spectrum, 148, 228
elastic constants, 24
elastic tensor, 23
elastic wave equation, 24
Euler approach, 2
Eulerian (Almansi) finite strain tensor, 5
exterior problem, 67
exteriror problem, 63

factorization of the far-field operator, 187
factorization of the operator, 197
far-field pattern, 239
fast algorithm, 177, 178
fluctuation of the wavefield, 168–170, 180, 182, 188
fluctuations of the Lamé parameters and mass density, 186, 199
Fourier integral transform, 46
Fourier transform, 214
Fourier-Bessel integral transform, 215
Fourier-Hankel and its inverse transforms, 215
Fourier-Hankel transform, 90, 100, 115
Fourier-Hankel transform for an elastic wavefield, 101–103
free index, 204

Gauss divergence theorem, 19, 20
generalized Fourier transform, 169, 171, 172, 177
governing equation, 24
gradient operator, 35
Green theorem, 21
Green's function, 39, 44
Green's function for an elastic half-space, 88, 97, 99, 107, 115
Green's function for the elastic wave equation, 52
Green's function for the scalar wave equation, 46
Green's function for traction, 59

Green's function in the frequency domain, 46
Green's function in the time domain, 46
Green's function in the wavenumber domain, 99, 106, 219

Hankel and inverse Hankel transforms, 215
Hankel function, 109, 110
Heaviside unit step function, 217
Helmholtz decomposition, 31
Helmholtz equation, 31
Hooke's law, 23

improper eigenfunction, 228
incident angle, 78
infinitesimal strain tensor, 6, 7, 9
inhomogeneous wave, 81, 82
interior problem, 59, 65
inverse Fourier transform, 214
inverse Fourier-Hankel transform for an elastic wave field, 102
irrotational vector field, 38
isotropic rank-2 tensor, 209
isotropic rank-4 tensor, 209

Jordan's lemma, 48

Krylov subspace, 178

Lagrange approach, 2
Lagrangian (Green) finite strain tensor, 5
Lamé constants, 24
Laplace operator, 21, 32
Lippmann-Schwinger equation, 169, 170, 172, 177
longitudinal wave, 42
LPG tank, 165

material derivative, 13
metric tensor, 206
monopole Green's function, 198
MUSIC algorithm, 186, 198, 241

Navier operator, 24
nodal point, 163
normal strain, 8

operator theory, 139
orthogonal decomposition of the S wavefield, 43

P wave, 71
P wave velocity, 42
P wavefield, 42
particle in a continuum body, 2
permissible sheet, 110, 111
plane wave, 33, 73
point spectrum, 148
point-like scatterers, 186, 193, 195, 199
Poisson ratio, 24
pseudo-projection, 186, 187, 193

Rayleigh function, 85
Rayleigh wave, 72, 86
reciprocity, 26
reciprocity of Green's function, 58
reflection angle, 75
representation of the solution, 59, 63, 64
representation theorem, 59, 63, 65, 67, 157
resolvent, 145
resolvent kernel, 142, 219
Reynolds transport theorem, 13
Riemann sheet, 91, 94
root of the Rayleigh function, 117
rotation operator, 36

S wave velocity, 42
S wavefield, 42
S-P wave, 133, 135
saddle point, 124
scalar field, 19
scalar potential, 38
scalar product, 202
scalar wave equation, 31
Schwartz inequality, 143
self-adjoint operator, 141
self-adjointness, 140
SH wave, 71
shear strain, 8

singular integral equation, 162
Snell law, 76
Sommerfeld integral, 91, 126
Sommerfeld radiation condition, 67, 68
Sommerfeld-Kupradze radiation condition, 61, 158
Sommerfeld-Kupradze radiation conditions, 66
spectral family, 146
spectral representation, 140, 146
spectral representation of Green's function, 155
spectrum, 145
spherical wave, 34
steady state, 26
steepest descent path, 124
steepest descent path method, 123
Stone theorem, 146
strain tensor, 5
stress tensor, 18
summation convention, 204
surface vector harmonics, 101
SV wave, 71
symmetry operator, 141

tensor field, 19
transient state, 28
transverse wave, 42
triple scalar product, 208

vector field, 19, 35
vector potential, 38
volume integral equation, 169
volumetric strain, 10, 11

wavenumber space, 47
wavenumber vector, 74
Weyl integral, 90

Young's modulus, 24